Engineering Design

Engineering Design

Rudolph J. Eggert
Boise State University

PEARSON
Prentice
Hall

Upper Saddle River, New Jersey 07458

Library of Congress Cataloging-in-Publication Data on File

Vice President and Editorial Director, ECS: *Marcia J. Horton*
Editorial Assistant: *Andrea Messineo*
Vice President and Director of Production and Manufacturing, ESM: *David W. Riccardi*
Executive Managing Editor: *Vince O'Brien*
Managing Editor: *David A. George*
Production Editor: *Rebecca Homiski*
Director of Creative Services: *Paul Belfanti*
Art Director: *Jayne Conte*
Cover Designer: *Bruce Kenselaar*
Art Editor: *Greg Dulles*
Manufacturing Manager: *Trudy Pisciotti*
Manufacturing Buyer: *Lynda Castillo*
Senior Marketing Manager: *Holly Stark*

© 2005 Pearson Education, Inc.
Pearson Prentice Hall
Pearson Education, Inc.
Upper Saddle River, NJ 07458

Pearson Prentice Hall® is a trademark of Pearson Education, Inc.

Microsoft® Word, Microsoft® Excel, Microsoft® PowerPoint®, and Microsoft® Project are registered trademarks of Microsoft Corporation in the United States and/or other countries. Mathsoft™ Mathcad® is a registered trademark of Mathsoft Engineering & Education, Inc.

Printed in the United States of America

10 9 8 7 6 5 4 3 2 1

Pearson Education Ltd., *London*
Pearson Education Australia Pty. Ltd., *Sydney*
Pearson Education Singapore, Pte. Ltd.
Pearson Education North Asia Ltd., *Hong Kong*
Pearson Education Canada, Inc., *Toronto*
Pearson Educación de Mexico, S.A. de C.V.
Pearson Education—Japan, *Tokyo*
Pearson Education Malaysia, Pte. Ltd.
Pearson Education, Inc., *Upper Saddle River, New Jersey*

Contents

APPENDICES

Index   **389**

Preface

Engineering design is the set of activities that lead to the manufacture of exciting new products such as aircraft, automobiles, home appliances, and hand tools as well as the construction of new facilities such as refineries, steel mills, and food processing plants. It is a pursuit that challenges our analytical abilities and our knowledge of mathematics, the sciences, and manufacturing to find solutions that work better, last longer, and are easy to maintain or repair. Just as important, however, is that engineering design is fulfilling and personally satisfying in that we can use our creative abilities to develop ideas, that they become real, and that they meet the needs of our fellow human beings.

This text is for students at the undergraduate or first-year graduate level. It is for design and nondesign faculty who wish to learn more about modern design concepts and methods. It is for practicing engineers who wish to enhance their knowledge and skills. The book is *not* a comprehensive, intellectual report of engineering design research. Rather, the text *is* a practical, down-to-earth presentation of engineering design integrating topics about the business of manufacturing. It draws upon the author's engineering and managerial experiences at organizations such as General Electric, Fisher Price Toys, Wurlitzer, and the New York State Energy Research and Development Authority. Preliminary versions of the text have been successful in semester-long design courses that include team-based design projects.

This book brings together the best design methods and practices in a consistent, clear, and orderly presentation. Readers are introduced to concepts and methods developed by the best engineering design authors in the world including Dixon and Poli, Dieter, Otto and Wood, Pahl and Beitz, and Ullman. Using a "just-in-time" philosophy of learning, topics are presented in an orderly fashion, progressively building design methods and terminology. The topics are presented and used, rather than stored away for use at a later date. Exercises at the end of each chapter *use* the knowledge and methods presented. Many different examples are presented to highlight the similarities and differences of various design methods. An extensive summary of key terms is included in a glossary to delineate the important subtleties that are embodied in design and manufacturing terminology. In addition, self-quiz exercises are included at the end of each chapter for immediate reinforcement of the reader's understanding.

A number of other important features are included. Concurrent engineering and teamwork, particularly the role of communication and project management, is emphasized throughout the text. The interdependence of product function, materials, and manufacturing processes is examined with respect to the product realization process in separate chapters on materials and process selection. The voice of the customer is interwoven in chapters on design problem formulation and includes using customer satisfaction measures to evaluate design alternatives. The subject of human factors is presented along with tables and illustrations in such detail so as to provide the reader with the necessary tools for preparing preliminary designs that incorporate human abilities and limitations. Instructors will find a variety of materials on the accompanying Instructor's CD ROM, including PowerPoint® slides, solutions to exercises, syllabi suggestions, design project write-ups, and electronic files of key figures from the text. To see current errata or contact the author, please visit <http://coen.boisestate.edu/reggert/>.

I wish to thank all the practicing engineers, students, and faculty who provided suggestions for the development of this book. In particular, I wish to acknowledge the following colleagues for their helpful review and advice:

Stephen H. Carr—Northwestern University
Tom Grimm—Michigan Tech
Raghu Echempati—Kettering University
John Lamancusa—Penn State
D. C. Anderson—Purdue University
Holly Ault—Worcester Polytechnic Institute
Thomas R. Kurfess—Georgia Tech

Finally, I wish to thank my son Randy and wife Linda for their many hours of editorial assistance and for their unwavering support during the preparation of the manuscript.

CHAPTER 1

Getting the Big Picture

LEARNING OBJECTIVES

When you have completed this chapter you will be able to

- Characterize analysis problems and the process used to solve them
- Characterize design problems and the process used to solve them
- Explain how the form and function of a product are interrelated
- Characterize the five phases of design
- Describe how ideas are realized as products
- Explain the economic life cycle of a product
- Explain various engineering roles in manufacturing enterprises
- Explain how and why employees are organized into functional departments
- Depict the distinguishing features of concurrent engineering
- Relate how business teams can be as successful as professional sports teams

1.1 INTRODUCTION

Engineering design is exciting, challenging, satisfying, and rewarding. Engineering design is exciting because we compete in world markets by engineering products such as motorcycles, airplanes, spacecraft, artificial hearts, automated assembly equipment, machine tools, and home appliances. It is challenging because we have to find solutions that are better, faster, less expensive, lighter, and safer. It is personally satisfying in that we can use our creativity to synthesize new ideas and can use our knowledge and skills in mathematics, sciences, and manufacturing to predict how well our new designs will behave before they are built. Engineering design is rewarding because we can see how our hard work leads to realization of new products that satisfy the needs of our fellow human beings.

Engineering design is a part of a bigger picture, the **product realization process** (PRP), which includes other activities of a firm such as sales, marketing, industrial design, manufacturing, production planning, distribution, service, and ultimate disposal. These activities require us to work with many different people both within and without our company. Often, we are a member of a product development team. And, as that team develops new and revised products we make many decisions that commit the company to invest large sums of money for design, prototyping, testing, and manufacturing ramp-up. The stakes can be big insomuch as the financial and human resource investments may be substantial. The risks can be large insomuch as we may not have perfect knowledge of our markets or new technologies. And the competition may be fierce, always challenging us to develop better products in less time with fewer resources.

The main goal of this book is to help us become effective participants in the product realization process. We recognize that our main contributions to our firm are not the engineering sketches that we prepare or the computer programs we run or the engineering tests we conduct. These activities merely produce information. Rather, our primary contributions are the decisions that we make using that information. If we do our jobs well, our decisions result in successful products that satisfy our customers and produce profits to keep our company healthy.

Making decisions as a member of a team is different than making them by oneself. We need to become skilled at systematic methods to resolve group conflicts and improve our group-decision-making abilities. This book, in essence, will examine procedures and methods that can provide us with a framework for effective, logical decision making.

This chapter will first examine engineering analysis problems and differentiate them from design problems. Next, we will examine how engineering analysis is part of the overall design process. Then we will examine how engineering design is part of the manufacturing enterprise's product realization process. Finally, we will discuss product development in the context of concurrent engineering and team-based decision making.

1.2   WHAT IS ENGINEERING DESIGN?

Engineering design is the set of decision-making processes and activities used to determine the form of an object given the functions desired by the customer. Whether we are designing a component, product, system, or process, we gather and process significant amounts of information. For example, we have the task of deciding which customer needs are important, including necessary product functions and desirable product features. We try to determine desirable levels of performance and establish evaluation criteria with which we can compare the merits of alternative designs. We consider the technical, economic, safety, social, or regulatory constraints that may restrict our choices. We use our creative abilities to synthesize alternative designs incorporating varied shapes, configurations, sizes, materials of composition, or different manufacturing processes.

We utilize knowledge and methods from the basic sciences, mathematics, and engineering sciences to predict or simulate the performance of each alternative before it is built, thereby avoiding the time and expense of tinkering. We thereby

Form ever follows function

The function of a product is what the product is expected to perform. Basic functions include: control, hold, move, protect, and store. We use verbs to characterize functions. We use the term "behavior" to describe how the product actually performs.

Form, on the other hand, is what the product looks like, what materials it is made of, and how it is made. We use nouns to characterize the form of a product. Form characteristics include: shape, size, configuration, material, and manufacturing processes used to make the product.

Form is intimately related to function. As the famous architect Luis Sullivan once said, "Form ever follows function." The form of an object, in other words, usually depends upon the function it will perform. Consider the screwdriver for example. Its tip has a shape that drives the screw into the hole. The handle is configured to conform to the human hand, permitting the application of torque and thrust. The injection molding process used to make the handle permits cost-effective manufacture. Finally, the steel shank material safely transmits the torque and thrust from the handle to the tip.

The essence of design is to determine a form to satisfy the required function, as shown in Figure 1.1.

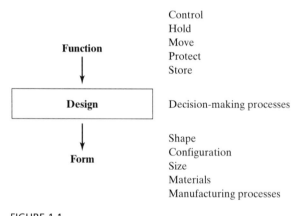

FIGURE 1.1

Design is the set of decision-making processes used to determine the form of an object given the functions desired by the customer.

consider how alternative "**forms**" influence each alternative's "**functions**," as shown in Figure 1.1. And we communicate our decisions as a set of detailed manufacturing specifications sufficient to fabricate and assemble the desired object.

We note that design can have a number of meanings. Sometimes we refer to an artifact or thing as a "design," as in "the design preformed well." At other times we refer to the decision-making activities or processes, as in "the team designed the new system in less than a year."

In the next sections we examine engineering analysis problems and the process used to solve them. We connect analysis to design by describing the character of design problems and discussing a simplified model of the design process. We also discuss an example shaft design problem to illustrate the five phases of design.

1.2.1 Engineering Analysis

Analysis problems have a number of characteristics in common. Consider the following "analysis" problems:

- Given an object of mass m, that has an applied force, F, acting on it, we determine its acceleration a, using Newton's second law of motion $a = F/m$.
- Given the cross-section geometry of an aluminum airplane wing, we determine the lift it produces by conducting wind tunnel experiments.
- Given the diameter and height of a welded, cylindrical tank, we calculate its volume and surface area using simple geometry relations.
- Given a cast steel engine block drawing, we estimate its weight.

In each of these examples we are given some information about an object and then asked to predict its "behavior." Having studied the basic sciences and mathematics, we are familiar with how things work and can model behavior as a function of input data. For example, we can model an object's acceleration (the predicted behavior) according to $a = F/m$ (an analytical equation), which, of course, depends upon the mass (form) and force (input data). Further, if we lack an equation, we might be able to conduct a number of engineering experiments to develop an empirical relation. Note too, that for analysis problems, the given information usually pertains to the form of an object such as shapes, configurations, sizes, material compositions, and or manufacturing processes. Therefore, when solving an *analysis* problem we *predict the behavior* or function of an object, based on its given form, using analytical equations or experimental methods.

> Predicted behavior is the solution to an analysis problem.

To help us better understand the differences between the analysis process and the design process, let's examine the analysis problem-solving process in detail. The analysis process can be further described as having three main stages: formulating, solving, and checking.

Formulating   As we formulate an analysis problem we are essentially trying to understand the problem and plan its solution. After gathering related information, we might draw a sketch or diagram using an approximate scale for lengths and angles. We would show pertinent geometry (shapes, dimensions, and angles) and illustrate the relevant physics, such as mechanical forces and moments. We would record all the "given" information and state what we are to "find." Next we would recall related problems that we had modeled in the past, listing appropriate principles and formulas. We might even break up the problem into smaller subproblems. We would then transform or interpret our problem into the model we knew how to solve, noting any and all assumptions between our "real" problem and the "ideal" model. At this point in the problem-solving process, we have a pretty good understanding of the problem and a set of modeling equations. We prepare a rough plan how to solve the equations whether by calculator, computer, or even analog experiments.

Solving   During this stage we logically try to determine the unknown(s) by solving an equation, or system of equations, using standard mathematical methods. We systematically set up the equations observing appropriate units and conversions. We verify that the number of unknowns is less than, or equal to, the number of equations. We might start by solving an equation that has more known quantities than unknowns, substituting our results into succeeding equations. Assuming that the equations are solvable, we find our solution and label it along with appropriate units.

Checking   Once we examine the validity, accuracy, and precision of our solution, we interpret our solution to determine whether it "makes sense." Does it violate laws of nature or mathematics? We evaluate the answer with regard to the number of significant digits of precision. We might also solve the problem using an alternative method such as a graphical technique or computer simulation. Typically, the independent solution methods give similar results, unless we did something wrong in our calculations or made a wrong assumption. In other words, there is usually just one answer to an analysis problem. We reread the initial problem to assure that we really answered the problem.

1.2.2   Engineering Design

Design problems, however, differ from analysis problems. Consider the following "design" problems:

- Given that the customer desires to manually lift a 500-pound engine block three feet above a given truck frame, determine a rope-and-pulley combination, including materials, sizes, and methods of construction.
- Given that the customer wishes to pump fuel oil from a 5,000-gallon storage tank to a boiler, select a satisfactory pump-and-motor combination.
- Given that the customer desires to store 350 gallons of gasoline, determine the shape and size of a storage tank.
- Given that the customer desires to toast slices of bread, develop a product that will perform this desired function.
- Given that the customer wishes to fasten together two steel plates, select appropriate sizes for the bolt, nut, and washer.

In the examples above, we are given information relating to the desired function and then asked to determine its form. For example, the customer desires to lift (function) an engine block, and determine a rope-and-pulley combination (form: pulley size and type, rope materials and size, configuration of pulley block, whether the pulley is welded or cast, and so on).

As in an analysis problem, we need to process input information. However, note that in a design problem the "given" information is somewhat fuzzy or ill-defined. For example, in the toaster design problem: how many slices of toast, how dark, what types of bread? In other words, in design problems we usually do "problem finding" before we do "problem solving." Design problems are also "open-ended." Since there are many possible design solutions that can provide the desired function, we see that design problems generally have more than one "solution."

Design problems involve consensus building and group decision making such as in determining customer needs and evaluation criteria. Unlike the routine simultaneous solution of a system of equations in an analysis problem, there are few, if any, structured procedures to follow that will guarantee a "solution." Design problems are therefore said to be ill-structured as well as open-ended.

Form is the solution to a design problem.

How does solving a design problem differ from solving an analysis problem? Let's look at a simplified model of the process used to solve design problems and compare it to the analysis process. We recognize that each business may use different decision-making methods or procedures that are tailored to its company or trade. The simplified model shown in Figure 1.2, however, incorporates four essential stages: formulating, generating, analyzing, and evaluating.

Formulating   As we formulate a design problem we try to understand the problem and plan its solution. We gather information about the customer such as necessary product functions, acceptable levels of performance, and desirable performance targets. We also try to determine relevant constraints regarding economic, technical, legal, and/or safety considerations. Most important, we record our findings as detailed engineering design specifications to help guide our future decisions. We develop a design project plan to coordinate "what" tasks will be completed by "whom" on our team and "when." In other words, **formulating** is the set of activities and decision-making processes used to understand a design problem and prepare a plan for its solution.

Generating   During this stage we synthesize, or generate alternative designs that might satisfy the customer. We might use creative methods such as brainstorming and Synectics to arrive at an initial concept design, then develop it into a layout or configuration, each alternative "form" having different shapes, configurations, sizes, materials, or made with different manufacturing processes.

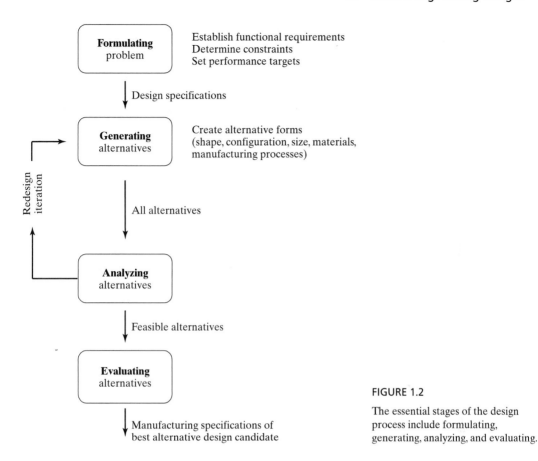

FIGURE 1.2

The essential stages of the design process include formulating, generating, analyzing, and evaluating.

Generating is the set of activities and decision-making processes used to create alternative design candidates for later analysis and evaluation.

Analyzing   Using knowledge from the basic sciences and computational skill from mathematics, we prepare engineering models to predict the performance of each design alternative. Briefly stated, **analyzing** is the process of predicting the performance or behavior of a design candidate. The candidate designs that fail to satisfy the constraints are reiterated. That is, new values of "form" are chosen and the redesigned candidate is reanalyzed. If no feasible candidates can be found, the problem may be overconstrained and may need to be respecified to relax some of the constraints. Redesign is shown as the solid reiteration loop in Figure 1.2.

Evaluating   **Evaluating** is the process of comparing the predicted performance of each feasible design candidate to determine the "best" design alternative. The merit of each design candidate is estimated using evaluation criteria laid out in the engineering design specification. Typical evaluation criteria include performance measures such as speed, size, reliability, maintenance intervals, power, weight, and cost. Also, optimal design methods, employing computerized numerical techniques, can be used in some cases to automatically regenerate new candidate designs to improve expected performance and overall quality.

Perhaps the most significant aspect about engineering design, compared to engineering analysis, is that engineering design includes analysis! All engineering design activities relate to developing the best product to satisfy customer needs. Design is the reason why we do what we do. Analysis is just one part of the bigger picture. In particular, the reason why a manufacturing firm pays an engineer to do analysis is because it contributes to the quality of the product and ultimately to the satisfaction of the customer, who really pays his salary.

When solving engineering design problems we often have to deal with *multiple evaluation criteria*, or multiple figures of merit. For example, our customer would like a stronger, heavy-duty transmission for his application. But a stronger transmission is heavier and more expensive. Which is more important, cost or strength? How should we decide? Like most design problems, this example exhibits *trade-offs*, or compromises, wherein one performance attribute improves while the other degrades.

In engineering design, we have *logical decisions* in addition to numerical analyses. Since a product development team will make many of the decisions as a team, various group decision-making methods will replace equation solving, as in analysis.

1.2.3 Design Phases

A product design evolves over time in **design phases** from the identification of a customer need to the realization of the finished product. Let's examine the following situation:

> Our customer owns and operates a custom-built machine tool used in her factory. The tool has a drive shaft that cannot be stopped other than by turning off the power and waiting about 90 seconds for the shaft to spin-down. The customer considers this a safety hazard and would like to actuate some device that would bring the shaft to a quicker stop. The shaft is 8 inches in diameter, weighs 1,000 pounds, is made of 4330 steel, and spins at 3,000 rpm. We have been hired to design and fabricate a solution to accomplish the rapid stopping of the shaft.

1. Initially, we explore the stated customer need and develop a list of objective design specifications. For example, we obtain details of the existing shaft geometry and operating conditions in the factory, including available power. Additionally, we try to clarify the type of performance that would satisfy the customer. For example, how quickly should the shaft stop? Would the customer be fully satisfied with 5 seconds and unsatisfied if longer than 20 seconds? We might do some simple calculations, using simple relations from physics, to better understand the overall mechanical forces and/or torques of the system. Our activities in this initial phase focus largely on problem *formulation*. In other words, we are not trying to solve the problem, only understand it.

2. During the **concept design** phase we synthesize a variety of candidate working principles or concepts and their abstract embodiments. For example, we might consider the following three alternative concepts:

  a. air friction as the working principle and fan blades as the embodiment,
  b. opposing magnetic fields as the working principle and an electric generator as the embodiment, and
  c. surface friction as the working principle and a disk-and-caliper brake as the embodiment.

Then, after developing a list of evaluation criteria, we select one of the concepts for further development.

3. Then, we generate a variety of *configurations* including the arrangement of individual components. For example, if during concept design we select a disk-and-caliper brake, during configuration design we consider alternative shaft locations, and geometries for the disk and caliper to select the best configuration. **Configuration design**, therefore, is the design phase when alternative configurations are generated, analyzed, and evaluated.

4. During the **parametric design** phase, we determine values for the controllable parameters, called design variables, identified as unknown during the configuration phase. Design variables deal with the form of a design, that is, its shape, configuration, size, material, and manufacturing process. For example, we determine specific values for: rotor diameter (outer), rotor thickness, and brake pad width. We also select pad material type, operating pressure of the hydraulic piston, and rotor manufacturing processes. Then, we analyze the performance using conventional machine design formulas, computer programs, or physical experiments. We check the analysis results to assure that all the constraints are satisfied and that an optimal performance is obtained. If not, we *iterate*, or **redesign** by generating new candidate designs with new values for the design variables. Then we reanalyze, evaluate, and so on.

5. During the **detail design** phase, we determine the remaining product specifications such as the surface finish, pad bonding resin, and assembly procedures. We might also fabricate and test a number of critical parts. We also prepare a complete package of *manufacturing specifications* including detail and assembly drawings, bill of materials, manufacturing process recommendations, prototype performance test results, and product specifications such as height, width, depth, weight, expected performance.

A number of terms are used in industry and academia to describe various design phases. For example, *preliminary design* is often referred to the *collection of activities* relating to concept design, configuration design, and parametric design. Also, *embodiment design* refers to those design activities relating to configuration design and parametric design activities.

Engineering design researchers have proposed similar models of the design phases. Pahl and Beitz (1996) proposed four phases including: (1) task clarification, (2) concept design, (3) embodiment design, and (4) detail design. Dixon and Poli (1995) presented a model that splits embodiment design into configuration and parametric design phases. Dieter (2000) included product architecture as an additional phase before configuration design.

In this text, we will closely examine a five-phase model as proposed in Figure 1.3. The five phases include: (1) formulation, (2) concept design, (3) configuration design, (4) parametric design, and (5) detail design. The only difference between the proposed model and Dixon and Poli's model is that formulation is explicitly emphasized as a crucial phase of design.

Engineering design and product development, in general, rarely proceed through all phases of design in a systematic, linear fashion as described in Figure 1.3. For example, revisions to products may need only parametric and detail design work, while a new technological breakthrough may require configuration, parametric, and detail design efforts. In another case, a candidate concept design may finally reveal itself during configuration design to be unmanufacturable, thereby requiring additional concept design.

Product development teams that generally follow the five phases of design benefit in three major ways:

1. *The teams focus on the right subproblem at the right time.* For example, team members will not be trying to establish machining tolerances during concept design, when the part dimensions are not even known. Or, in another case, alternative configuration design candidates will not be evaluated before the evaluation criteria are established during formulation.

2. *The teams efficiently gather and process the right type of information.* In formulation, for example, customer needs are typically abstract and ill-defined, as compared to parametric design when more specific information is necessary.

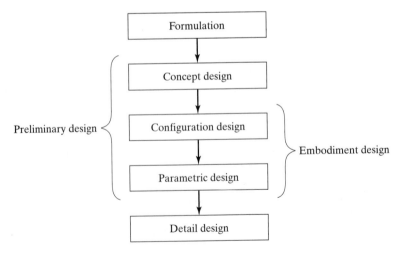

FIGURE 1.3

Five phases of design, emphasizing the crucial nature of design problem formulation.

3. *The teams focus on alternative methods pertinent to the phase.* For example, a variety of creative methods can be explored during concept design. In another case, we might investigate alternative CAD software packages during parametric design.

Successful design teams recognize the value of doing tasks appropriate to the design phase.

1.3   HOW DOES ENGINEERING DESIGN FIT INTO THE PRODUCT REALIZATION PROCESS?

In the previous sections we examined how engineering design activities translate customer needs into product manufacturing specifications. But, how are manufacturing specifications turned into real products? Are we finished with our work once we "turn over" the manufacturing specifications? Are the manufacturing specifications ever really finished?

We recognize that most parts and components are mass-produced in today's modern design and manufacturing environment. As a consequence, the parts must be made to be interchangeable with each other. We accomplish interchangeability and mass production using intricate systems composed of people, procedures, and machinery. Even when only a few products are "manufactured," as in the production of a nuclear-powered aircraft carrier, we use similar systems involving many people, systematic procedures, and expensive machinery.

In the next sections we will consider how engineering design is an important part of the product realization process and how a product contributes to the life of the business enterprise.

1.3.1   Product Realization Process: The Big Picture

The product realization process is the means by which a customer need is transformed into a realized product. Alternatively, Dixon and Poli define the product realization process as a complex set of interrelated activities, both cognitive and physical, involving the whole firm, by which new and modified products are conceived, produced, brought to market, serviced, and disposed of. The process involves physical activities and decision-making activities (cognitive), including sales, marketing, industrial design, engineering design, production design, manufacturing, distribution, service, and disposal, as shown in Figure 1.4.

A customer need for a new or improved product can originate from almost anywhere in the firm. Often the service department is the first to become aware of an existing product's shortcomings. In some cases we might discover inadequate product packaging during distribution shipments to retail outlets. But the majority of new or revised product ideas usually originate from the *sales* or *marketing* groups.

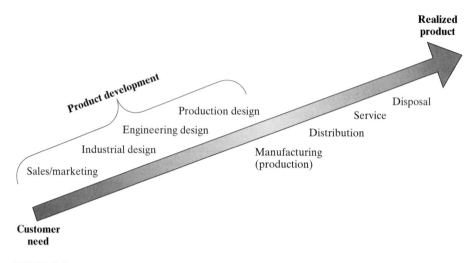

FIGURE 1.4

The product realization process is the means by which a customer need is transformed into a realized product (adapted from Dixon and Poli).

Industrial design activities focus on how the new or revised product idea is compatible with the customer's anatomical limitations and/or aesthetic trends in the marketplace. Often the industrial design group will prepare an artistic rendering or a physical model that illustrates basic product form, color, texture, and intended functionality.

Engineering design activities result in recommended manufacturing specifications that satisfy the customer's functional performance requirements and manufacturing constraints.

Production design activities involve the design, fabrication, and installation of production equipment such as jigs, fixtures, machine tools, quality control instrumentation, and material-handling equipment. In some cases, it might involve the construction of a new factory. Production design also considers manual and automated assembly equipment. Even as the first units come off the production lines, the production design group may be making changes to reduce material waste and rework.

Manufacturing activities relate to fabrication, assembly, and testing. They also include training, scheduling, and supervising production employees. Significant coordination between engineering design, production planning, and manufacturing is necessary during ramp-up as bugs in the product design and manufacturing processes are worked out. Others sometimes describe "Manufacturing" as *all* the activities of a manufacturing enterprise, often referred to as the big "M." Note that in this text manufacturing will be denoted as fabrication, assembly, and testing, usually referred to as the little "m" manufacturing.

Distribution activities involve shipping the product in wholesale-sized lots to distribution centers located around the country or world. In some cases, railcar-full shipments are packed. Most often, containerized freight trailers are used.

Some companies ship directly to retailers or the customer, as in personal-computer mail ordering.

Service activities for consumer products usually relate to repair or replacement at the factory. However, large appliance manufacturers will train repair persons for home service such as washing machine or dryer repair. For some industrial products such as commercial refrigeration systems, service persons actually do the equipment installation, as well as routine maintenance, at the customer's site.

Disposal activities involve the removal, elimination, and/or recycling of hazardous chemicals or scarce materials such as in nuclear power plant fuel rods, automotive oil, or printer cartridge recycling.

Indirectly, the product realization process involves administrative activities such as accounting, finance, personnel, strategic planning, legal, and management. These behind-the-scenes activities help to coordinate the whole organization and their costs are often referred to as *indirect expenses.*

We refer to **product development** as the collection of activities leading up to, but not including, production. Therefore, we can see that product development is more than engineering design. It also involves postdesign activities such as production planning and the coordination of activities relating to ramping-up of production. Even after the product is launched into the market, engineering design may be involved in making minor improvements for improved performance, safety, or cost reasons.

We can aggregate some of the decisions made by the various departments and explicitly incorporate how the customer will "use" the product. This results in a simplified model of the product realization process having four main stages that represent the whole life of a product: design, manufacture, use, and retire, as shown in Figure 1.5. As we look over the four stages let's take special note of what engineering fields are involved.

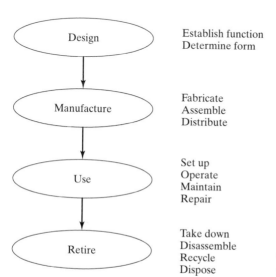

Design — Establish function / Determine form

Manufacture — Fabricate / Assemble / Distribute

Use — Set up / Operate / Maintain / Repair

Retire — Take down / Disassemble / Recycle / Dispose

FIGURE 1.5

The four stages in the life of a product include design, manufacture, use, and retire.

Early in the life of a product, sales engineers, applications engineers, field service engineers, and others identify customer needs. This information leads to new product ideas, which are subsequently developed into mature product designs by industrial designers, design engineers, materials engineers, and test engineers as they determine the form of the new product.

Then, industrial, manufacturing, and quality engineers convert the design into a final product for manufacture. During this phase they plan, organize, and fabricate new tooling and fixtures. They also construct facilities and equipment to assemble and distribute the product.

Upon delivery, the consumer receives and sets up the product. This typically involves unpacking, cleaning, and in some cases assembling components. During and after operation, the product is *maintained* by cleaning, lubricating, and adjusting. If necessary, the product is repaired by replacing worn or broken parts. Field service engineers are sometimes responsible for installation, maintenance, and repair activities for large or complex industrial and commercial products.

Once the product is retired, it is removed from service, disconnected, and relocated. Large assemblies are taken down and disassembled into smaller, more manageable units. Materials are recycled when possible and nonrecyclable parts are disposed of.

Thus, we can conclude that engineering design is just one part of the big picture, called the product realization process, that involves many departments and employees during the life of a product.

1.3.2 Economic Life Cycle of a Product

A business will typically devote significant financial and manpower resources to the development of a new product, as shown in Figure 1.6. Unlike annual operating expenses, product development expenditures are usually considered long-term investments. Furthermore, the business will expect to receive returns on that investment over the economic life of the product.

Initially, as the product is introduced to the market, its sales begin to grow slowly (Crawford and DiBenedetto, 2003). As more and more customers learn about the product, sales begin to increase during a *growth phase*. Then, at *maturity*, sales income levels off. Once the market is *saturated*, few new sales occur and sales income *declines*.

The time over which these phases occur can be years or months. The basic hammer, for example, has been around more than a hundred years. In some cases, technological improvements make a product obsolete, as in the case of the beta-format video recorder. Products live and die because conditions change, customer preferences change, regulations change, and material costs and availability change.

As a product declines, the business will have a new product to take its place, as shown by the dashed line in Figure 1.6. The "birth" and ultimate "death" of a product is often called the **product life cycle**. Thus, as one product approaches maturity a new product emerges from product development. Successful companies usually have many new product ideas in various stages of development, such that their annual volumes continue to increase each year.

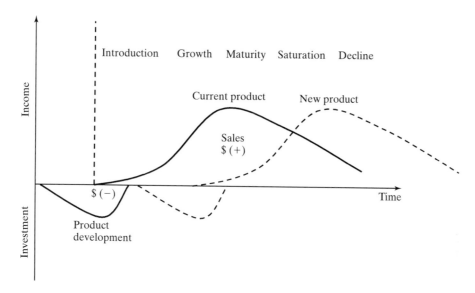

FIGURE 1.6

Economic life cycle of a product. As the current product declines, a new product is introduced and thereby keeps the company healthy.

1.4 THE MANUFACTURING ENTERPRISE

As we learned from the product realization process and the economic life of a product, significant resources both human and financial are involved. A couple of questions naturally come to mind, such as, who is entrusted with the resources of the enterprise to make the right decisions? Or, how does a business enterprise coordinate the decisions made by so many employees?

In the next sections we will examine the various roles that engineers play in the product realization process, how a manufacturing enterprise is organized into an effective structure, and how modern product development practice makes use of the concurrent engineering approach.

1.4.1 Engineering Roles

Manufacturing employment directly accounts for about one half of the engineering jobs in the United States. The Bureau of Labor Statistics reported that a total of 1,465,000 engineers were employed in 2000 (U.S. Department of Labor, 2000) and that approximately 50 percent of all engineers worked in manufacturing companies, 27 percent worked in engineering services, 12 percent worked for government, and 3 percent were self-employed. Approximately 8 percent were classified as "other."

Even when engineers are not working directly for a manufacturing business, they often interact with manufacturing businesses, providing engineering services or purchasing their goods. For example, engineering-service firms often design systems and facilities for manufacturing companies. Also, engineers in government positions will specify and purchase equipment made by manufacturing companies, such as tanks, airplanes, and ships.

Recognizing that we will be working in or with a manufacturing business, we might now ask ourselves about the roles and responsibilities that we might have.

Engineers are deployed throughout the product realization process, making full use of their technical, mathematical, and/or problem-solving abilities, as shown in Table 1.1. Note that the titles and responsibilities are representative of those used in industry and do not represent an exhaustive list.

In the sales and marketing department we find *sales engineers* meeting with customers, determining needs, presenting product offerings, preparing advertising campaigns, negotiating terms of sale, and closing the sale. General Electric, for example, may have two or three sales engineers assigned to the "sale" of a $100 million gas turbine power plant. *Applications engineers* assist sales and marketing personnel by solving technical issues with respect to the use of product. An applications engineer assigned to a turbine power plant sale, for example, would investigate its operating costs and benefits, and the integration of the plant with the customer's site. *Field service engineers* are usually responsible for installing, maintaining, and repairing equipment, often at customer sites.

In the research and development department we often find industrial designers, design engineers, materials engineers, and test engineers. *Industrial*

TABLE 1.1   Engineering Roles in the Product Realization Process

Department/function	Job title	Responsibilities
Sales and marketing	Sales engineer	Meets customers, determines needs, presents product offerings, negotiates terms of sale
	Applications engineer	Assists sales and marketing solving technical issues with respect to the use of product
	Field service engineer	Installs, maintains, and repairs equipment at customers' sites
Research and development	Industrial designer	Establishes essential product appearance, human factors
	Design engineer	Decides part or product form, including: shape, size, configuration, materials, and manufacturing processes
	Materials engineer	Investigates and develops improved materials
	Test engineer	Designs and conducts performance and safety tests
Manufacturing	Industrial engineer	Designs fabrication, assembly, and warehousing systems
	Manufacturing engineer	Develops manufacturing tools and fixtures
	Quality control engineer	Establishes and maintains raw materials and finished goods quality controls
Processing/operations	Plant engineer	Designs and maintains processing plant facilities
Miscellaneous	Project engineer	Coordinates project work tasks, budgets, and schedules

designers establish the essential product appearance as it relates to human factors and aesthetic concerns. *Design engineers* translate the function desired by the customer into part or product form including: shape, size, configuration, materials, and manufacturing processes. *Materials engineers* investigate and develops improved materials. *Test engineers* are responsible for designing and conducting performance and safety tests at various stages in the product development.

In the manufacturing department we often employ industrial engineers, manufacturing engineers, and quality control engineers. *Industrial engineers* are usually responsible for designing fabrication, assembly, and warehousing systems. They try to optimize the system of workers, machines, and materials handling. *Manufacturing engineers* develop manufacturing tools and fixtures used to fabricate and assemble parts. *Quality control engineers* establish and maintain instrumentation and documentation systems to control the quality of raw materials and finished goods.

In processing companies we find plant engineers and project engineers. *Plant engineers* design and maintain processing plant facilities. For example, a chemical plant producing thermoplastic polymers may have $500 million invested in pumps, tanks, chemical reactors, boilers, and piping. In addition to making improvements from time to time, the plant engineers are responsible for keeping the plant operating 24 hours a day, seven days a week.

Project engineers coordinate project work tasks, budgets, and schedules. Projects are commissioned to complete a variety of activities, including new equipment installations, new product development, the design of a new factory, and replacement of automated assembly equipment.

1.4.2 Organization

Most manufacturing enterprises are formed as a corporation owned by stockholders. Stockholders provide most of the funds that a business uses to purchase land, buildings, equipment, and other assets for its operations. Stockholders elect the board of directors to oversee the activities of the president and other officers of the company. The board of directors provides advice to the president and officers regarding broad policies and long-term strategies, while the president, as chief executive officer and as principal trustee for the stockholders, has the responsibility and authority to make day-to-day decisions committing financial and human resources.

An example *organization chart* is shown in Figure 1.7. In this example, the president's chain of command includes five subordinate managers responsible for specific functions of the business. These positions often carry the title of vice president or manager.

Businesses organize their employees according to their specialized functions called departments, such as: sales, marketing, finance, purchasing, manufacturing, and research and development. Employees in the same department perform similar types of work. Consequently, the department manager can be most effective in hiring, training, developing, and supervising his or her employees. A company's **organization structure** establishes areas of responsibility and

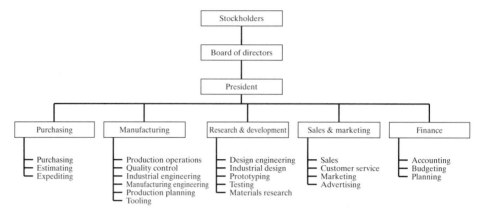

FIGURE 1.7

Functional organization chart of a typical manufacturing company.

authority facilitating the management of employees and coordination of its operations (Badiro and Pulat, 1995).

Businesses also *process* raw materials or agricultural products such as iron ore to steel, steel ingots to sheet metal, petroleum to plastics, potatoes to French fries, or cattle to beef products. These businesses also organize their operations into similar *specialized functions* that facilitate the nature of their business, as shown in Figure 1.7.

Since different businesses may have more or less departments, employees, and/or managers, we find many variations of the basic organization presented in Figures 1.7 and 1.8. However, regardless of the size of the business, the basic functions have to be performed. In smaller companies this is often accomplished by having one department or manager responsible for many functions. A typical example would be the combining of manufacturing activities with research and development activities. Another example would be the grouping of purchasing with finance.

1.5 CONCURRENT ENGINEERING

With so many people in the typical enterprise, with so many responsibilities associated with the product realization process...how is product development conducted? Product development is usually assigned to a team of employees representing the various functions or departments of the enterprise.

Modern product development practice takes advantage of increased communication, coordination, and employee motivation by using the concurrent engineering approach (Wesner et al., 1995). **Concurrent engineering** is a team approach to product design, in which team members, representing critical business functions, work together in the same office and are coordinated by one senior manager. The aim of concurrent engineering, also called simultaneous engineering, is to achieve superior product designs.

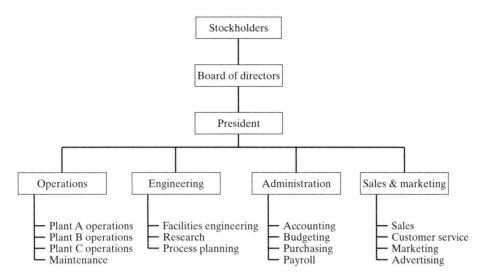

FIGURE 1.8

Functional organization chart of a typical processing company.

Prior to concurrent engineering, customer requirements information was handed over to the design group. Product design specifications were prepared and then handed over to the manufacturing group for production. Little interaction occurred during product development, leading to many miscommunications and production delays. This approach is referred to as "over the wall" product development.

An example of a concurrent engineering product development team is shown in Figure 1.9. A concurrent engineering team is composed of members representing the primary business functions. We call this cross-pollinating mix of team members a *cross-functional* team composition.

Concurrent engineering teams are also *colocated* (Carter and Stilwell-Baker, 1992). Placing team members next to each other in the same office area increases

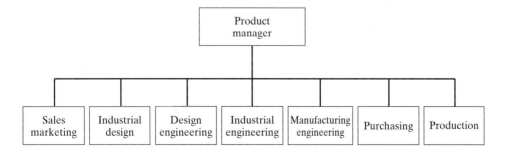

FIGURE 1.9

Concurrent engineering team. A collection of cross-functional team members, colocated in the same office area and coordinated by a high-level manager.

the quantity and quality of communication. Members also tend to develop a sense of teamwork, loyalty, and camaraderie.

Concurrent engineering teams are *coordinated*. A product development manager, at a high level in the organization structure, has sufficient authority and responsibility to make timely decisions and commit resources effectively.

1.6   PRODUCT REALIZATION: A PROFESSIONAL TEAM SPORT

Many people, including employees, consultants, customers, and vendors, contribute to the realization of a new or revised product. We appreciate that we cannot design, manufacture, and distribute mass-produced products by ourselves. We must work with the team of employees who make up the company. We also appreciate that teamwork is essential for successful product realization.

Companies involved in product realization are similar to teams in professional sports. Let's examine the characteristics of professional teams, such as professional football teams. Then, we can extend the analogy to suggest ways in which we might improve the teamwork within our own company. It is worth noting that companies often hire distinguished coaches to present in-plant seminars on improving teamwork.

Competitors, Companies, and Customers. Professional teams compete against one another with the primary goal of winning games for the pleasure of their paying customers. Companies compete against one another attempting to satisfy customer needs with the goal of earning profits for their investors.

Owners and Investors. Professional teams and businesses are owned by individuals or corporations that provide the initial funds to set up and operate the franchise or business. Owners and investors expect to receive some returns on their investments in the form of dividends or increases in equity.

Professional Players, Employees, and Teams. In both businesses and professional team sports, the participants are paid professionals. Employees and players are members of the big leagues and earn professional salaries and bonuses. As they join the serious ranks of a professional team, professional players leave behind the jovial, friendly game of "touch football." Similarly, as a member of a department, project, or product development team, we have serious roles and responsibilities to perform. The competition is exciting. The job is challenging. And, in many cases, it is personally satisfying and financially rewarding.

Coaches and Managers. Professional team coaches select, guide, and direct their players. Managers hire, train, and supervise employees in the performance of their duties. Coaches and managers also provide strategic advice, expertise, or direction to guide their teams on the competitive playing field.

Referees, Umpires, Lawyers, and Judges. Professional team sports have rules that are enforced by referees or umpires. Manufacturing enterprises are regulated by codes, standards, and other local, state, and federal laws that are enforced by lawyers and judges.

Communication and Coordination. Coaches and players communicate during the game. Signals are spoken clearly and loudly so that every player knows the specific actions he or she is to take. As we have often seen on television, games can be lost by players who "miss the call." Communication is similarly essential in business so that everyone in the product realization process knows what to do, when to do it, and how to do it. Therefore, in addition to being skillful in the modes of communication, we must also know the jargon used in product realization.

Fumbles, Injuries, Risk, and Uncertainty. Competition and product realization is filled with uncertainty, such as weather changing the playing conditions or new regulatory laws changing competitive conditions. A player can drop or fumble the ball, as well as get injured on the field. Employees can similarly make mistakes or get injured on the job. A certain amount of risk is inevitable in both cases. Imperfect knowledge and risk are just part of business and sports and must be accommodated. Decisions need to be made and actions need to be taken. Note, however, that successful teams plan for and prepare to overcome these difficulties.

Individual Skills. Winning players are *good at the position they play*. Initially, they prepare themselves in physical training such as weight lifting, jogging, sprinting, and other physical exercises, to develop individual abilities of speed and strength. Then they learn and memorize all the "plays" that the team may execute during a game. Then during practices, they hone and coordinate their physical abilities with knowledge of the plays, to improve execution and timing. Similarly, successful employees recognize the importance of individual contributions to their department, project, or product development team. Initially they prepare themselves by developing technical knowledge and skills in the basic sciences, mathematics, engineering sciences, manufacturing processes, and design. They gain experience in design methods, computational tools, prototyping, and testing. They develop communication skills for effective listening, speaking, reading, interpreting, writing, sketching, and drawing. They learn and use "the jargon." They develop a professional interpersonal "style" embracing empathy, tolerance, honesty, trust, and personal integrity.

Team Skills. Winning teams have outstanding players and coaches. They routinely practice offensive and defensive plays, so that they can flawlessly execute them during games. They also have effective, alternative, and adaptive game strategies. If our company is to be successful, we need to have outstanding employees and managers. We need to know and fulfill our roles and responsibilities well. We need to adopt sound alternative business strategies. We need to learn and employ team-based decision-making methods and practices. We need to be able to resolve conflicts equitably and efficiently. We must be able to communicate well to establish clear and effective coordination.

Whether we participate as a sales engineer, applications engineer, industrial designer, design engineer, manufacturing engineer, or field service engineer, our knowledge of the product realization process and our abilities in making decisions will make us valuable team members and effective employees.

1.7 SUMMARY

- Engineering design is exciting, challenging, satisfying, and rewarding.
- The solution to an engineering analysis problem includes predicting the behavior of an object.
- Engineering analysis activities include formulating, solving, and checking.
- Engineering design is the set of decision-making processes and activities used to determine the form of an object given the functions desired by the customer.
- Form includes shape, configuration, size, materials, and manufacturing processes.
- The stages of the design process include formulation, generation, analysis, and evaluation.
- The five phases of design are problem formulation, concept design, configuration design, parametric design, and detail design.
- Designing requires sound decision-making skills.
- The product realization process is the means by which a customer's needs are transformed into a realized product.
- Businesses are organized along functional lines, facilitating the coordination of operations.
- Engineers perform various functional roles in a manufacturing enterprise's product realization process.
- Concurrent engineering teams include employees who are colocated, cross-functional, and coordinated.
- Department, project, or product development teams may improve their effectiveness by studying the characteristics of successful professional sports teams.

REFERENCES

Badiro, A. B., and P. S. Pulat. 1995. *Comprehensive Project Management*. Englewood Cliffs, NJ: Prentice Hall.

Carter, D. E., and B. Stilwell-Baker. 1992. *Concurrent Engineering: The Product Environment for the 1990's*. Reading, MA: Addison-Wesley.

Crawford, C. M., and C. A. DiBenedetto. 2003. *New Products Management*, 7th ed. New York: McGraw-Hill.

Dieter, G. E. 2000. *Engineering Design*, 3d ed. New York: McGraw-Hill.

Dixon, J. R., and C. Poli. 1995. *Engineering Design and Design for Manufacturing*. Conway, MA: Field Stone Publishers.

Pahl, G., and W. Beitz. 1996. *Engineering Design*, 2d ed. New York: Springer-Verlag.

U.S. Department of Labor Web Site: http://stats.bls.gov/.

Ullman, D. G. 1997. *The Mechanical Design Process*, 2d ed. New York: McGraw-Hill.

Wesner, J. W., J. M. Hiatt, and D. C. Trimble. 1995. *Winning with Quality*. Reading, MA: Addison-Wesley.

KEY TERMS

Analyzing	Engineering design	Organization structure
Concept design	Evaluating	Parametric design
Concurrent engineering	Form	Product development
Configuration design	Formulating	Product life cycle
Design phase	Function	Product realization process
Detail design	Generating	Redesign

EXERCISES

Self-Test. Write the letter of the choice that best answers the question.

1. _____ That which a product should do or perform is called:
    a. form
    b. function
    c. configuration
    d. process

2. _____ What a product looks like, what materials it is made of, and how it is made is called:
    a. process
    b. configuration
    c. form
    d. function

3. _____ Which of the following is not one of the five phases of design:
    a. configuration design
    b. preliminary design
    c. a concept design
    d. formulation
    e. detailed design

4. _____ Most sales income occur in which part of the product life cycle:
    a. introduction
    b. growth
    c. maturity
    d. saturation

5. _____ The stages in the life of a product include all the following except:
    a. design
    b. manufacture
    c. integrate
    d. retire

6. _____ The president of a corporation oversees the activities of:
   a. board of directors
   b. stockholders
   c. R&D
   d. all of the above

7. _____ The field service engineer of a company:
   a. investigates and develops approved materials
   b. coordinates project work tasks
   c. establishes product quality controls
   d. installs and maintains equipment at customers' sites

8. _____ Developing manufacturing processes is the responsibility of the:
   a. mechanical engineer
   b. manufacturing engineer
   c. quality control engineer
   d. applications engineer

9. _____ The job title of the person who designs fabrication assembly and warehousing systems is:
   a. design engineer
   b. industrial engineer
   c. project engineer
   d. applications engineer

10. _____ The team-based approach to product development using engineers from different business functions is called:
    a. concurrent engineering
    b. production engineering
    c. spontaneous engineering
    d. design engineering

2: c, 4: c, 6: c, 8: b, 10: a

11. Describe the four stages of the design process.
12. Differentiate engineering analysis from design.
13. What are the five principal categories of form?
14. Describe how design differs from analysis.
15. What types of decisions are made during conceptual design?
16. What types of decisions are made during configuration design?
17. What types of decisions are made during parametric design?
18. What is the difference between preliminary design and embodiment design?
19. Why is tinkering not engineering?
20. List five different functions (hint: verb) and give a different example form (hint: noun) for each function (e.g., function-fasten, form-staple).

21. Describe how company funds flow in or out during a product's life cycle.

22. How do companies avoid going out of business as their product's sales revenues decline?

23. List the four phases in the life of a product and the activities that occur in each phase.

24. What specific aspects would an outboard motor design engineer consider with regard to its "use" phase?

25. What specific aspects would a nuclear power plant design engineer consider with regard to its "retire" phase?

26. Connect to the Bureau of Labor Statistics site at http://stats.bls.gov/oco/ocos033.htm. How does the Bureau describe the work of a mechanical engineer and what is the average starting salary of an ME?

27. Briefly describe the responsibilities performed by the following engineers:
    a. sales engineer
    b. applications engineer
    c. field service engineer
    d. industrial designer
    e. design engineer
    f. materials engineer
    g. test engineer
    h. industrial engineer
    i. manufacturing engineer
    j. quality control engineer
    k. plant engineer
    l. project engineer

28. What job title appeals to you and why?

29. Why do businesses organize their employees along functional lines?

30. List the five principal business functions of a manufacturing company.

31. List the four principal business functions of a processing company.

32. What are the roles of the stockholders, board of directors, and president?

33. List the types of team members on a concurrent engineering team?

34. Why should concurrent engineering teams be
    a. cross-functional?
    b. colocated?
    c. coordinated by a high-level manager?

35. What is "over-the-wall" product development?

36. Connect to the Web site of a company you would like to work for. Do they have any job listings for engineers that appeal to you?

CHAPTER 2

Defining and Solving Design Problems

When you have completed this chapter you will be able to

- Characterize different types of design problems
- Describe product and process plant components
- Decompose and diagram a product's components
- Characterize types of design
- Select and apply design problem solution strategies

2.1 INTRODUCTION

What characteristics distinguish one design problem from another? Let's consider the following design problem examples.

Longer-life lightbulbs. Our R&D department indicates that a newly developed material will extend the life of the current product by 20 percent. Should we modify our existing product line?

Safer toaster. Our customer service department reports that many customers have complained about toaster oven, model #453, blowing circuit breakers in their home wiring. Is a manufacturing defect the cause or is a new design required to fix the problem? Is it ethical to continue shipping the toaster ovens without fixing the problem?

Lower-emissions lawn mower. Due to new Environmental Protection Agency regulations, we must select a new engine for our current product line of power lawn mowers. What factors will we consider?

Lightweight canoe paddle. Marketing studies indicate that our current line of solid wood paddles is too heavy for the growing market of women paddlers. Should we invest in the development of a new paddle using lighter materials?

Special-duty robot welder. The U.S. Navy has contacted our company to design and manufacture a new remotely controlled underwater welding robot. If only six units are ordered, what manufacturing processes would be economical?

Fewer broken potato chips. We have been asked to redesign our company's packaging equipment to reduce the number of broken potato chips per bag. Can a modification to the existing equipment fix the problem?

Lower-cost car seat. Our customer, a leading automobile manufacturer, has contacted our company to lower the total cost of a line of passenger seats that we make for them. What compromises will our customer be willing to make? Cost for comfort? Cost for safety?

We readily see that some design problems involve *improvements* to existing products, such as the lightbulb and car seat. Others require the development of a new product, something that never existed before, such as the robot welder. A *design problem*, therefore, can be defined as a product deficiency that needs resolution, or a product opportunity that needs consideration.

Some design problems involve relatively simple, *one-piece products*, shaped from a single material such as a toothpick or baseball bat. Others, such as an automobile and a commercial jet airplane, on the other hand, are examples of very *complex products* that include thousands of components made with different materials or manufacturing processes.

Some design problems deal with knowing the difference between the *customer* and the *consumer*. For example, redesigned car seats are bought by the customer and then sold to the consumers who actually use them. Will satisfying the customer's desire for a lower cost be acceptable to the consumers if it means a trade-off in comfort or safety?

Technology readiness is also used to characterize design problem types (Ulrich and Eppinger, 1995). Although R&D may have developed a new light bulb filament material in the laboratory, is it manufacturable in the factory? Will the new technology be cost-effective? Will it work reliably?

Finally, design problems can also involve different *quantities of production*, such as the *mass-produced* toothpick or the *one-of-a-kind* potato-chip packaging machine.

In the remainder of this chapter, we describe the basic anatomy of products and process plants to establish a nomenclature of component terminology. We then use the new terms to examine strategies for solving design problems.

2.2   PRODUCT AND PROCESS PLANT ANATOMY

Products are made of one or more fundamental components arranged in structured assemblies. Process plants, factories, and facilities are also designed to function with multiple components. Products and process plants have structures, like the parts and systems of the human body that work together. Much like medical students, we can benefit from the study of the "anatomy" of products and process plants. In sections 2.2.1 and 2.2.2 we define a number of important terms relating to the anatomy of products and process plants.

2.2.1   Product Anatomy

A **product** is an item that is purchased and used as a unit (Dixon and Poli, 1995). Table 2.1 lists a small portion of the many thousands of products manufactured each year. Note that each item listed can be purchased and used as is. In other words, no further fabrication or assembly is typically required.

Some products are simple and others are complex depending on the *number, type, and function of their components*. For example, the paper clip, canoe paddle, and toothpick are single-component products. The penlight, bicycle, and toaster oven are somewhat more complex. The refrigerator, automobile, and commercial jet airplane are very complex.

Products are composed of components that include parts and assemblies. A **part** is a single piece requiring no assembly, sometimes called **piece-part**. An **assembly** is a collection of two or more parts or subassemblies. A **subassembly** is an assembly that is incorporated into another assembly or subassembly.

A **standard part** is a common interchangeable item, having standard features, typically mass-produced, used in various applications. Examples include: screw, bolt, nut, washer, v-belt, rivet, key, gasket, shear pins, and lubricants. A **special-purpose part** is a part designed and manufactured for a specific custom application. Examples include: engine housing, control link, access cover, support bracket, and washing-machine tub.

A **standard assembly** or subassembly is one that is commonly sold and used in various applications. Examples include pumps, electric motors, brake calipers, heat exchangers, ball bearings, clutches, valves, small gasoline engines, and electric

TABLE 2.1   Example Products

army helicopter	milling machine
automobile	paper clip
bag of potato chips	paper cup
baseball bat	penlight
bicycle	portable CD player
canoe paddle	power lawn mower
carton of milk	refrigerator
coffee maker	garage door opener
commercial jet	steam boiler
fishing reel	toaster oven
incandescent light bulb	toothpick
inflatable kayak	vacuum cleaner
laser printer	welding robot
leaf rake	wrench

switches. They are sometimes referred to as modules. Companies that use standard components purchased from one or more suppliers are called original equipment manufacturers, or OEMs. A **special-purpose assembly** is an assembly that is designed for a specific application: for example, a garden tractor transmission that incorporates a custom housing, drive shafts, standard bearings and gears. A variety of examples for each type of component are presented in Table 2.2.

Unfortunately, whether the component is a part or an assembly, it is sometimes referred to as a "part." For example, a bill of materials will list the components of a product, by part number, part name and quantity. In some cases, these "parts" will be assemblies of other "parts." Usually, the reader or listener must interpret whether the component is a part or an assembly by the context or the way that the term is used. Sometimes this can lead to misunderstandings that waste time and effort among fellow workers. Therefore, it is important for us to know and appreciate the subtle differences between a part and a "part."

The process of categorizing the components of a product into parts and assemblies is called **product-component decomposition**. Product component decomposition is used to identify the existence of individual parts or assemblies and the corresponding implications of multiple manufacturing processes, inventories, and materials handling, including manual or automated equipment to mate and fasten individual components. A product component decomposition diagram illustrates the basic structure of the components, or the **product anatomy** of the product. Since the "form" of a component usually denotes its "function," product component decomposition diagrams assist the development team in exploring the relationships between component types and functional performance. Decomposition diagrams are especially useful during **reverse engineering** when a development team physically disassembles an existing competitive product to learn how each component contributes to the product's overall performance (Otto and Wood, 2001).

TABLE 2.2   Examples of Each Type of Product Component

Standard parts	Standard assemblies	Special-purpose parts
bolt	pump	equipment housing
nut	valve	access cover
washer	electric motor	control link
rivet	clutch	support bracket
shaft key	chain	washing machine tub
gasket	heat exchanger	automobile windshield
v-belt	brake caliper	hood ornament
gear blank	ball bearing	motorcycle handlebar
shear pin	power screw	flashlight case
lubricant	gasoline engines	
sprocket	electric switches	

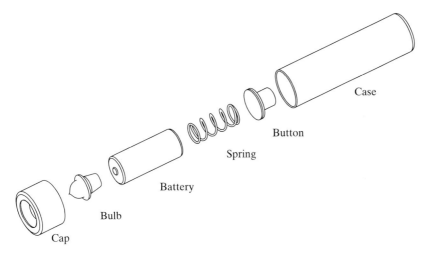

FIGURE 2.1

Assembly drawing of penlight showing standard and special-purpose components.

Example

Decompose the penlight shown in Figure 2.1 into its constituent components. Identify whether the components are parts or assemblies and in particular whether they are standard or special purpose. Then prepare a product component decomposition diagram to illustrate the basic anatomy of the penlight.

Looking at the assembly drawing from left to right, we see that the penlight is an assembly of a cap, bulb, battery, spring, button switch, and case. The cap is a special purpose part made specifically to hold the bulb. The bulb is a subassembly composed of a glass lens, filament, and base. Since the bulb is mass-produced and made in standard sizes it is a standard subassembly. The spring is a part that is usually purchased as a standard part. The special-purpose button part acts as a switch by pushing the battery toward the bulb to connect the circuit. The battery is a standard subassembly including an anode, cathode, electrolyte paste, and plastic cover. The metal case is a special-purpose part, cylindrical in shape and electrically conducting.

A product component decomposition diagram is shown in Figure 2.2. Note that the diagramming scheme uses an oval to represent a product and a rectangle for a subassembly. Individual parts are shown as plain text. Also note that solid lines are used to show the hierarchy of the parts and subassemblies.

The decomposition diagram readily illustrates that the product is an assembly of two subassemblies and four parts. We note that the function of the bulb is to convert electricity to light. The function of the battery is to store electrical energy, and the case supports the components as well as conduct electricity from the battery to the bulb.

The decomposition reveals that the penlight assembly depends upon selecting and purchasing standard components such as the bulb, battery, and spring. It also depends on

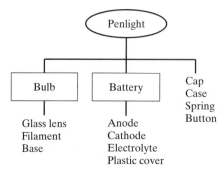

FIGURE 2.2

Product component decomposition diagram of a penlight.

the sound design and fabrication of the special cap, case, and button, each of which will necessitate decisions regarding appropriate materials and manufacturing processes.

Example

Prepare a product component decomposition diagram of an electric space heater. The heater has: (1) a special-purpose metal enclosure subassembly composed of a housing and an open-faced grill, (2) a standard blower subassembly composed of a fan and electric motor, (3) an electric special purpose module subassembly of a blower switch, heater switch, safety tip-over switch, and power cord, (4) a special-purpose heating element, which uses a nickel cadmium wire wrapped around a ceramic frame and six machine screws that fasten the subassemblies.

The space heater includes four subassemblies fastened by machine screws as shown in Figure 2.3. The blower unit is a standard subassembly. The enclosure, control module, and heating element are special-purpose subassemblies.

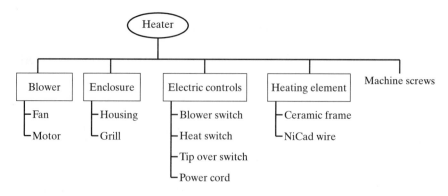

FIGURE 2.3

Product component decomposition diagram of an electric heater.

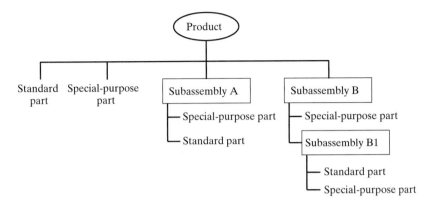

FIGURE 2.4

Product component decomposition diagram of a product having parts and subassemblies, both standard and special-purpose.

In general, a product can be decomposed into its components including subassemblies or parts, which may be standard or special-purpose in nature, as shown in Figure 2.4.

2.2.2 Process Plant Anatomy

Design engineers sometimes design process plants. A **process plant** is a combination of systems used to process energy or materials (both organic and inorganic). Examples include: meat and vegetable processing, iron ore processing, petroleum plants, stamping plants, hydroelectric power plants, and steel plants. A process plant is usually a one-of-a-kind design, customized to meet specific needs. A frozen-vegetable processing plant is shown in Figure 2.5.

The plant includes four systems: washing, blanching, freezing, and packaging. A **system** is two or more pieces of equipment used to perform a set of processes. Other examples include: waste treatment, auxiliary power generation, materials handling, HVAC.

Each system is composed of pieces of **equipment** that perform simpler specialized processes. To **process** means to mechanically or chemically treat matter so as to change its properties. Examples include: refrigerating, heating, separating, distilling, refining, spraying, chilling, evaporating, melting, homogenizing, freezing, heat-treating, cleaning, inspecting, and sorting.

Equipment refers to machines or apparatus designed to perform a process or portion thereof. Examples include: feed water pumps, boilers, condensers, electroplating tanks, and paint sprayers.

In general, a process plant is an integrated arrangement of systems, and pieces of equipment, as shown in Figure 2.6.

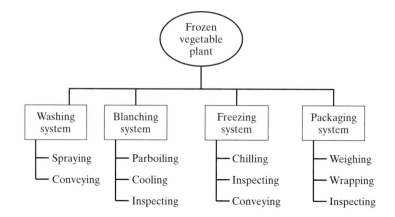

FIGURE 2.5

Decomposition diagram of a frozen-vegetable processing plant.

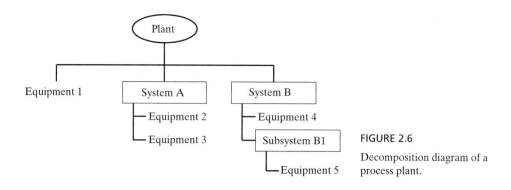

FIGURE 2.6

Decomposition diagram of a process plant.

Although this book will most often refer to the design of products, the design methods and techniques presented are commonly used to design larger systems and process plants, as well.

2.3 TYPES OF DESIGN

We can characterize design in a variety of ways. The following terms are commonly used to describe different types of design or design activities. Take note how the nomenclature and anatomy of products and process plants is used to describe the following types of design.

Variant Design. **Variant design** seeks to modify the performance of an existing product by varying some of its design variable values or product parameters such as size, or specific material, or manufacturing processes (Pahl and Beitz, 1996). Examples would include: modifying the length of a lever to increase mechanical advantage, or using aluminum for a part rather than steel to reduce weight, or using die casting rather than sand casting to reduce processing costs. Note, however, that the fundamental working principle or concept is usually maintained. For example, the size of the gasoline engine is revised rather than switching to an electric motor.

Adaptive Design. **Adaptive design** is when we adapt a known solution to accomplish a new task. Examples of adaptive design would include: adapting the ink-jet printer concept to spray a glue to bind powders in layers as a rapid prototyping method, adapting the cell phone concept to include personal digital assistant functions, and adapting the positive displacement pump concept to be used as an internal combustion engine (Pahl and Beitz, 1996).

Original Design. **Original design** refers to conceiving and embodying an original, innovative concept for a given task (Pahl and Beitz, 1996). Ullman (1997) describes original design as the development of a new component, assembly, or process that had not existed before.

Selection Design. In **selection design** we match the desired functional requirements of a component with the actual performance of standard components listed in vendors' catalogs. For example, if we were designing a belt-and-pulley drive, we would determine the belt type and size, or the pulley shaft-bearing types and sizes. Or, if we were designing a new lawn tractor, we would have selection design problems relating to the size and type of gasoline engine or the size and type of wheels.

Part, Assembly, Product Design. Design may be characterized with respect to whether a part, assembly, or product is being designed. For example, the following types of design generally increase in component complexity: standard part design, special-purpose part design, standard assembly design, special-purpose assembly design, and product design.

Concept Design, Configuration Design, Parametric Design, Detail Design. Design may also be characterized by the timing and type of information being processed, as in the design phases such as: formulation, concept design, configuration design, parametric design, or detail design. Configuration design of a special-purpose part, for example, requires determining the number, type, and approximate arrangement of geometric features, whereas concept design of a prime mover would examine alternative working principles such as electric motors, steam engines, or gasoline engines.

Redesign. Much of our working career will be devoted to the improvement of existing products. To obtain the improvements we usually modify parts, or subassemblies, or combinations thereof, by changing their shapes, sizes, configurations, materials, and manufacturing processes. Since design is determining "form," whenever we improve an aspect of form we are essentially redesigning.

Artistic Design. **Artistic design** deals with an object's appearance, such as "designer" clothing or furniture. Since engineering design calls for the application of science and mathematics to predict the behavior of a candidate design before it is made, artistic design is not engineering design, however.

Tinkering as "Design." Throughout history we find examples of products that have not been engineered including: pots, pans, cutlery, chairs, sofas, and beds. Manufacturers of such "designs" employ new materials or try different fabrication methods without regard to the underlying sciences, and if successful, adopt them in their future products. As in artistic design, **tinkering** is not engineering design.

2.4 STRATEGIES FOR SOLVING DESIGN PROBLEMS

Is there an obvious strategy, or plan of action, for solving a particular design problem? Does the solution strategy depend upon the problem?

Let's reconsider the examples presented in Section 2.1. The *toaster* has a defect that needs to be remedied. The situation is annoying and potentially dangerous to the consumer. The toaster has a part or subassembly that is not functioning, causing the product to malfunction. If the company does not act promptly, it will likely suffer significant financial losses. The current *lawn mower engine* pollutes the air and will no longer be legally acceptable. This too is a problem that needs immediate solving, similar to the toaster problem. The *potato-chip-packaging equipment* works. It currently functions. However, it might be able to work better. It represents an opportunity that would likely lead to improved customer satisfaction and company profits. Similarly, the *lightbulb, canoe paddle, car seat*, and *robot welder* are opportunities to improve performance, enter a new market, reduce costs, or serve a new customer. While problems need to be fixed, opportunities may be deferred. Design problems, therefore, exhibit varying amounts of urgency or necessity.

In addition to the urgency or necessity of solving a design problem, we might consider other factors. For example, we may not have enough time available to redesign a part or subassembly because the product's life cycle is too short. Or, we may not have enough manpower or other resources available in the company. We may have to settle for a quick fix. The technical challenges may be too risky. Or, it may be better to withdraw a poor product from the market than to poorly redesign a substitute. For example, a tire manufacturer might withdraw a specific brand of off-road tires because delaminating tread failures appear to result in rollovers of some sports utility vehicles. In this case, the company decided that the best strategy was "not solve the problem."

A number of approaches or strategies can be used to solve a design problem. Let's reconsider the toaster design problem to illustrate some plausible strategies. Assume that we know that a defective part causes 5 percent of the manufactured toasters to fail during normal use. We might consider the following alternative strategies:

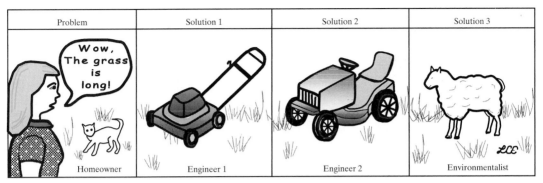

Problem	Solution 1	Solution 2	Solution 3
Homeowner	Engineer 1	Engineer 2	Environmentalist

Quite often, there are many alternative solutions to a design problem. Each of the three solutions will "cut" the grass. Solution 1 requires the user to walk behind the power lawn mower. Solution 2 permits the user to ride. Solution 3 requires no gasoline, walking or riding. But then again, . . . other waste products *are* generated.

A. Change the part's thickness, length, or material

B. Reconfigure the part by rearranging some of its geometric features,

C. Select and purchase a similar part from a reputable supplier

D. Redesign the subassembly eliminating the defective part

E. Replace the part with one that uses a different working principle

F. Pursue combination of above

G. Discontinue the product, or

H. Do nothing.

We see that alternative A relates to revising a part's size. We can call this strategy choice a variant design or parametric design strategy. Alternative strategy B relates to changing a part's configuration, which we call configuration design. Choice C can be called a selection design strategy. Choice D can be called a redesign strategy. And similarly, choice E can be called a concept design strategy. We see that there are many viable technical approaches to solving the defective toaster part design problem. It may not be possible, given limited time and resources, to determine the root cause of the defect. Therefore, discontinuing the product may be the right business strategy.

We can also conclude that a design problem is different than the solution strategy. The toaster design problem was that it was defective. A defective part caused product failures. And moreover, we can use any one of a number of strategies to "solve" the problem.

Depending on the circumstances, a design problem may have a self-evident solution strategy. For example, if the toaster part was a standard part, and determined to be improperly fabricated, we would likely pursue alternative C: Select and purchase a similar part from a reputable supplier. Choosing an

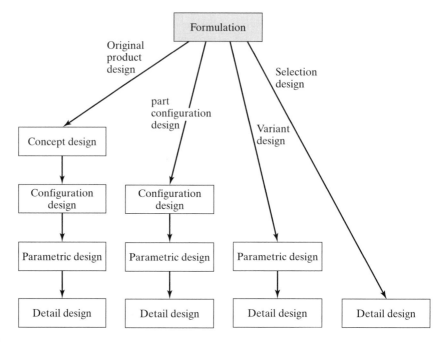

FIGURE 2.7

Formulation initiates all solution strategies.

appropriate strategy is usually easier when we have more information, such as knowing the root of the problem, or the significance of the lost customer satisfaction. However, we rarely have all the information! Therefore, self-evident strategies are rarely obvious, especially when we consider more complex components such as special-purpose subassemblies.

We cannot know for certain all the information that would lead to a perfect decision. It usually costs too much and takes too much time to gather. Businesses, however, make decisions on limited information all the time. Otherwise, the competition would pass them by.

One key to success in selecting the right, initial solution strategy is getting the right information. This is a major objective in design problem formulation. Figure 2.7 emphasizes that a given design problem may be solved in a variety of ways. But take particular note how formulation starts them all. We examine design problem formulation in the next chapter.

2.5   SUMMARY

- Design problems are product deficiencies that require resolution or product opportunities that require consideration.

- Products may be composed of parts and subassemblies, which may be standard or special-purpose.
- A process plant is an integrated arrangement of systems and pieces of equipment used to process energy or materials.
- Design types include redesign, selection design, variant design, adaptive design, original design, artistic design, and tinkered design.
- Alternative strategies can be used to solve a design problem.

REFERENCES

Dixon, J. R., and C. Poli. 1995. *Engineering Design and Design for Manufacturing.* Conway, MA: Field Stone Publishers.

Otto, K. P., and K. L. Wood. 2001. *Product Design: Techniques in Reverse Engineering and New Product Development.* Upper Saddle River, NJ: Prentice Hall.

Pahl, G., and W. Beitz. 1996. *Engineering Design*, 2d ed. New York: Springer-Verlag.

Ullman, D. G. 1997. *The Mechanical Design Process*, 2d ed. New York: McGraw-Hill.

Ulrich, K. T., and S. D. Eppinger. 1995. *Product Design and Development.* New York: McGraw-Hill.

KEY TERMS

Adaptive design	Process plant	Special-purpose assembly
Artistic design	Product	Special-purpose part
Assembly	Product anatomy	Standard part
Equipment	Product-component	Standard (sub)assembly
Original design	decomposition	Subassembly
Part	Reverse engineering	System
Piece-part	Selection design	Tinkering
Process	Solution strategy	Variant design

EXERCISES

Self-Test. Write the letter of the choice that best answers the question.

1. _____ A collection of two or more parts is called a:
    a. subassembly
    b. component
    c. product
    d. assembly

2. _____ A single-piece component requiring no assembly is called a:
    a. part
    b. piece-part
    c. both a and b
    d. neither a nor b

3. _____ An item that is purchased and used as a unit is called a:
   a. part
   b. piece-part
   c. product
   d. assembly

4. _____ Component complexity refers to all of the following except:
   a. types of components
   b. number
   c. color
   d. function

5. _____ All of the following would be considered subassemblies for a penlight except:
   a. bulb
   b. battery
   c. spring
   d. reflector

6. _____ Two or more pieces of equipment used to perform a set of processes is called a:
   a. plant
   b. assembly
   c. system
   d. function

7. _____ Which of the following is not a system of a frozen-vegetable plant:
   a. washing
   b. inspecting
   c. freezing
   d. blanching

8. _____ Designing a part or subassembly by matching desired performance requirements with actual performance of components listed in vendor catalogs is called:
   a. adaptive design
   b. variant design
   c. selection design
   d. artistic design

9. _____ Modifying the performance of an existing product by changing parameter(s) is:
   a. selection design
   b. variant design
   c. adaptive design
   d. artistic design

10. _____ Which of the following design types is not engineering design?

    a. configuration design

    b. original design

    c. selection design

    d. tinkering

2: c, 4: c, 6: c, 8: c, 10: d

11. How does a *design problem* relate to deficiencies and opportunities and give an example of each.
12. Sketch a product component decomposition diagram of a toy that uses an electric motor.
13. Sketch a product component decomposition diagram of a wheelbarrow.
14. Sketch a product component decomposition diagram of a refrigerator.
15. What is the difference between a special-purpose part and a standard part?
16. Why is a penlight battery not a product? What is it?
17. How does component complexity influence design difficulty?
18. What does the anatomy of a product tell us?
19. For the products listed below, give an example of the following: standard part, special-purpose part, special-purpose subassembly, standard subassembly.

    a. bicycle

    b. passenger jet

20. Describe whether the working principle or physical concept is changed in variant design, adaptive design, and original design?
21. Why is artistic design and tinkering not engineering design?
22. Describe a situation when a design problem should not be solved?
23. Select and describe three strategies for designing a new product?

Formulating a Design Problem

When you have completed this chapter you will be able to

■ Understand the complexities and subtleties of design problems
■ Describe the overall process of formulating a design problem
■ Determine customer and company requirements
■ Describe and use sources of product and customer information
■ Prepare an engineering design specification
■ Understand and implement the house of quality
■ Establish a consensus among the team members and management

3.1 INTRODUCTION

Formulating a design problem is an exciting and challenging activity. Like a detective when solving a complicated homicide case, we investigate leads, uncover hidden facts, and determine motives. We pursue some dead-ends and have to backtrack. We may find our data to be false, or misleading, or uncertain. Conditions we once thought fixed may change during the investigation.

Often the initial description of a design problem lacks the type of information necessary for successful design and manufacturing. Whatever information there is may be too general, or incorrect, or not technically specific. As we begin we try to get an overall understanding of the situation. But then, as we become familiar with the problem, we probe deeper into various sources to obtain more details. Piece by piece, as the puzzle begins to take shape, we systematically and doggedly pursue the facts to build a solid foundation for solving the problem.

The process of formulating a design problem, as shown in Figure 3.1, usually includes four primary activities: seeking information, interpreting, gaining consensus, and obtaining management approval.

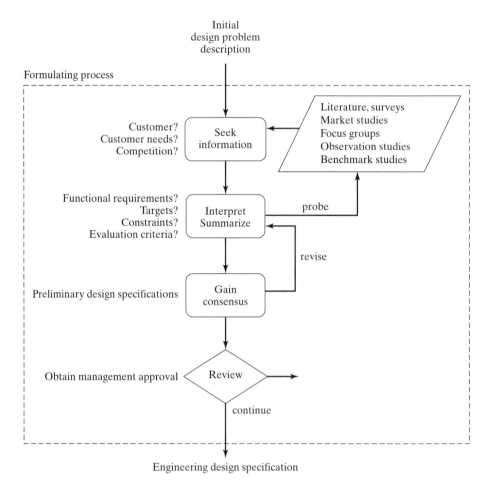

FIGURE 3.1

The process of formulating a design problem includes seeking and interpreting information, gaining team consensus on the basics of the problem, and obtaining approval from management to proceed.

Seeking Information. We gather data from various information sources, including surveys and studies, to obtain a detailed understanding of the customer, his or her specific needs, and the competition.

Interpreting. As we translate the raw data, we detail the specific functional requirements of the part, assembly, or product, desired performance targets, necessary constraints, and evaluation criteria. If the data are too general or vague, we revisit information sources and probe deeper.

Gaining Consensus. The prior activities are diverging in nature and lead to an information explosion. To gain consensus, team members discuss their interpretations of the data, revising their conclusions and reprobing the data sources, if necessary, to resolve conflicts. Then they prepare a preliminary engineering design specification, which is a written "summary" of the problem.

Obtaining Management Approval. To keep management informed and to obtain their approval, the team presents the preliminary engineering design specification in a design review meeting, or in a memorandum or technical report.

The formulating process shown in Figure 3.1 is no doubt an oversimplification of how individual teams formulate design problems in their companies. However, the figure does illustrate that a product development team processes a significant amount of information. The figure also emphasizes the need for a common understanding among team members and management.

In the next sections of this chapter we examine the specific types of information to be processed and methods that facilitate building consensus.

3.2   OBTAINING A DETAILED UNDERSTANDING OF THE DESIGN PROBLEM

The quantity and quality of information we obtain often depends upon whether the design problem pertains to a part, subassembly, or product. Imagine, for example, the original design of a high-performance motorcycle, versus the variant design of a multicylinder internal-combustion-engine subassembly, versus the selection design of a ball bearing. Let's examine these specific situations to identify the types of information we might search for and acquire in any given design problem.

High-Performance Motorcycle Design Example. Motorcycle manufacturers such as Harley-Davidson, Honda, Kawasaki, and Suzuki design and manufacture millions of motorcycles each year, including off-road trail bikes, touring bikes, street bikes, and high-performance bikes. The high-performance bike is fast and maneuverable and consequently demands the utmost in engineering design skill. The team would normally seek information regarding:

- How quickly should the cycle accelerate to 60 miles per hour?
- What should the top speed be?
- How maneuverable should the motorcycle be?
- Is fuel consumption less important than acceleration?
- What riding comforts are expected?
- Is an electric starter desired?
- Will the customer tolerate a liquid cooling system?
- Will the customer care about aesthetics?
- Which is more preferred: low-end torque or high-end speed?
- What is the target cost of manufacture?
- What will the maximum and minimum air temperatures and humidity be?
- What is the anticipated production run quantity?
- What types of instruments are preferred, digital or analog?

- What fuel capacity is desired?
- What are the desired service intervals?
- What are the average rider statistics for weight, height, arm length, leg length, age?

Multicylinder-Engine Design Example. Multicylinder internal combustion engines can be purchased by original equipment manufacturers (OEMs) for use in products such as: riding lawn mowers, auxiliary electric power generators, and powerboats. While most engine manufacturers have standard lines of 2-, 4-, 6-, or 8-cylinder engines, they will custom-design engines for special OEMs. To complete a custom design, however, the team would consider requirements such as:

- What fuel will the engine burn?
- What horsepower and torque are required at what speeds?
- Are there any height, width, and or depth size constraints?
- Is there a preferred output shaft size?
- Will the engine run at full speed all the time?
- What features or performance issues are important?
- Will the engine run at idle speed for long periods?
- Is an electric starter preferred?
- What will the air temperatures and humidity be?
- How long should the oil and air filters last before replacement?
- Will an integral fuel tank be required?
- What is the desired life expectancy in operating hours?
- What is the desired service interval for valve clearance and timing?
- What safety standards will apply?
- Will components be made in-house or purchased?
- What is the desired specific fuel consumption (gal/hr per hp)?
- What new tooling will be required?
- What is the target cost of manufacture?
- What is the anticipated production run quantity?

Ball-Bearing Selection Design Example. Ball bearings are standard subassemblies made by a number of manufacturers. Each subassembly includes: an outer ring, inner ring, balls, and ball separator. Optional features include shields and seals. When we design a ball bearing we select a type and size from a manufacturer's catalog based on rated load capacity. The type and size depend on a number of requirements, including:

- What radial and thrust loads will the bearing need to support?
- How many rotations should 90 percent of the bearings survive at the rated loads?

- What will the operating speed of the rotating ring be?
- Will the bearing need to be self-aligning?
- Will the bearing need to be shielded from dirt and dust?
- Will the bearing need to be sealed against liquids?
- How will the bearing be lubricated?
- Will the bearing be subjected to impact?
- How will the bearing be mounted?
- What will the range of air temperatures and humidity be?
- How many units will be purchased?
- What type of warranty does the supplier provide?

As we see from these examples, during formulation we actively search for, acquire, and interpret essential details of the design problem involving: what product functions are desired, how well the functions are performed, the operating environment that the product is subjected to, as well as marketing, manufacturing, and finance concerns. These can be broadly classified into **customer requirements** and **company requirements**.

3.2.1 Customer Requirements

Fundamentally, products are bought to satisfy needs and wants. Products have value. They provide utility to the customer. Value can be defined as the ratio of benefits to costs. The greater a product's benefits, in relation to its costs, the more the value it has and, usually, the greater the customers' satisfaction. If we listen carefully to the "voice of the customer," and endeavor to identify important customer requirements, we will most likely maximize our customers' satisfaction.

Function and Performance. The most important requirement of a product is that it should work. In other words, it should perform one or more functions. The principal function of a motorcycle, for example, is to transport a person. However, if we decompose, or subdivide, the main function into subfunctions we might understand our customer a little better. For example, a motorcycle performs a number of subfunctions, including: transport rider(s) fast, steer bike easily, support rider(s) comfortably, absorb road shocks, and start engine quickly.

Products or parts produce or react to forces and moments. They also provide motion (kinematics), convert types of energy, or control matter or energy. For example:

- A screwdriver drives a screw by pushing and twisting,
- a coffeemaker converts electricity to heat water for brewing,
- a wall switch connects house current to control overhead light,
- a pencil transmits hand force to graphite tip to mark paper,
- a ship propeller converts rotational shaft torque to thrust force,

- a hammer drives a nail by converting kinetic energy to an impact force, and

- a thermostat senses temperatures to control furnace and blower.

Note how the functions generally relate to energy, matter, and signal (i.e., control). We will discuss these and other functions and subfunctions in more detail in Chapter 4, "Concept Design."

After we determine the primary functions that our customer desires, we try to ascertain how important each function is to the customer. For example, the customer often describes more important functions as those that *must* be included versus those that *should* be included. Pahl and Beitz (1996) describe a "must" function as a demand and a "should" function as a want. Both sets of descriptors are in common use. In addition to using words as importance descriptors, design teams sometimes estimate importance using quantitative **importance weights**, as shown in Table 3.1. Note that the total adds up to one, or 100 percent. As we will see in later chapters, importance weights are measures that can be used to evaluate alternatives. An alternative to importance weights is the use of importance ratings, or measures that use ordinal numbers scales, such as "5" for most important, "3" for important, and "1" for not important.

Operating Environment. The operating environment for a product includes maximum and minimum air temperatures, humidity, and pressure. A motorcycle operating in a desert climate, for example, will experience extremely high air temperatures and low humidity, as compared to cooler mountain climates with lower air pressures at higher elevations. We might also consider dirt, mud, dust, corrosive gases (e.g., chemical vapors/pollutants) or liquids (e.g., salt spray), shock, and vibration.

Safety. Products must not injure anyone or damage anything during their installation, use, or retirement. Manufacturers are ethically and legally responsible to ensure that their products are free from hazards caused by defects of design and manufacture.

Economic. Customers may have specific economic requirements, including: the initial price they pay, installation fees, delivery charges, financing expenses, operating expenses (e.g., fuel and maintenance), useful life

TABLE 3.1   Customer Importance Weights by Subfunction

Subfunction	Weight
transport rider(s) fast	50%
steer bike easily	20%
support rider(s) comfortably	10%
absorb road shocks	5%
start engine quickly	15%
total	100%

before replacement, salvage value (e.g., resale or trade-in value), and disposal costs. Taken as a whole, these are often called the life-cycle cost of ownership. Customers typically scrutinize the life-cycle costs of expensive consumer products such as appliances and vehicles. Business customers, too, "justify" company purchases using life-cycle cost/benefit analyses.

Geometric Limitations. The product may need to fit in a space limited by height, width, and or depth. Or it may need to be connected at certain locations or have some angular requirements. Note that this category should not be used to set specific sizes or configurations of a product. It should rather consider only limitations on the sizes or configurations.

Maintenance. Maintenance requirements include how and when the product is cleaned, adjusted, and/or lubricated for proper working condition. The need for special training and/or tools to perform the maintenance should also be considered.

Repair. The costs and convenience associated with repairs may be of importance, especially for unexpected breakdowns or damage due to accidents.

Retirement. A product may need to be disassembled at the end of its useful life for remanufacturing or recycling some of its components. Some components may have scrap value. Others may have disposal fees and require regulatory procedures.

Reliability. Customers expect their products to work all of the time. Reliability is a measure of the likelihood that a product and all of its components work.

Robustness. A robust product performs in spite of variations in its material properties, how it was manufactured, the operating environment, or how it is used. For example, customers might be somewhat dissatisfied when a product's moving parts jam because of thermal expansion.

Pollution. The customer may require specific considerations of air, water, and noise pollution.

Ease of Use. Will the product require special controls to operate effectively (e.g., air-conditioner thermostat, automatic garage door opener)? Will the customer need to be trained before using the product (e.g., private airplane, all-terrain vehicle, computer numerical control (CNC) machine, or firearm)? In other words, will the product be user-friendly?

Human Factors. When a product is used, controlled, or operated by a person, human factors must be considered, such as the user's ability to apply forces and torques, body size and range of motion (i.e., hands, feet, arms, and legs), and ability to sense (i.e., touch, hear, or see).

Appearance. Does the customer have any specific requirements as to color, shape, surface finish, texture, and/or aesthetics?

3.2.2   Company Requirements

Successful products must also satisfy company requirements as well as customer requirements. As mentioned in Chapter 1, the company will make a significant investment during the product development phase and the manufacturing ramp-up, before the product provides any sales revenues. Considering that it may take months, if not years, of continuous sales revenues to pay for these early investments, the company will have requirements pertaining to marketing, manufacturing, and financing.

Marketing   **Customers** purchase products. **Consumers** use them. Grandparents, for example, are customers who buy toys for their grandchildren consumers. As toy designers we would want to examine whether the customer might consider safety over and above entertainment value.

Knowing whether the customer is an individual, group, business, or government agency can make a big difference, too. Some products are bought by two or more people, such as an automobile or family home appliance. In such cases the evaluation of individual products and the decision to purchase is made by two or more people. Similarly, businesses buy products such as photocopy machines, milling machines, personal computers, and welding robots. Businesses usually establish rigorous criteria and purchasing procedures that involve significant rational decision making. Finally, perhaps at the highest level, government agencies buy products, such as helicopters and tanks. Obtaining an understanding of the federal agency's evaluation criteria, and procedures it will use, is instrumental for the company's success.

Other marketing requirements include understanding what the competition is and what it will be doing during the time it takes to get the product to the market, otherwise called the "time to market." The company may decide to pursue a strategy to serve a segment of the market, such as large discount chains, or develop a product for a "high-end" niche market. The company will also try to estimate the necessary advertising resources and the annual volume that is to be produced and sold.

Manufacturing   Manufacturing requirements pertain to purchasing, fabrication, assembly, warehousing, and distribution. Since the final cost of a product relates to labor, materials, and manufacturing processes used, each company may have special advantages or limitations to take into consideration.

Perhaps the most important factor is the expected annual production volume (number of units made). Mass-produced food products, chemicals, automobiles, light bulbs, and refrigerators use continuous production processes. Raw materials, parts, and subassemblies flow through the factory plant without being stored. Batch production, which fabricates and inventories parts and subassemblies in batches, is often used to make bicycles, helicopters, inflatable kayaks, and milling machines. A custom-design machine tool is an example of "one-off" production. High-production-volume manufacturing processes such as injection molding and die casting would not be considered for it.

Financing   The product development team will also obtain information that can be used to prepare estimates of the capital expenditures necessary to purchase and install manufacturing equipment or modify existing facilities. Similarly, projections of sales revenues, expenses, profits, and return on investment will be estimated. These estimates are usually prepared and updated throughout the product development project. The figures are periodically reviewed by senior management to determine whether to continue or terminate the project.

Other   In some cases, existing patents may block the product design. If a license agreement cannot be negotiated with the patent holder, a totally new design concept may have to be considered. Similarly, legal and voluntary regulations, standards, and codes need to be examined.

A summary of fundamental customer and company requirements is presented in Table 3.2.

3.2.3   Engineering Characteristics

Next we select **engineering characteristics**, or measures that can quantify how well a product performs each requirement. For example, what does a motorcycle customer mean when she says that she wants to go fast? We might consider quantifying the "go fast" requirement by selecting top speed measured in miles per hour, or maximum acceleration measured in feet per second per second, or both. Similarly, we might quantify "start quickly" by selecting cranking time measured in seconds.

TABLE 3.2   Fundamental Customer and Company Requirements

Customer requirements	Company requirements
Functional performance	Marketing
Motions/kinematics	Customer/consumer
Forces/torques	Competition
Energy conversion/usage	Strategy
Control	Time to market
Operating environment	Pricing
Air temp., humidity, pressure	Advertising
Contaminants	Sales demand/targets
Shock, vibration	Manufacturing
Safety	Production quantity
Economic	Processes, materials
Geometric limitations	New factory equipment
Maintenance	Warehousing and distribution
Repair	Financial
Retirement	Product development Investment
Reliability	Return on investment
Robustness	Other
Pollution	Regulations, standards, codes
Human factors	Patents/intellectual property
Appearance	

Engineering characteristics are essential to the design process and therefore should be objective and not subject to interpretation. First, they can be can be used to impartially assess how well an *existing product* satisfies its customer requirements. Second, since engineers use these same measures in formulas to predict the physical behavior of objects, we can use engineering characteristics to assess how well a *new product* might perform. The phrase "engineering requirements" is sometimes used in industry to describe these measures. Requirements, however, are somewhat different from characteristics. A "requirement" is more of a desirable or necessary target value for an engineering characteristic. A "characteristic" is really the means to establish a "requirement." Therefore, in this text we refer to these quantitative measures as engineering characteristics.

We also designate appropriate units and limits for each engineering characteristic. Limits refer to minimum or maximum values that customers often demand, such as: needing *at least* 10 miles per gallon of fuel. A few engineering characteristics for the motorcycle example are given in Table 3.3.

3.2.4   Constraints

Restrictions on function or form are called **constraints**. Constraints limit our freedom to design. Therefore, maximum or minimum performance limits relating to the desired functions or subfunctions are appropriately called constraints. Specific limitations regarding shape, size, configuration, materials, or manufacturing processes are also called constraints. Constraints also originate from economic or legal considerations. Finally, constraints emanate from the laws of nature. The conservation of energy, and static equilibrium for example, are often used to develop constraints.

During formulation, we take careful note of all explicitly stated constraints. For example, "injection molding must be used," "the customer requires aluminum components," "a round tank is required." "the length must be less than 5 feet," or "a left-handed thread is required." Implicit constraints, on the other hand, are those restrictions that are implied, though not directly expressed. For example, the customer implicitly wants a safe product. Safety, however, implies that all the parts will not buckle, fracture, melt, corrode, and so on.

TABLE 3.3   Engineering Characteristics, Units, and Limits

Subfunction	Engineering Characteristic	Units	Limits
start engine quickly	cranking time	seconds	≤6 sec
support rider(s) comfortably	cushion compression	inches	
transport rider(s) fast	acceleration	feet/sec^2	≥32 ft/sec^2
	top speed	mph/kph	≥90 mph
	0–60 mph	seconds	≤6 sec
steer bike easy	steering torque	pound-ft	
	turning radius	feet	
absorb road shocks	suspension travel	inches	>5 inches

> We should consider all design constraints, both explicit and implicit.

When a part (or product) satisfies all the constraints it is called a feasible design. Those alternative designs violating any constraints are said to be infeasible. During the parametric design phase we will generate a number of alternative designs, predict their performance, and check that the constraints are satisfied for each design. Only those designs that are determined to be feasible will be developed further.

3.2.5   Customer Satisfaction

Finally, contemporary product design teams try to qualify or quantify how satisfied a customer is at various levels of performance, that is, a measure of the utility a product provides (Badiru and Pulat, 1995; Dixon and Poli, 1995; Hazelrigg, 1996; Keeney and Raiffa, 1976; Siddall, 1982; Stub et al., 1994). In other words, they try to estimate customer satisfaction with respect to key engineering characteristics. For example, would a 90-miles-per-hour top speed for a high-performance motorcycle be unsatisfactory, moderately satisfactory, or highly satisfactory? What about 120 or 150 miles per hour?

Customer satisfaction is extremely important. First, we need the customer to tell us what a good or excellent design is. We should not impose our values of good or bad. A high-performance motorcycle customer might value a 90-miles-per-hour top speed bike as unsatisfactory, even though we might value it as good or excellent. Second, we should try to establish qualitative or quantitative levels of satisfaction. We could then measure how good an existing product is, or how good a competitive product is or a new design might be.

What performance would the customer consider as unsatisfactory, poor, fair, good, or excellent? Continuing with the motorcycle example, we might find that various top speeds would result in the qualitative levels of satisfaction shown in Table 3.4. Note that the table uses qualitative words such as not satisfied, somewhat satisfied, and moderately satisfied.

A quantitative means to express customer satisfaction is to use numerical values. For example, when the customer is not satisfied, we quantify it with the number 0 (i.e., 0 percent satisfied). When the customer is most satisfied, we quantify his/her satisfaction value as 1 (i.e., 100 percent satisfied). Similarly, other values can be established for levels in between, as shown in Table 3.5. Although somewhat subjective, quantitative satisfaction values can be used to evaluate the "goodness" of existing products or new design candidates.

Graphing provides a third means to provide satisfaction values as a function of a performance variable. Let's continue with the high-performance motorcycle. Using focus groups and other customer satisfaction assessment information, we might decide that our customers would be totally unsatisfied with a bike having a top speed of less than 90 miles-per-hour. Further, we decide

TABLE 3.4 Qualitative Satisfaction Levels as a Function of Top Speed

Top speed (mph)	Satisfaction
$0 < $ speed $ \leq 90$	Not satisfied
$90 < $ speed $ \leq 100$	Hardly satisfied
$100 < $ speed $ \leq 110$	Somewhat satisfied
$110 < $ speed $ \leq 120$	Moderately satisfied
$120 < $ speed $ \leq 135$	Very satisfied
$135 < $ speed $ \leq 150$	Most satisfied
speed $ > 150$	Most satisfied

Satisfaction level tables such as this help us to evaluate the merits of alternative designs.

TABLE 3.5 Quantitative Satisfaction Values Based on the Amount of Customer Satisfaction

Satisfaction	Value
Most satisfied	1.0
Very satisfied	0.9
Moderately satisfied	0.8
Somewhat satisfied	0.5
Hardly satisfied	0.3
Not satisfied	0.0

that the customer would be totally satisfied with any speed greater than 150 miles per hour. Let's also assume that the customers would be proportionately satisfied in between those speeds. Let's use the value of 0.0 for not satisfied and 1.0 for totally satisfied. Rather than using a table, we graph the *satisfaction versus "top speed"* curve as the dashed line shown in Figure 3.2.

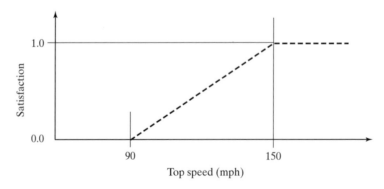

FIGURE 3.2

Satisfaction as a function of top speed (mph). Note that this is an example of a "more-is-better" relationship (i.e., more top speed results in more customer satisfaction).

We could similarly decide the customer's satisfaction with respect to cushion compression and cranking time, as shown in Figure 3.3 and Figure 3.4.

Using the above **customer satisfaction curves** for cushion compression and cranking time, we can readily determine that the customer will be unsatisfied with cushions too soft or too stiff, and that a cranking time greater than 6 seconds is unacceptable. Having that knowledge, in a quantifiable manner, will help us decide which motorcycle designs are "better," that is, satisfy our customer the most.

In general, satisfaction curves fall into three categories: (1) more-is-better, (2) target-value-is-better (i.e., nominal-is-better), and (3) less-is-better. These same categories are illustrated in Figures 3.2–3.4. Note that the shape of the curve between minimum and maximum satisfaction values is arbitrary and may

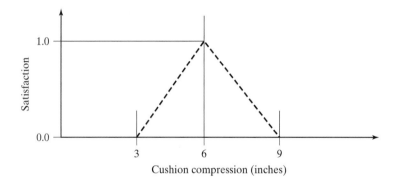

FIGURE 3.3

Satisfaction versus cushion compression (in.). Note that this is an example of a "target is best" relationship (i.e., greatest customer satisfaction occurs at target value).

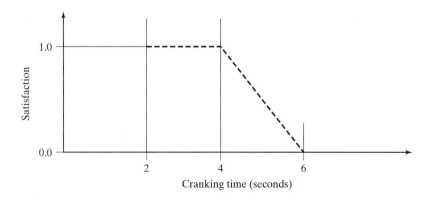

FIGURE 3.4

Satisfaction versus cranking time (seconds). This is an example of a "less-is-better" relationship.

be curved rather than linear. The curve is shown as a dashed line to emphasize that it is an estimate of the customer's satisfaction. More examples will be discussed later in Chapter 8, "Parametric Design."

Most important, we recognize that even though the shape of the curve is arbitrary and even though the end points are only estimates, if we understand how our customer appreciates different levels of performance, qualitatively or quantitatively, we will make better design decisions.

3.3   INFORMATION SOURCES

As we formulate our design problem we will search out and examine information from a variety of sources, including: surveys, market studies, literature, focus groups, observation studies, and benchmark studies.

Surveys. Perhaps the first place to look is inside the company. Many companies have established customer feedback systems that include: service reports, customer support reports, and warranty claims. Customer surveys provide a major source of information on existing and potential customers. Customers can be queried using phone, mail, and e-mail surveys in addition to in-person interview surveys. Some surveys are available from trade associations, also.

Market Studies. Trade associations, and/or government agencies, publish market studies that provide information about general customer trends. The U.S. Department of Commerce, for example, provides annual production statistics on thousands of companies in dozens of industry classifications.

Literature. Literature sources include reference handbooks, monographs, technical journals, trade journals, and general periodicals, in addition to electronic sources such as compendiums and Internet searches.

Focus Groups. A **focus group** can be employed wherein a group of prospective customers are gathered together in one location to discuss a new product (Crawford and DiBenedetto, 2003). The participants are usually compensated for their time and a highly trained moderator often leads the discussion.

Observation Studies. Customer use patterns can be obtained by observation studies wherein company representatives or paid consultants observe customers as they use a product, without being seen themselves. For example, we might observe a shopping mall parking lot, to investigate how people use their car trunk to stow parcels.

Benchmark Studies. Assessments of competitive products, often called **benchmark studies**, are also informative. A number of competitive products are identified, purchased, disassembled, and evaluated with respect to their common and unique features. Much can be learned about how each product attempts to satisfy customer and company requirements.

How much information is enough? Which sources of information are more reliable? The answers depend on the time and money available to the team, and the consequences of false, misleading, or insufficient information. To help us answer these difficult questions, we should consult company experts and upper management for their experience, wisdom, and guidance.

Note that the information gathered involves many departmental functions from across the company. It is no wonder, therefore, that successful product design teams have representatives from sales and marketing, engineering, finance, and manufacturing. In addition, recall that the industrial designer on the team will focus on desired aesthetics, while the design engineer will concentrate on the desired functions.

In the next section we examine **quality function deployment** (QFD) and the **house of quality** for product planning. The house of quality, in particular, is an outstanding method that systematically structures and develops the design problem information.

3.4   QUALITY FUNCTION DEPLOYMENT/HOUSE OF QUALITY

As we consider the entire product realization process, from identifying customer requirements to delivering the finished product, we recognize that thousands of decisions by many different people using various evaluation criteria are involved. Hopefully our company will appoint a concurrent engineering product development team that is colocated, cross functional, and coordinated by a high-ranking manager. But will the product have the quality that the customer is expecting?

3.4.1   What Is Quality?

Consumers were surveyed by *Time* magazine (1989) about what a **quality product** is? The most frequent responses were: (1) works as it should, (2) lasts a long time, and (3) is easy to maintain. An earlier work by Garvin (1987) corroborates the survey's findings by identifying the following characteristics of quality: (1) performance, (2) features, (3) reliability, (4) durability, (5) serviceability, (6) conformance to conventions/standards, (7) aesthetics, and (8) perceived quality/reputation of manufacturer.

Since a product is as good as its parts, a quality *product* is made of quality *parts*, which are made by high-quality processes. Consequently it will function or perform as expected (reliable), last a long time (durable), and be easy to maintain (serviceable), among other things. But, which department in the company is responsible for quality? Is it sales and marketing? Production? Engineering design? Other?

3.4.2   Quality Function Deployment (QFD)

Every department contributes to the quality of the product, and is therefore responsible. Just like the responsibility for financial matters is a business function that can be assigned to a group of employees in the "finance" department, so can quality. But everyone in the company is responsible for quality. How can we assign or "deploy" quality throughout our company? We can't call every depart-ment the quality department. Yes, there is the quality control "department." But that group is usually responsible for only a limited set of raw material and fin-ished goods tolerance checking activities. But, don't we need to have all the de-partments focus on quality?

Quality function deployment (QFD) is a team-based method that draws upon the expertise of the group members to carefully integrate the voice of the customer in all activities of the company. The method makes use of focused dis-cussion groups to systematically address product, part, process, and production quality. Group discussions are summarized in four house of quality diagrams that structure (1) product, (2) part, (3) process, and (4) production information. Since representatives from all corners of the company are involved in the deci-sion making, the method achieves a high level of consensus, and consequently results in high-quality products. In other words, quality, as defined by the cus-tomer, is deployed throughout all the other functions of the company. Let's ex-amine the first of the four houses, the house of quality for product planning.

3.4.3   House of Quality for Product Planning

The "house of quality" (HoQ) for product planning is a systematic graphic rep-resentation of product design information organized as a matrix of "rooms," "roof," and "basement." The house of quality is a useful and illustrative summary of product information. The three other houses of quality (part design, process planning, production planning) will be discussed later in this chapter.

The true value of the HoQ is not the diagram. Rather, the true value lies in the group decision making, which requires that the team discuss and ultimately obtain a common understanding of the design problem.

Recall that during problem formulation the team tries to obtain a detailed understanding of the design problem. It gathers and evaluates information re-lating to: customer and company requirements, their importance weights, engi-neering characteristics, and competitive products used as benchmarks. Some of this work will be done in groups. But, a lot of the work will be done individually. If discussion and summary are a group activity, however, the team is likely to obtain a consensus of opinion on many aspects of the product design require-ments.

The HoQ for product planning, shown in Figure 3.5, systematically struc-tures the following information:

1. customer requirements
2. customer importance weights
3. engineering characteristics
4. correlation ratings of requirements and characteristics
5. benchmark satisfaction ratings
6. coupling between engineering characteristics

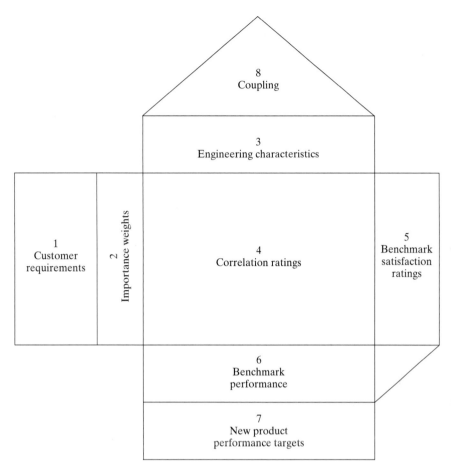

FIGURE 3.5

Rooms of the product planning house of quality.

7. benchmark performance values
8. new product design target values

1. *Customer Requirements (Room 1).* Customer requirements are summa-rized as rows in the first column. A clear list of functions and subfunc-tions focuses on important needs of the customer. Customer wording or terminology is frequently used to express the "voice of the customer." The list should contain only the more important requirements and rarely exceeds 25 items.

2. *Customer Importance Weights (Room 2).* Adjacent to the requirements column is the importance weights column. Using values between 0.0 and 1.0, the weights establish how important the customer considers each re-quirement with respect to the other requirements. The importance weights sum to 1.0.

3. *Engineering Characteristics (Room 3).* Along the top row, underneath the roof triangle, is a list of quantitative performance parameters and their associated units, arranged in a row vector. An engineering characteristic can be used to quantify the amount of satisfaction of each customer requirement.

4. **Correlation Ratings Matrix** *(Room 4).* At the intersection of a row and column is a cell that is used to indicate the amount of correlation between a customer requirement and an engineering characteristic. Each cell is given six of three correlation rating numbers: 1 (low), 3 (medium), or 9 (high) for positive correlation and −1, −3, −9 for negative correlation. The cell is left blank for no discernible correlation. The numbers 1, 3, and 9 are frequently used in practice, although other number schemes have been used.

    Note that poorly correlated customer requirement/engineering characteristic pairs are not good indicators or predictors for the team. If all we have are uncorrelated or poorly correlated engineering characteristics, we have no real measure of the satisfaction of a customer requirement. How will we ever know whether our new product design will satisfy the customer if we don't have any correlated measures?

5. *Benchmark Satisfaction Ratings (Room 5).* To the right of the correlation matrix we list customer ratings for competitive products used as **benchmarks**. First, the team rates its own current product (CP), if it has one, as to how well it satisfies the customer on each requirement. Then, the team rates each competitive product. This, of course, requires the team to consider what the customer thinks about each alternative product. Market research data can be used too. A new product idea can also be added to the benchmark section, to be rated against the competition.

6. *Benchmark Performance (Room 6).* Below the correlation matrix we indicate the performance of each benchmark product, using the measurement units designated for each engineering characteristic. This section, therefore, is an arrangement of statistics gathered on the competitive products as to how well each product performs or "measures up."

7. *New Product Targets (Room 7).* Below the performance statistics, in the basement, we list **performance targets**, or desirable goals for the new product.

8. *Coupling Matrix (Room 8).* The triangular roof of the HoQ, called the **coupling matrix**, is a matrix of values that estimate the amount of coupling, or interaction, between engineering characteristics. Rating numbers such as 1 (low), 3, and 9 (high) are given for positive coupling and −1, −3, and −9 for negative coupling. Uncoupled engineering characteristics can be optimized, one by one, without affecting other engineering characteristics. Inversely correlated characteristics indicate that compromises will need to be made, or "trade off." In other words, if we improve one characteristic we will worsen the other.

Example

A product development team has been formed to design a new electric pencil sharpener for use in homes or offices. The team has obtained market research data and, combined with their own surveys, have come to a consensus about the major customer requirements including: doesn't slide when using, needs little insertion force, requires little insertion torque, operates when pencil is inserted, collects pencils shavings well, empties shavings easily, plugs into wall socket easily, cord is long enough, grinds pencil to sharp point, and needs only one hand two operate. Through a process of voting they also determined approximate importance weights for each.

After considerable discussion they agree upon the following engineering characteristics: slides (yes/no); friction factor, start switch force (lbf.), insertion force to sharpen (lbf.), hold force required (lbf.), grasp torque (in.-lbf.), shavings storage volume (cu. in.), number of steps to empty, standard 120 VAC (yes/no), cord length (ft.), point cone angle (degrees), number of hands to operate, weight (oz.), and point roughness (micro in.).

Then, systematically, they discussed correlation ratings, customer satisfaction ratings, and coupling ratings and establish new product targets. They summarized their understanding of the product design problem in a house of quality for product planning, shown in Figure 3.6.

3.4.4 Downstream Houses of Quality

The house of quality for product planning is the first of four houses that can be constructed. As we wish to deploy quality throughout the whole organization, we need to involve people responsible for part design, manufacturing process planning, and production planning.

Quality function deployment is a systematic method that includes all the houses of quality, not just the product planning house. As shown in Figure 3.7, the product planning house of quality connects customer requirements to engineering characteristics. Then, the part design house of quality connects or relates the engineering characteristics to part characteristics. The process planning house of quality is used to determine process characteristics and/or parameters based on part characteristics. Finally, the production planning house of quality relates process characteristics and production characteristics. The group decision making that goes into the development of the cascading houses produces a logical sequence of product engineering characteristics, part characteristics, process parameters, and production characteristics.

Further information on quality function deployment and the house(s) of quality can be found in Clausing (1994); Cohen (1995); Hauser and Clausing (1988); King (1989); Summers (1997); and Urban (1993).

3.5 PREPARING AN ENGINEERING DESIGN SPECIFICATION (EDS)

In the early phases of the formulating process, we gather, examine, and evaluate information regarding customer requirements, company requirements, engineering characteristics, constraints, and customer satisfaction.

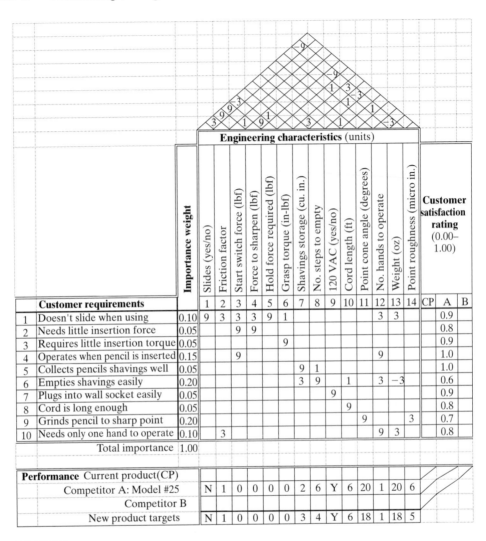

FIGURE 3.6

House of quality for an electric pencil sharpener.

At this juncture, we usually devote some effort to summarizing the design requirements in a document commonly called an **engineering design specification**, or EDS. The information is fresh in our memories and we need to consolidate the notes of our individual team members. More important, some members may have different interpretations of the data, which need to be resolved. This may be especially true if we did not prepare a house of quality for product planning. Therefore, discussing and writing the engineering design specification is more than recording the team's findings. It is a useful process that can correct misunderstandings and clarify terminology among team members from other

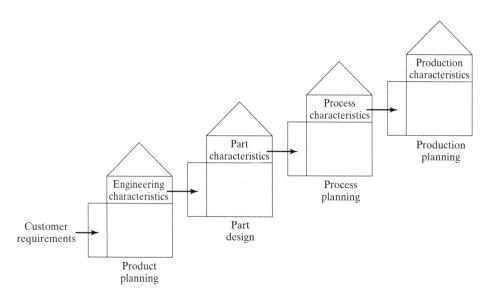

FIGURE 3.7

Quality function deployment method uses cascading house of quality diagrams to structure product, part, process, and production information.

departments. It is a homogenizing process that usually results in a common understanding of the customer's needs and priorities.

Consequently, the engineering design specification (EDS) is a single document that captures the whole team's understanding of the *specific* details of
the design problem (Dieter, 2000; Dixon and Poli, 1995; Haik, 2003). It includes
information on customer requirements, company requirements, engineering
characteristics, constraints, and customer satisfaction. An example template,
which lists some of the more important categories, is shown in Table 3.6. In
some cases it is only one page long and in others it may be dozens of pages long.
Of course, the contents of an EDS will be tailored to the item being designed,
i.e., part, subassembly, product, system, or plant. Note that the EDS is sometimes called a product design specification (PDS) (e.g. Pugh, 1991). This step is
equivalent to preparing a "statement of the problem" that we find in general
problem-solving methods.

Not all of the design problem details will be known at this time. Therefore,
we should consider the engineering design specification to be a work-in-
progress and dynamic in nature. As we develop the product, more information
and knowledge develops. Therefore, we can fine-tune the EDS as we proceed,
even up to the point of production engineering. Of course, the more we know
ahead of time, the fewer the engineering design changes we have to make. That
will save us time and effort since in a manufacturing enterprise, changes are
documented by engineering change notices that need to be circulated through
the company for approval signatures.

TABLE 3.6   Engineering Design Specification Template

Cover page	Retirement
Title	Reliability
Stakeholders	Robustness
Date	Pollution
Introduction	Ease of use
Design problem description	Human factors
Intended/unintended uses	Appearance
Special features	**Company requirements**
Customer requirements	Marketing
Functional performance	Manufacturing
Operating environment	Financial
Safety	Other
Economic	**Appendices**
Geometric limitations	Site visit data
Maintenance	Sales/marketing data
Repair	House of quality

Note that the term "product specifications" is different from the EDS. Product specifications refer to printed information about the finished product, usually given to the customer in the owner's manual or "instructions" sheet. For example, a product specification sheet for a water pump might list items such as: weight 5 pounds, 120 volts, and pumps 5 gallons per minute. Also, note the absence of the word "design" in the phrase. Engineers design equipment, systems, plants, and facilities in addition to just products. Therefore, the phrase "engineering design specification" is perhaps more appropriate because it includes "design" and covers more artifacts than just products.

Engineering design specification = Product *design* specification

Engineering design specification ≠ Product specification

As the team prepares the EDS, draft sections are circulated among team members for review and comment. Frequent discussions may result as team members resolve specific issues. A team meeting is often convened to discuss its contents. The finished version, however, usually integrates different viewpoints into a common understanding of the design problem. In other words, the process of writing the EDS establishes a team consensus on the important customer and company requirements.

The engineering design specification provides a convenient mechanism to communicate the team's findings to all stakeholders. In some cases, the EDS is presented to upper management in a design review meeting to obtain their approval to continue the product development efforts.

Example

Smart Kitchens, Inc. makes a variety of kitchen appliances for residential use. The company is interested in expanding their product line into electric coffeemakers. The typical Smart Kitchens customer wants a coffeemaker that can brew about 8 cups of hot, delicious coffee. Company management formed a product development team that gathered and interpreted pertinent data. The team then summarized their findings in a preliminary engineering design specification, shown in Tables 3.7–3.8 and Figures 3.8–3.10.

TABLE 3.7   Example Engineering Design Specification for a Coffeemaker

Title: New Coffee Maker for Smart Kitchens, Inc., May 2003

Introduction
Design problem: home kitchen coffeemaker
Intended purpose or use: brew coffee
Unintended purpose: heat water for tea or hot chocolate
Special features:
Control switch should have on/off indicator light,
Means to keep coffee warm after brewing

Customer Requirements
Functional performance
- Water should be heated to temperatures between 135° and 175° F
- Brewing time should be less than 6 minutes
- Drip brewing method is required rather than percolation
- Input electricity must be 110–120 volts AC
- Power consumption should be less than 400 watts

Operating environment
- Residential temperatures 50°–125°F and humidity 10–100%
- Pot and basket should be dishwasher safe
- Minimal dust

Economic
- Should have economic life of more than 5 years
- Should not require any routine servicing other than cleaning

Geometric limitations
- Compact size is desired
- Height, width, and depth less than 15 in. by 10 in. by 10 in.
- Pot must contain a minimum of 48 oz (eight 6-oz cups) of brewed beverage
- Brew chamber should accommodate up to 4 cu. in. of coffee grounds

Maintenance, repair, retirement
- The coffeemaker casing should be easy to clean
- No repairs should be required during economic life
- No special disposal efforts should be required

Reliability, robustness
- No failures should occur during economic life
- Will accommodate variations in water, coffee ground quality, supply power voltage

---

TABLE 3.7 *Continued*

Safety
- Will not burn or electrocute user
- Will not combust or catch fire during normal use

Pollution
- Will not create noise >40 db

Ease of use
- Simple to fill water, add/remove grounds and filter paper
- One switch to turn on/off
- Simple to remove basket and place in dishwasher

Human factors
- No large forces or torques required to operate
- Pot handle to fit 5–95th percentile females and males
- Switch to have obvious mode of operation
- Removable parts should be graspable and not slippery

Appearance
- Color scheme to match current/popular appliance trends
- Surface finish should be very smooth to facilitate cleaning
- Shape should be consistent with current trends

Company Requirements
Marketing
- Retail price should be less than $30

Manufacturing
- Production run quantity is estimated at about 25,000 units
- A beta prototype should be ready for testing in 12 months
- Components must be made with injection-molding processes available in current plant

Financial
- Development costs should be paid back in three years

Other
- Production prototype must be UL-approved

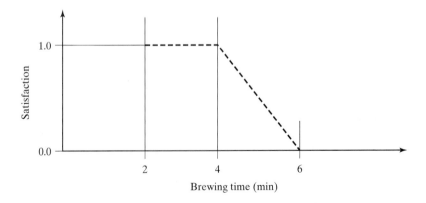

FIGURE 3.8

Customer satisfaction versus brewing time in minutes for example coffeemaker.

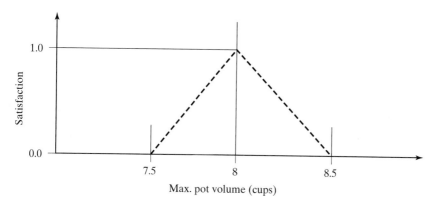

FIGURE 3.9

Customer satisfaction versus maximum pot volume (in 6-oz cups) for example coffeemaker.

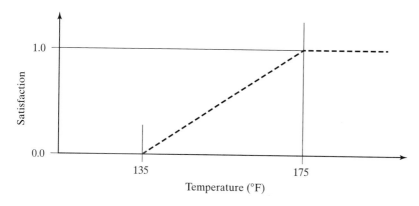

FIGURE 3.10

Customer satisfaction versus beverage brewing temperature (in degrees Fahrenheit) for example coffeemaker.

TABLE 3.8   Customer Importance Weights for Selected Engineering Characteristics of the Example Coffeemaker

Engineering Characteristic	Importance wt.
Brewing time	0.20
Max. pot volume	0.30
Beverage temperature	0.50
Total	1.00

3.6 ESTABLISHING CONSENSUS AMONG STAKEHOLDERS

Others in the company, including various layers of management, along with the development team members will be involved. Most everyone will want to contribute in some way or another.

To obtain a consensus of opinion, however, each stakeholder must have a chance to communicate his support for, or objections to, aspects of the project. Most projects will not proceed unanimously. Critical individuals should be given the opportunity to express their doubts or objections as minority positions. However, to move the project forward, everyone recognizes that some command decisions will have to be made. But, by letting stakeholders participate in the deliberations, they will at least understand why, even if they don't always agree.

This chapter has presented a number of activities that provide opportunities to build consensus including:

Obtaining a detailed understanding. A product links the customer to the product realization process. To fully understand customer and company requirements, team members from departments such as sales, marketing, engineering, manufacturing, purchasing, and finance need to actively participate and communicate.

Preparing the engineering design specification. This joint activity requires the amalgamation of team member findings. As disagreements are disclosed, they can be openly resolved.

Preparing a house of quality. Preparing the house of quality is a group-problem-exploration process. It helps to structure the team's interpretation of the design problem. It homogenizes their knowledge of customer and company requirements, engineering characteristics, importance weights, competitive benchmarks, and design targets.

Developing a project plan. As the team members jointly develop a design project workscope, organization chart, budget, responsibilities table, and schedule they will gain a common understanding of where the project is headed and how they will get there. These aspects are covered in Chapter 14, on design projects.

Formulating is not just learning about a problem. Formulating is not just becoming familiar with the circumstances. Formulating is not just "going through the motions," so that we can begin solving the problem. Thorough and systematic design problem formulation is the *secret* to a successful solution.

The *secret* to a successful solution is a sound *formulation.*

3.7 SUMMARY

- Careful problem formulation leads to successful problem solution.
- The first step in design problem formulation is to obtain a detailed understanding of the problem, including customer and company requirements, importance weights, engineering characteristics, constraints, and customer satisfaction.
- Engineering characteristics are used to quantify how well an existing product, or future product design, fulfills a given customer or company requirement.
- Customer satisfaction curves illustrate the acceptable upper and lower limits of an engineering characteristic and the incremental satisfaction for a given change in performance.
- A variety of sources are used to gain a better understanding of the problem including: surveys, market studies, literature, focus groups, observation studies, and benchmark studies.
- The engineering design specification summarizes critical product design requirements.
- Team member consensus can be developed using the concurrent engineering approach while completing the EDS and the HoQ.
- The house of quality for product planning is a systematic and graphic representation of product design information organized as a matrix of rooms, roof, and basement.
- Quality function deployment is a method that utilizes houses of quality to systematically deploy the "voice of the customer" throughout the whole company.

REFERENCES

Badiru, A. B., and P. S. Pulat. 1995. *Comprehensive Project Management.* Englewood Cliffs, NJ: Prentice Hall.

Clausing, D. 1994. *Total Quality Development.* New York: ASME Press.

Cohen, L. 1995. *Quality Function Deployment.* Reading, MA: Addison-Wesley.

Crawford, C. M., and C. A. DiBenedetto. 2003. *New Products Management,* 7th ed. New York: McGraw-Hill.

Dieter, G. E. 2000. *Engineering Design,* 3d ed. New York: McGraw-Hill.

Dixon, J. R., and C. Poli. 1995. *Engineering Design and Design for Manufacturing.* Conway, MA: Field Stone Publishers.

Garvin, D. A. 1987. "Competing on Eight Dimensions of Quality." *Harvard Business Review,* November–December, pp. 101–109.

Haik, Y. 2003. *Engineering Design Process.* Pacific Grove, CA: Brooks/Cole.

Hauser, J. R., and D. Clausing. "The House of Quality," *Harvard Business Review,* May–June 1988, pp. 63–73.

Hazelrigg, G. A. 1996. *Systems Engineering: An Approach to Information-Based Design,* Upper Saddle River, NJ: Prentice Hall.

Keeney, R. L., and H. Raiffa. 1976. *Decisions with Multiple Objectives.* New York: John Wiley & Sons.

King, B. 1989. *Better Designs in Half the Time,* 3d ed. Methuen, MA: GOAL/QPC.

Pahl, G., and W. Beitz. 1996. *Engineering Design*, 2d ed. New York: Springer-Verlag.

Pugh, S. 1991. *Total Design*. Reading, MA: Addison-Wesley.

Siddall, J. N. 1982. *Optimal Engineering Design: Principles and Application*. New York: Marcel Dekker.

Stub, A., J. F. Bard, and S. Globerson. 1994. *Project Management: Engineering, Technology and Implementation*. Upper Saddle River, NJ: Prentice Hall.

Summers, D. C. S. 1997. *Quality*. Upper Saddle River, NJ: Prentice Hall.

Urban, G. L., and I. R. Hauser. 1993. *Design and Marketing of New Products*, 2d ed. Englewood Cliffs, NJ: Prentice Hall.

KEY TERMS

Benchmark	Customer	Focus groups
Benchmark studies	Customer requirements	House of quality
Company requirements	Customer satisfaction	Importance weights
Constraints	curves	Performance targets
Consumer	Engineering characteristics	Quality function
Correlation ratings matrix	Engineering design	deployment
Coupling matrix	specification	Quality product

EXERCISES

Self-Test. Write the letter of the choice that best answers the question.

1. _____ The most important category of customer requirements is:

    a. constraints

    b. functional performance

    c. appearance

    d. reliability

2. _____ All of the following are types of customer requirements except:

    a. robustness

    b. strategy

    c. ease-of-use

    d. geometry

3. _____ The performance of a product in spite of variations in its material properties, how it is manufactured, or how it is used is called:

    a. reliability

    b. maintenance

    c. robustness

    d. human factors

4. _____ The secret to the successful solution of a design problem is a sound:

    a. project plan

    b. concept design

    c. parametric design

    d. formulation

5. _____ Assessments of competitive products are called:
   a. focus groups
   b. observation studies
   c. market studies
   d. benchmark studies

6. _____ Measures that help us quantify how well a product performs customer and company requirements are called:
   a. importance weights
   b. constraints
   c. engineering characteristics
   d. variables

7. _____ Restrictions on the function or form of a design candidate are called:
   a. importance weights
   b. variables
   c. engineering characteristics
   d. constraints

8. _____ Customer satisfaction curves tend to fall into the following types except:
   a. more is better
   b. target is better
   c. benchmark is best
   d. less is better

9. _____ An estimate of how well an engineering characteristic fulfills a customer's expectations is called:
   a. goodness unit
   b. constraint
   c. customer satisfaction
   d. parameter

10. _____ According to the *Time* magazine survey, a quality product it one that: works as it should, lasts a long time, and is:
   a. reliable
   b. easy to maintain
   c. durable
   d. low-cost

11. _____ An engineering design specification includes the following except:
   a. constraints
   b. engineering characteristics
   c. demographics
   d. importance weights

2: b, 4: d, 6: c, 8: c, 10: b

12. Briefly explain the difference between an explicit constraint and an implicit constraint.

13. When designing a new electric coffeemaker, we learn that our customers require the coffeemaker to "produce hot, delicious coffee." Define an engineering characteristic to quantify this customer requirement. What units could it have? What limits might be imposed?

14. Assume that a motorcycle design team is debating whether to spend a significant proportion of their development budget on designing a better suspension to reduce road shock transmitted to the rider. How could you use information from Table 3.1 to settle the issue?

15. Assume that our development team is designing a new bicycle frame. Give a specific example for each category of company requirements:

   a. marketing

   b. manufacturing

   c. financial (e.g., manufacturing—New design must use welding and machining processes)

16. Assume that for a high-performance motorcycle design, acceleration has been selected to measure the "go fast" customer requirement. Use the number of G's for units (i.e., the multiples of 32 ft./sec./sec.). Sketch a possible customer satisfaction curve for this engineering characteristic. Include units and limits. Explain the shape of your curve.

17. Sketch a possible customer satisfaction curve for the price tag of a new automobile. Include units and limits. Explain the shape of your curve.

18. Consider the design of a new pocketknife for hiking and camping.

   a. Define three possible customer requirements (use verbs to express functions) (e.g., cuts easily),

   b. Establish a reasonable set of importance weights for the three requirements,

   c. Define one or more engineering characteristics for each of the requirements. Include units and limits,

   d. On a separate sheet of paper sketch a customer satisfaction curve for each engineering characteristic.

19. The house of quality for product planning has eight "rooms." Briefly describe the contents of each room with a simple phrase.

20. Prepare a simple house of quality for the pocketknife information developed in the previous question. Use a Swiss Army knife as a benchmark. Estimate values.

21. Briefly explain how the quality function is deployed throughout the whole company using cascading houses of quality.

22. Briefly describe how a project team can develop a consensus of opinion.

CHAPTER 4

Concept Design

When you have completed this chapter you will be able to

- Distinguish alternative design concepts as different abstract embodiments of physical principles, materials, and geometry
- Clarify the functional requirements of a design
- Explain and use activity analysis
- Describe and apply function decomposition diagrams
- Generate alternative design concepts using various methods
- Analyze concepts for feasibility and/or preliminary performance
- Evaluate feasible concepts using Pugh's method
- Evaluate alternative concepts using weighted-rating method
- Describe and select alternative means to protect intellectual property

4.1 INTRODUCTION

Concept design is a phase in the evolution of product when alternative design concepts are generated, evaluated, and selected for further development. But what is a design "concept?" What differentiates one concept from another? A nonengineer would likely define a concept as an idea or thought. A design concept is *not* the same thing. As engineers, we need to be more specific in our use of the term. Let's examine the two examples below to get a better understanding.

First, consider the desired function of stopping a spinning shaft. Three alternative concepts might include a fan brake, a regenerative brake, and a disk brake, as shown in Table 4.1. Each concept uses a different physical principle to achieve the stopping action. The concepts differ in geometric and material aspects also.

TABLE 4.1 Alternative Concepts for Slowing and Stopping a Spinning Shaft

Alternative	Physical principle	Abstract embodiment
1	fluid friction	fan blade on shaft
2	magnetic field	regenerative brake
3	surface friction	disk-and-caliper brake

Second, consider the required function to fasten sheets of paper. A few alternative concepts are shown in Table 4.2. A paper clip is a loop of solid, elastic material that when separated causes a clamping force. The paper clip is an embodiment that behaves according to the "spring force" physical principle described by Hooke's law, which states that a force in an elastic body is proportional to its elongation.

As we can readily conclude from the two examples, a **design concept** is an alternative that includes at least physical principles and abstract embodiments. But there is more.

Design concepts exhibit a number of similarities:

1. concepts work by some physical principle, phenomenon, or principle,
2. physical principles act on some surface or location,
3. concepts exhibit geometric properties, or shapes,
4. concepts have an abstract embodiment,
5. concepts imply relative motion of surfaces, or objects, and
6. concepts imply general material types.

We define a **physical principle** as the means by which some effect is caused, or produced. Physical principles are often described by analytical or empirical relationships that couple the causes and effects, such as Hooke's law $F = k\Delta L$, or Coulomb friction, $F = \mu N$. A list of some physical principles is given in Table 4.3.

The physical principle acts on a working material. The *working material* has mechanical, physical, and chemical properties. It may be a solid, liquid, or gas, having inherent mechanical properties such as hardness, ductility, coefficient of friction, modulus of elasticity, and yield strength.

TABLE 4.2 Alternative Concepts for Fastening Sheets of Paper

Alternative	Physical principle	Abstract embodiment
1	spring force	paper clip
2	bent clamp	staple
3	bendable clamp	cotter pin
4	adhesion	glue

TABLE 4.3   Representative Physical Principles

Conservation of energy	Archimedes' principle	Ohm's law
Conservation of mass	Bernoulli's law	Ampere's law
Conservation of momentum	Boyle's law	Coulomb's laws of electricity
	Diffusion law	Gauss' law
Newton's laws of motion	Doppler effect	Hall effect
Newton's law of gravitation	Joule-Thompson effect	Photoelectric effect
	Pascal's principle	Photovoltaic effect
Coriolis effect	Siphon effect	Piezoelectric effect
Coulomb friction	Thermal expansion effect	
Euler's buckling law		
Hooke's law	Newton's law of viscosity	
Poisson effect/ratio	Newton's law of cooling	
	Heat conduction	
	Heat convection	
	Heat radiation	

The physical principle acts on the **working geometry** composed of surfaces and motions. A brake disk (rotor), for example, is a flat, circular surface. Physical principles can act at a point, line, and/or area surface. Motions can be rotational, translational, or nonmoving. Motions can also vary with time and magnitude. The successful embodiment of a physical principle on a working geometry in a working material has been defined as a **working principle** (Pahl and Beitz, 1996).

Example

Prepare a sketch of a disc brake concept. Show the disc (rotor) and label the physical principle, working geometry, motion, and material.

When we step on a brake pedal it pushes a rod that compresses hydraulic fluid in the master cylinder. The fluid pressure, approximating 1,000 psi, is transmitted to the brake's caliper piston causing it to expand. The piston forces the brake pads to clamp an annular portion of the rotor surface, causing the frictional braking force.

The solid disc rotates as shown in Figure 4.1 The friction force, F_f, acts on the planar surface that is perpendicular to the axis of rotation.

In concept design we purposely delay making decisions about specific shapes, configurations, sizes, materials, or manufacturing processes. In concept design we do not determine any sizes or configurations. We allow the concept to be an abstract embodiment. For example, we do not select a rotor diameter or a pad thickness for the disk brake; and we do not determine any configuration details for the paper clip. In other words, an abstract embodiment will allow us the freedom to generate many alternative configurations, which will be analyzed and evaluated in the configuration design phase, discussed later in the text.

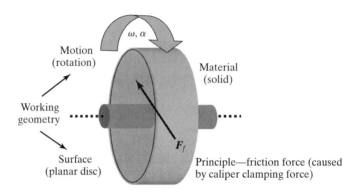

Note: abstract sizes and shapes

FIGURE 4.1

The disc-brake concept is the abstract embodiment of a physical principle (friction force) acting on the working geometry (rotating planar disc) of a material (solid).

A design concept is the abstract embodiment of a physical principle, material, and geometry.

The concept design phase begins with a review of the engineering design specifications and related documents, and concludes with one or more concepts to be developed further, as shown in Figure 4.2. During concept design we participate in a number of decision-making activities. We clarify functional requirements, generate alternative concepts, and analyze the concepts to determine if they are feasible. We reject infeasible concepts and iterate. Then, we evaluate the feasible concepts to select the best ones for further development. We examine each of theses activities in the remaining sections of this chapter.

4.2 CLARIFYING FUNCTIONAL REQUIREMENTS

The engineering design specification will usually provide information on customer and company requirements. However, it may lack sufficient details on specific functions and/or subfunctions. Therefore, the following three methods may be used to help us clarify the product's functional requirements: (1) activity analysis, (2) component decomposition, and (3) functional decomposition.

4.2.1 Activity Analysis

Activity analysis can be used to learn how the customer will use and ultimately retire the product. Customer activity categories are shown in Table 4.4.

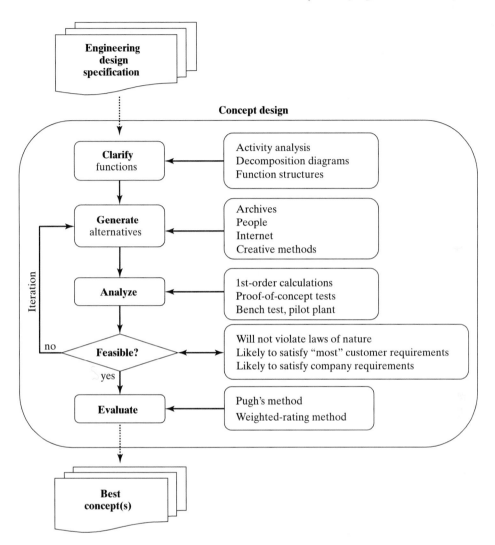

FIGURE 4.2

Concept design decision-making activities.

Let's examine the customer activities associated with using and retiring a rechargeable electric shaver, shown in Table 4.5. We often consider just the daily-use pattern, activities 7–15. But maintenance activities 16–18 must also be considered, as well as the setup activities 1–6.

We readily see that an activity analysis helps us to understand all the required functions, not just those during daily use. We also learn how the product interacts with the environment. In this case, the shaver is recharged and stored in a drawer. In essence, the activities lead to subfunctions that must be accommodated in the final design.

TABLE 4.4   Customer Activities Relating to Using and Retiring a Product

Use	set up
	operating
	maintain
	repairing
Retire	take down
	disassemble
	recycle
	dispose

TABLE 4.5   Activity Analysis List for Electric Shaver Use and Retirement

	Setup	1. open package
		2. examine shaver, cord, travel case, and cleaning brush
		3. read instruction booklet
		4. fill out warranty card
		5. plug in shaver to charge batteries
		6. put shaver, case, cord, brush in bathroom cabinet drawer
	Daily use	7. remove charged shaver from drawer
		8. trim hair
		9. shave face or legs
		10. remove cutter blade cover
		11. brush cutter blade
Use		12. replace cover
		13. repeat step 5
		14. store shaver in drawer
		15. repeat steps 7–14 until blades need replacing
	Replace blade	16. remove cutter blade cover
		17. install new cutter blade
		18. replace cutter cover
	Daily use	19. repeat steps 7–13 until batteries need replacing
	Replace batteries	20. install new rechargeable batteries
	Daily use	21. repeat steps 17–19 until shave becomes unrepairable
Retire	Dispose of shaver	22. throw out shaver and auxiliaries

4.2.2   Product Component Decomposition

When competitive products are available, they can be disassembled into their respective components, and a **component decomposition** diagram can be drawn. The product component decomposition diagram is a block diagram of the parts and subassemblies that make up the product. Penlight and coffeemaker product component decomposition diagrams were presented in Chapter 2. An additional example, that of a coffeemaker, is shown in Figure 4.3.

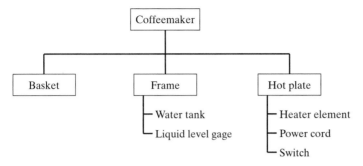

FIGURE 4.3

Product component decomposition diagram of a coffeemaker.

Component decomposition diagrams illustrate the hierarchical structure of component forms, not functions. As we subdivide individual subassemblies into their constituent components, however, we can obtain a better overall understanding of how individual components interact with each other and ultimately contribute to the overall product function.

4.2.3 Product Function Decomposition

A product **function decomposition** subdivides the major functional requirement into its respective subfunctions and subsubfunctions. The function decomposition diagram is a hierarchical structure of functions, not forms. The diagrams help us to identify whether functions are connected, and where the interface connections might be. For example, let's examine a function decomposition diagram of a coffeemaker, as shown in Figure 4.4.

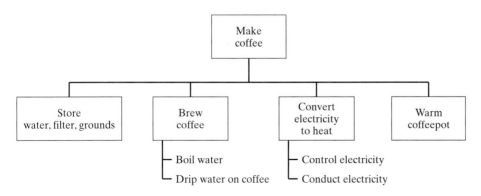

FIGURE 4.4

Function decomposition diagram of a coffeemaker.

The first block describes the overall product function, such as make coffee. Then, the product function is further decomposed into subfunction blocks, such as brew coffee and convert electricity to heat. The decomposition is continued so as to successively refine the subfunctions and understand the logical changes in energy, materials, and signal.

Function decomposition diagrams help us separate what functions need to be performed versus how it gets done (form). Later, when we investigate alternative concepts, we will have the freedom to select any form to meet the desired functions. For example, we could choose to warm the pot by hot-air convection rather than a conducting hot plate. Or, we could boil the water in the pot, then pump it into a brewing chamber.

Functions are usually expressed as verbs and act on objects or entities expressed as nouns. In the example above, some verb noun pairs are: brew coffee, warm coffeepot, store water, and convert electricity. A number of other functions are given in Table 4.6.

Noun-objects can be classified into energy, material, and signal categories. When preparing a function decomposition diagram, we should carefully consider how energy, material, and signals are changed or acted on by the function. For example, the cold water is heated, resulting in hot water. Electric energy is transformed to thermal energy. A "switch" signal controls the electric power. The general process is shown in Figure 4.5.

TABLE 4.6   Fundamental Functions

amplify	dissipate	protect
change	fasten	release
channel	heat	rotate
collect	hold	separate
conduct	increase	store
control	join	supply
convert	lift	support
cool	lower	transform
decrease	move	translate

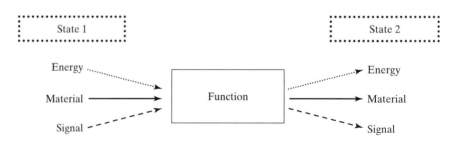

FIGURE 4.5

Function changes the state of energy, material, and information.

Upon examining the completed function decomposition diagram, the team members may find that they can remove, combine and/or reorganize some subfunctions. For example, they may decide not to warm the pot, that is, remove the subfunction. Other functions might be combined, such as store coffee grounds and brew coffee, as in a basket. During the process of removing, combining, or reorganizing subfunctions, we will likely produce more diagrams. But that is all right. Our goal is to understand the desired product function as best we can.

Detailed function structure diagrams can be created by combining changes in energy, materials, and signals with function decomposition and activity analysis. For further information on function structure diagrams see Hundal (1997), Otto and Wood (2001), and Ullman (1997).

4.3   GENERATING ALTERNATIVE CONCEPTS

To be selective, we need a selection. In other words, if we are to select the best concepts for further development, we need to systematically generate a lot of alternatives. We will be looking for concepts that will potentially satisfy the product subfunctions. Alternative concepts will differ primarily in physical principle or in abstract embodiment. Note that the process of generating alternatives is sometimes called **synthesis**. Along the way we will investigate archives, talk to people, connect to the Internet, and use creative methods.

Archives.   University, public, and corporate libraries should be first on the list of places to look for alternative concepts. In addition to design catalogs, reference handbooks, encyclopedias, and specialized monographs, libraries also maintain collections of periodicals, including technical journals and trade magazines. Some companies also maintain file cabinets full of current and past design information. While most archives are paper-based, other media used include microfiche, microfilm, and computer-based or electronic databases. A systematic search of these sources should trigger a number of alternative concepts.

People.   Starting with our co-workers, we can confer, one on one, with people knowledgeable in the field. We might also contact our local university engineering professors. Then, current vendor representatives and professional society acquaintances should be contacted. Finally, we might hire a consultant.

Internet.   The Internet should also be searched including: the U.S. Patent Office, Thomas' On-Line Register, vendor Web site catalogs, professional societies, and trade organizations.

Existing Products.   If available, competitive products can be purchased and dissected. Although patents may protect the product, the examination often reveals opportunities for new designs.

Creative Methods.   By researching archives, people, and the Internet, we will likely find many alternative concepts that have fulfilled previously defined subfunction requirements. When we are looking for new solutions, however, we might consider the following innovative methods:

Brainstorming. **Brainstorming** is an iterative group method that takes advantage of team members' diverse skills, experience, and personalities to generate innovative ideas. A group of participants is gathered in a room with an easel or blackboard/whiteboard. After having the "problem" presented, participants suggest ideas that are then transcribed to the easel or board. No criticism of alternatives or ideas is permitted. Wild and crazy ideas are encouraged. The ideas, written on the board, act to stimulate participants. Ideas are transcribed until no new ideas are stimulated.

A variant of the brainstorming technique uses 3-by-5 cards or Post-it notes. Each participant is asked to write down three ideas. The cards are collected and anonymously transcribed to a large whiteboard for everyone to see. After everyone has had a chance to read through the whole list, each member fills out another 3-by-5 card, building on the first set of alternatives. Again, no criticism of ideas is permitted. The cards are again collected, and information transcribed to the larger board. The cycle can be repeated until no more new alternatives are generated. Verbal brainstorming, without cards, is not as good because some members may dominate discussion or telegraph criticism, which can quash creative thinking. The cards, on the other hand, are available for a permanent record of the meeting, too.

Method 6-3-5. Method 6-3-5 is a refinement of the brainstorming method, and was developed by Rohrbach (1969). A group of about six members gather. Each member writes down three ideas on a sheet of paper. Each sheet of paper is circulated to a neighbor. After reading the ideas, the neighbor writes down three more ideas. The sheets of paper are circulated five times.

Synectics. **Synectics** is a method that requires the problem solver to view the problem from four perspectives: analogy, fantasy, empathy, and inversion (Gordon, 1961). Viewing similar problems in nature by looking for an **analogy** stimulates idea creation: a tree as an analogy to a structure; a piping network as an analogy to an electrical wiring network. **Fantasy** asks us to imagine the impossible: for example, using an antigravity belt to commute to work. **Empathy** asks us to imagine being a product, and how we would perform each function. **Inversion** asks us to take an "inverted" or reverse point of view, inside versus outside, black versus white, quiet versus noisy, and so on.

Checklists. Companies and individuals have prepared some checklists that stimulate creative thinking. For example, Osborn (1957) proposed a checklist of nine starter questions: Substitute? Combine? Adapt? Magnify? Modify? Put to other uses? Eliminate? Rearrange? Reverse?

4.4   DEVELOPING PRODUCT CONCEPTS

During concept design we attempt to generate many alternative concepts for each subfunction. A product concept variant, on the other hand, is a development or combination of specific concepts. For example, let's assume that for an

TABLE 4.7   Morphological Matrix

		Alternative Concepts				
		1	2	3	. . .	n
Subfunctions	SF_1	C_{11}	C_{12}	C_{13}		C_{1n}
	SF_2	C_{21}	C_{22}	C_{23}		C_{2n}
	SF_3	C_{31}	C_{32}	C_{33}		C_{3n}
	. . .					
	SF_m	C_{m1}	C_{m2}	C_{m3}		C_{mn}

arbitrary product, we generate two concepts for subfunction SF_1 and three concepts for subfunction SF_2. We could designate the alternative concepts as $C_{i,j}$, where i represents the subfunction and j represents the alternative concept, resulting in SF_1: $\{C_{11}, C_{12}\}$ and SF_2: $\{C_{21}, C_{22}, C_{23}\}$. The following six product concept variants are theoretically possible: (1) C_{11}, C_{21}, (2) C_{11}, C_{22}, (3) C_{11}, C_{23}, (4) C_{12}, C_{21}, (5) C_{12}, C_{22}, (6) C_{12}, C_{23}.

Sometimes, the combinations are not compatible or realizable, in which case that product concept variant would be eliminated from further consideration. In other cases we may find that a concept can perform or share two functions, like the screwdriver. It can push and turn.

We can list the subfunctions in a column of a matrix, and the alternative concepts for each function in adjacent rows, as shown in Table 4.7. This approach is called a **morphological matrix**. To stimulate alternative combinations, the design team selects one concept from any column, for each function, proceeding down the matrix. The total number of theoretically possible combinations is equal to the product of the number of concepts for each subfunction. In our example we have 2×3, or 6 combinations. Obviously, for complicated products the total number of possible combinations can be large. If a subfunction is independent of the others, it can be eliminated from the matrix, and considered separately. For example, the type of radio in an automobile has little or no connection to concept alternatives for drive train. Sometimes, to simplify the work effort, teams consider customer and company requirements as they develop the combinations, and thereby eliminate less prospective combinations.

Example

A design team is developing new concepts for a gasoline-engine-powered minbike. They decompose the primary functions into transmit engine power, steer, and brake. They select three concepts for transmitting power: chain drive, belt drive, and gearbox; two concepts for braking: disc and drum; and threes concepts for steering: handlebar, airplane control stick, and fly-by-wire. Prepare a morphological matrix and systematically develop alternative combinations.

We prepare the morphological matrix by listing the subfunctions in the leftmost column and the alternative concepts for each subfunction in the same row, as shown in Table 4.8.

TABLE 4.8

		Alternative Concepts		
		1	2	3
Subfunctions	Transmit	Chain	Belt	Gearbox
	Brake	Disc	Drum	
	Steer	Handlebar	Control stick	Fly-by-wire

Combining three alternatives for subfunction 1, two alternatives for subfunction 2, and three alternatives for subfunction 3, we have $3(2)3 = 18$ possible combinations. By systematically indexing row and column subscripts we can produce all 18 combinations, as follows:

$$C_{11}, C_{21}, C_{31}, \quad C_{12}, C_{21}, C_{31}, \quad C_{13}, C_{21}, C_{31},$$
$$C_{11}, C_{21}, C_{32}, \quad C_{12}, C_{21}, C_{32}, \quad C_{13}, C_{21}, C_{32},$$
$$C_{11}, C_{21}, C_{32}, \quad C_{12}, C_{21}, C_{33}, \quad C_{13}, C_{21}, C_{33},$$
$$C_{11}, C_{22}, C_{31}, \quad C_{12}, C_{22}, C_{31}, \quad C_{13}, C_{22}, C_{31},$$
$$C_{11}, C_{22}, C_{32}, \quad C_{12}, C_{22}, C_{32}, \quad C_{13}, C_{22}, C_{32},$$
$$C_{11}, C_{22}, C_{32}, \quad C_{12}, C_{22}, C_{33}, \quad C_{132}, C_{22}, C_{33},$$

Each of the 18 combinations is shown in Table 4.9. The 18 alternatives are then analyzed and evaluated to select the better ones for further development.

TABLE 4.9   Alternative Designs Created by Systematically Indexing the Rows and Columns of a Morphological Matrix

	Subfunction		
Alternative	Transmit	Brake	Steer
1	Chain	Disc	Handlebar
2	Chain	Disc	Control stick
3	Chain	Disc	Fly-by-wire
4	Chain	Drum	Handlebar
5	Chain	Drum	Control stick
6	Chain	Drum	Fly-by-wire
7	Belt	Disc	Handlebar
8	Belt	Disc	Control stick
9	Belt	Disc	Fly-by-wire
10	Belt	Drum	Handlebar
11	Belt	Drum	Control stick
12	Belt	Drum	Fly-by-wire
13	Gearbox	Disc	Handlebar
14	Gearbox	Disc	Control stick
15	Gearbox	Disc	Fly-by-wire
16	Gearbox	Drum	Handlebar
17	Gearbox	Drum	Control stick
18	Gearbox	Drum	Fly-by-wire

4.5 ANALYZING ALTERNATIVE CONCEPTS

Not every alternative that we generate will function or be manufacturable. We should screen out, or eliminate, those that are not feasible.

During parametric design we analyze alternatives using analytical relations and/or engineering tests to predict their performance. But, during the concept design phase, we do not have specific information about sizes, configurations, material properties, or manufacturing processes. Each concept variant is characterized only by its physical principle and abstract embodiment.

We can, however, get a rough idea of whether an alternative will function and/or whether we can manufacture it. For example, we can make a few assumptions and calculate a few performance estimates based on simple laws of motion, heat transfer, solid mechanics, and/or thermodynamics. These back-of-the-envelope calculations can rule out "impossible" ideas. We can perform benchtop experiments to see whether a physical principle will work in a specific application. We can confer with manufacturing personnel to confirm the manufacturability of a concept. And we can investigate whether a supporting technology is ready for the production line or should stay in R&D.

The screening criteria should focus on functionality and manufacturability, and should include:

1. Will the concept likely function?
2. Will the concept likely meet the customer's minimum performance requirements? (These are the "musts," not the "shoulds.")
3. Will the concept likely survive the operating environment?
4. Will the concept likely satisfy other critically important customer requirements?
5. Will the concept be manufacturable, (i.e., materials and manufacturing processes)?
6. Will the concept likely satisfy the financial and or marketing requirements?

During the analyzing and screening process, we will usually find concept variants that should be eliminated. Or, on the other hand, upon reexamination, some of those to be eliminated might be reconceived, to remove their deficiency. This is an option only if time and resources permit.

4.6 EVALUATING ALTERNATIVE CONCEPTS

Assuming that we have screened out those candidates that were not functional or manufacturable, those remaining can be evaluated to determine which should be developed further.

Two methods commonly used are **Pugh's concept selection method** and the **weighted-rating method**.

Pugh's Method for Concept Selection. The concept selection method developed by Pugh (1991) includes the following steps:

1. The team selects evaluation criteria, principally from the engineering design specification and other documents prepared during the formulation phase.

2. A matrix is prepared listing the evaluation criteria in the first column.
3. The concepts are identified in the remaining columns.
4. One concept is selected as they datum concept or reference concept and labeled as datum.
5. The team selects a concept to evaluate. For each criterion the team marks whether the concept is better ($+$), worse ($-$), or about the same (S) as the datum.
6. Each of the other concepts is similarly rated, using the same marking system.
7. All the $+$'s, $-$'s, and S's are summed and recorded at the bottom of the matrix.

An example matrix for a go-cart transmission is shown in Table 4.10. As the team discusses each entry, the team gains a greater understanding of the design problem, the alternative concepts, and the specified requirements. Also, new or revised concepts may originate. These may be added to the matrix for consideration. It would appear that the team favors the gears concept, in that there are more $+$'s and fewer $-$'s, as compared to the chain concept.

When the results are tallied, some concepts may appear as "strong" concepts in that they have more $+$'s than $-$'s. However, strong concepts may exhibit a few weaknesses that could be improved upon. Similarly, weak concepts might be strengthened. The method will often indicate the weakest concepts that should be eliminated from further development.

The method, as proposed by Pugh, has one drawback, however. Each criterion is assumed to have equal importance. There is no importance weight factored into the evaluation. Modified versions have been proposed that include an importance weight column and calculations that factor in the importance weight. A modified version for the transmission example is shown in Table 4.11.

Weighted Rating Method. The weighted-rating method uses a similar matrix layout as the modified Pugh's method. It is also called the weighted sum

TABLE 4.10 Pugh's Concept Selection Method

Criteria	Concept Alternatives		
	Gears	V-belts	Chain
high efficiency	$+$	D	$+$
high reliability	$+$	A	$+$
low maintenance	$+$	T	S
low cost	$-$	U	$-$
light weight	$-$	M	$-$
$\Sigma+$	3	NA	2
$\Sigma-$	2	NA	2
ΣS	0	NA	1

TABLE 4.11   Modified Pugh's Method

Criteria	Importance Wt. (%)	Concept Alternatives		
		Gears	V-belts	Chain
high efficiency	30	+	D	+
high reliability	25	+	A	+
low maintenance	20	+	T	S
low cost	15	−	U	−
light weight	10	−	M	−
	100			
Σ+		75	NA	55
Σ−		25	NA	25
ΣS		0	NA	20

method or the Pahl and Beitz method. The method is quite similar and includes the following steps:

1. Team selects evaluation criteria.
2. A matrix is prepared listing the evaluation criteria in the first column.
3. Importance weights are given for the criteria, usually as percentage points, adding to 100.
4. Concepts are identified in columns.
5. Team rates each concept as unsatisfactory, just tolerable, adequate, good, or very good using an ordinal scale such as 0, 1, 2, 3, or 4. Other scales have also been used.
6. Each concept rating is multiplied by its respective weight and summed to produce an overall rating for the concept.

An example weighted rating matrix for a go-cart transmission is shown in Table 4.12.

The process of analyzing and evaluating alternative concepts is naturally subjective, mainly because of the abstractness or fuzziness of the concepts. Some concepts will be obviously infeasible. A few, however, will be identified for further development. Concept design activities also provide an opportunity for the team to obtain a consensus, or common understanding of the important design issues.

4.7   CONCEPT DESIGN PHASE COMMUNICATIONS

A significant amount of information will be produced and processed by the team as concepts are generated, analyzed, and evaluated, including:

- photocopies of archival matter,
- printouts from the Internet,
- vendor catalogs and data sheets,

TABLE 4.12   Weighted Rating Method

Criteria	Importance Weight (%)	Gears		V-belts		Chain	
		Rating	Weighted Rating	Rating	Weighted Rating	Rating	Weighted Rating
high efficiency	30	4	1.20	2	0.60	3	0.90
high reliability	25	4	1.00	3	0.75	3	0.75
low maintenance	20	4	0.80	3	0.60	2	0.40
low cost	15	2	0.30	4	0.60	3	0.45
light weight	10	2	0.20	4	0.40	3	0.30
	100	NA	3.50	NA	2.95	NA	2.80

Rating	Value
Unsatisfactory	0
Just tolerable	1
Adequate	2
Good	3
Very Good	4

- preliminary test results,
- first-order calculations,
- patent abstracts,
- minutes of meetings,
- concept sketches,
- concept screening sheets,
- concept evaluation matrices,
- expert interview notes.

Each team member is usually required to maintain a project notebook to organize his personal notes. These notes, especially product ideas and sketches, may be used for patent disclosures, as well as trademark and/or copyright registrations. In addition, the team should establish a central information repository, such as a file cabinet, for the safe keeping of documents and other information that is pertinent and valuable to the project and company.

Sketching is also encouraged during this phase, as compared to CAD drawings. Hand-drawn sketches can quickly communicate form and function. Layout sketches are often used to illustrate the whole idea. Detail and assembly drawings, usually CAD-drawn, are not typically desired until further configuration information becomes available.

The team will also find that centralized word document files are valuable. Usually available on company file servers, these centralized files can serve as product data archives. Product data management systems can also be utilized as data sources.

Design review meetings minutes should be religiously maintained to record the team's findings as well as their reasoning. The team may find that reviewing these will often resolve misunderstandings. Also, subsequent design teams may find the records to be valuable.

4.8 INTELLECTUAL PROPERTY

Employees act on behalf of the stockholders or owners of a company. We are entrusted with the safekeeping of the company's property. Real property, for example, includes assets relating to real estate, like land, buildings, and land improvements such as bridges and parking lots. Personal property is another category of assets; it includes cash, marketable securities, vehicles, and office equipment. To protect these forms of property we use lockable safes, install building fire sprinkler systems and burglar alarms, and purchase insurance contracts.

Intellectual property is another form of property that includes ideas. We typically protect these with trade secrets, contracts, trademarks, trade dress, copyrights, and patents. Protection is especially important during the early stages of product design efforts. First and foremost we must recognize that our company is investing thousands, if not hundreds of thousands, of dollars in developing the new product. Automobile companies, for example, will invest millions of dollars in each new model. Second, we need to acknowledge that it is our responsibility, as trustees for the stockholders, to protect product development information. Let's examine the basic types of intellectual property protection.

Trade Secrets. Companies can protect secret intellectual property called **trade secrets** such as formulas, recipes, methods, processes, devices, and/or techniques from unauthorized use by anyone who obtained it by improper means or because of a confidential affiliation. The protection is afforded under most state laws and the Federal Economic Espionage Act of 1996 (ASME, 2001). Examples include the recipes for Coca-Cola, Coors beer, other brand-name food products, in addition to secret processes used to manufacture microcircuits. Two major provisions are that the secret have economic value and that the company maintain its secrecy, including taking steps to:

- identify specific sensitive information,
- notify those having access about its secrecy,
- require employees and visitors to sign confidentiality and nondisclosure agreements,
- restrict access to only those who need to know, and
- compartmentalize a sequence of processes among separate departments, to limit exposure.

A trade secret does not require an application or registration as in patenting, and can last an indefinite period of time. However, there is no protection from a third party independently discovering the secret on his own.

Contracts. A **contract** is a written or verbal agreement between two parties, such as between individuals and corporations. Examples include employment contracts having clauses that assign patent rights and copyrights to the employer, and also require the employee to nondisclose, or keep confidential, any proprietary information. Contracts have a finite or specific term, and do not have to be prepared by a lawyer to be valid. However, it is highly recommended and a simple contract may cost as little as $500 in legal fees.

Trademark. A **trademark** is a symbol, design, word, or combination thereof used by a manufacturer to distinguish its products from those of its competitors, principally to distinguish its source. Examples include IBM, GE, XEROX, Coke, and Pentium. Protection is provided against others from making, using, or selling products using trademarks under individual state laws and the Federal Trade Act of 1946. Trademark law also protects trade dress. Trade dress is a distinctive, nonfunctional feature (i.e., appearance) that distinguishes a manufacturer's goods or services from those of another. Trade dress involves product color, configuration, and packaging. Examples include the packaging for Wonder Bread, the tray configuration for Healthy Choice frozen dinners, and the color scheme of Subway sub shops as well as the Golden Arches, the International House of Pancakes' blue roof, and the Howard Johnson's orange roof.

Different from a patent, a trademark is protected from its first commercial use, whether it is registered or not. It need not be registered with the U.S. Patent and Trademark Office (USPTO) or the individual state. However, if registered with the USPTO, protection is afforded in all states by the one application. Registered trademark owners may also display the ® symbol. A trademark has a 20-year life and is indefinitely renewable.

Copyrights. Authors of creative literary, musical, or artistic works are afforded protection under the constitution and the U.S. Copyright Act of 1976. The law provides **copyrights** such that the author can exclusively publish and produce his works and/or sell the rights to his works. Examples include books, sheet music, photos, paintings, sculptures, dramas, sermons, movies, sound recordings, and computer software. The work must be creative, original, and fixed on paper, canvas, videotape, film, hard drive and/or disk. Under the "fair use" provision, small portions of the work may be reproduced for purposes of news reporting, research, scholarship, or teaching.

Works need not be registered to have some protection. The three requirements of copyright notice include the © symbol or the word "copyright," the name of the author, and the year that the work was first published. However, unless registered with the Copyright Office (a department of the Library of Congress), authors may not file a law suit in the United States. Registration includes a two-page application, copies of the work, and a modest filing fee. The term of a copyright is the author's life plus 70 years.

Patent. A **patent** is a document granting legal monopoly rights to produce, use, sell, or profit from an invention, process, plant (biological), or design. It is provided by U.S. laws dating back to 1790 and administered by the U.S. Department of Commerce—Patent and Trademark Office. **Utility patents** protect inventions such as Xerox copying, Polaroid photography, halogen light bulbs, and countless other machinery. Examples for process patents include polymer processes for Lexan, rayon, and Delrin. **Design patents** are granted for ornamental aspects of a product such as shape, configuration, and/or any surface decoration including toys, automotive trim products, and household products.

A number of conditions exist for a utility-patent invention including that it be (1) new, (2) useful, and (3) unobvious. The invention must also be completely and adequately described in the application. Finally, the invention must not be disclosed to the public more than one year prior to the patent application. A U.S. patent is granted only after a thorough review of the inventor's application by the USPTO.

Utility and process patents last for 20 years. The design patent lasts for 14 years. The filing fees range from about $500 to $1,100. However, the patenting process may cost $5,000 or more if a lawyer is used to expedite the application process. These costs are typically borne by an employer on behalf of the employee-inventor.

The basic types of intellectual property protection are summarized in Table 4.13.

TABLE 4.13 Summary of Methods to Protect Intellectual Property

	Protects	Length	Application Required	Registration Available	Costs
Trade Secret	formulas, recipes, processes	indefinite	no	no	some
Contract	items specified	length of contract	no	no	$500>
Trademark	graphical symbol or word	20 yrs renewable	no	yes	$>350
Copyright	literary, musical, or artistic works	author's life +70 yrs	no	yes	$>30
Utility Patent	function, process	20 yrs	yes	yes	$>1,100
Design Patent	appearance	14 yrs	yes	yes	$500>

4.9 SUMMARY

- A concept is an abstract embodiment of physical principle, material, and geometry.
- Function decomposition diagrams help to identify critical product functions and subfunctions.
- Alternative concepts can be generated using systematic searches and creative methods.
- Concept analysis examines whether a concept is functional and manufacturable.

- Concept evaluation can determine the relative worthiness of feasible alternatives using Pugh's method or the weighted-rating method.
- Effective design communication relies upon the proper use of adequate textual and graphical information.
- Trade secrets, contracts, trademarks, copyrights, and patents can protect intellectual property.

REFERENCES

ASME. 2001. *Intellectual Property: A Guide for Engineers.* New York: American Society of Mechanical Engineers.

Gordon, W. J. 1961. *Synectics: The Development of Creative Capacity.* New York: Harper & Row.

Hundal, M. S. 1997. *Systematic Mechanical Design.* New York: ASME Press.

Osborn, A. 1957. *Applied Imagination—Principles and Practices of Thinking.* New York: Charles Scribbner & Sons.

Otto, K. P., and K. L. Wood. 2001. *Product Design: Techniques in Reverse Engineering and New Product Development.* Upper Saddle River, NJ: Prentice Hall.

Pahl, G., and W. Beitz. 1996. *Engineering Design*, 2d ed. New York: Springer-Verlag.

Pugh, S. 1991. *Total Design.* Reading, MA: Addison-Wesley.

Rohrbach, B. *Kreatic Nach Regeln—Methode 635.* Absatzwirtschaft 12, 73–75, 1969.

Ullman, D. G. 1997. *The Mechanical Design Process*, 2d ed. New York: McGraw-Hill.

KEY TERMS

Activity analysis	Fantasy	Synthesis
Analogy	Function decomposition	Trade secret
Brainstorming	Intellectual property	Trademark
Concept design	Inversion	Utility patent
Component decomposition	Morphological matrix	Weighted-rating method
Contract	Patent	Working geometry
Copyrights	Physical principle	Working principle
Design concept	Pugh's concept selection	
Design patent	method	
Empathy	Synectics	

EXERCISES

Self-Test. Write the letter of the choice that best answers the question.

1. _____ The abstract embodiment of a physical principle, material, and geometry is called:

    a. synthesis

    b. activity analysis

    c. synectics

    d. design concept

2. _____ The means by which some effect is caused is called a:
   a. design concept
   b. motion principle
   c. physical principle
   d. function decomposition

3. _____ Diagrams that help to identify critical product functions and subfunctions are called:
   a. Pugh's method
   b. brainstorming
   c. function decomposition
   d. physical principle

4. _____ Examining how a customer will use and ultimately retire a product is called:
   a. geometry analysis
   b. physical analysis
   c. activity analysis
   d. decomposition analysis

5. _____ All of the following methods may be used to clarify the products functional requirements except:
   a. activity analysis
   b. function structures
   c. inversion
   d. functional decomposition

6. _____ All the following are functions performed by an electric coffeemaker except:
   a. store water
   b. convert electricity
   c. grind coffee
   d. warm coffeepot

7. _____ The process of generating alternatives is sometimes called:
   a. synectics
   b. synthesis
   c. inversion
   d. brainstorming

8. _____ A method that requires a problem solver to view the problem from the four perspectives—analogy, fantasy, empathy, and inversion—is called:
   a. brainstorming
   b. archives
   c. synthesis
   d. synectics

9. _____ As a function is performed, the following states may be changed except:

   a. matter

   b. energy

   c. time

   d. signal

10. _____ A group method that takes advantage of team members' diverse skills, experience, and personalities to generate innovative ideas is called:

    a. Internet

    b. archives

    c. brainstorming

    d. empathy

11. _____ A feasible alternative concept design will likely:

    a. satisfy customer requirements

    b. satisfy company requirements

    c. not violate laws of nature

    d. all of these

12. _____ The one drawback of the original Pugh's method is:

    a. criteria are not evaluated

    b. importance weights are not factored into the evaluation

    c. concepts are identified in columns

    d. the team is unable to obtain a consensus

13. _____ Concept evaluation methods to determine the relative worthiness of feasible alternatives include all of the following methods except:

    a. weighted rating

    b. Pugh's

    c. modified Pugh's

    d. Hooke's

14. _____ A matrix used to form combinations of concepts into an alternative design is called:

    a. manufactured matrix

    b. morphological matrix

    c. concept matrix

    d. mini matrix

15. _____ The method of intellectual protection that can last an indefinite period of time is called:

    a. trademark

    b. trade secret

    c. contract

    d. copyright

16. _____ A symbol, design, word, or combination thereof, used by a manufacturer to distinguish its products from those of its competitors, is called a:

    a. trade

    b. copyright

    c. trademark

    d. trade dress

17. _____ A document granting legal monopoly rights to produce, use, sell, or profit from an invention, process, plant (biological), or design is called a:

    a. trademark

    b. trade dress

    c. contract

    d. patent

18. _____ Authors of creative literary, musical, or artistic works have their work protected with a:

    a. contract

    b. copyright

    c. patent

    d. trade dress

19. _____ A written or verbal agreement between two parties is called a:

    a. trademark

    b. litigation

    c. deal

    d. contract

20. _____ Intellectual property can be protected by using all of the following except:

    a. contracts

    b. sprinkler systems

    c. copyrights

    d. trade secrets

2: c, 4: c, 6: c, 8: d, 10: c, 12: b, 14: b, 16: c, 18: c, 20: b

21. List five fundamental functions that act on matter, energy, and information.

22. List nine physical principles, laws, or effects.

23. On a separate sheet of paper, and using the format of Table 4.5, prepare an activity analysis for a power lawn mower (hint: Table 4.4).

24. On a separate sheet of paper, prepare a simplified functional decomposition for a wheelbarrow used for landscaping.

25. Consider the functional requirement to "fasten" a cover plate on an automobile transmission.

    a. Talk to at least five people to get "concepts." List them here.

    b. Connect to the Internet and do a search for "fasteners" or "fastening methods." List five of the better concepts found.

26. List five literature sources for "generating" concepts.
27. Connect to the U.S. Patent and Trademark Office at www.uspto.gov. Select a product type of interest to you. Examine some of the patent abstracts. Print out a patent diagram. Describe the basic principles of operation.
28. Examine the spreadsheet below. Note that the design team has decided upon a rating for each of the four alternatives. Complete the weighted-rating calculations. Remember to convert the percentages to decimals before multiplying. Sum the weighted ratings for each candidate.

Concept Evaluation Using Weighted Rating Method (Pahl and Beitz Method)

| | | Alternative Concepts / Embodiments | | | | | | | |
| | | Gears | | V-belts | | Chain | | Linkages | |
Criteria	Weight (%)	rating	wt. rating	rating	wt. rating	rating	wt. rating	rating	wt. rating
cost	20	3		4		4		4	
weight	15	2		4		3		4	
size	15	3		2		3		3	
reliability	9	4		3		4		4	
efficiency	8	4		3		4		3	
force	8	4		3		4		4	
integral braking	10	4		3		4		3	
customer appeal	15	2		1		4		3	
	100	na		na		na		na	

Unsatisfactory	0
Just tolerable	1
Adequate	2
Good	3
Very Good	4

a. Rank the alternatives based on their weighted rating.
b. Is there an obvious winner? Why?

CHAPTER 5

Selecting Materials

L E A R N I N G
O B J E C T I V E S

When you have completed this chapter you will be able to

- Explain the interdependency of product function, material, process, and geometry
- Describe fundamental material classes
- Characterize fundamental mechanical and physical properties
- Establish criteria for screening materials
- Parametrically rate alternative materials

5.1 INTRODUCTION

During the formulation phase of product development we determined the operating environment and the primary functions of the product. Then during conceptual design we selected physical effects and abstract embodiments that would be functional and manufacturable. The abstract embodiments included considerations of working geometries and materials. Finally, even though we did not need to choose specific materials at that time, we made some general assumptions as to their basic properties.

For the brake rotor, for example, we established that the selected material would need to be strong, resist thermal warping, not deflect or deform during use, and not wear out prematurely. From our general knowledge of metals, polymers, ceramics, and composites we assumed that metallic materials would be considered further.

As our product design develops during the configuration and parametric design phases, however, we need to select specific materials for each special-purpose part, and in the cases when optional materials are available from the supplier of standard parts.

97

During configuration design we will examine product configuration and part configuration. Product configuration includes the spatial arrangement and or connectivity of components, whereas part configuration considers the selection and arrangement of geometric features.

Parametric design is the last phase of embodiment design. During this phase we want to predict the brake's operating performance, which depends on the specific values of the design variables, such as rotor diameter and thickness. Therefore, we need to examine specific materials utilizing their properties in our estimates. In the brake rotor design example:

given the following material properties:	*we can estimate:*
coefficient of friction	brake torque
specific heat, thermal conductivity	maximum operating temperature
density	rotor weight
modulus of elasticity	deformations

Then, in detail design we confirm our material choices and their properties, rerunning our calculations with as-delivered material properties, and in so doing, validate our preliminary performance estimates.

Making material selection decisions is difficult, however. A successful product must work and must be manufacturable. For it to work, it must have the right geometry in relation to its function. For example, if the material is too weak, we can make the shape stouter. However, making a bigger part will likely increase material and manufacturing costs. In fact, as we shall see in Chapter 6, "Selecting Manufacturing Processes," some shapes are not compatible with some manufacturing processes. Material properties, manufacturing processes, product geometry, and product function are all interrelated (Dieter, 2000; Groover, 1996). This interdependence of product function, product geometry, material properties, and manufacturing processes is illustrated in Figure 5.1.

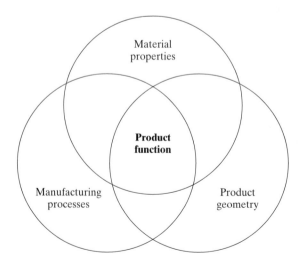

FIGURE 5.1

Interdependence of product function, material properties, manufacturing processes, and product geometry.

In the remaining sections, we will:

- define some fundamental mechanical properties and physical properties,
- examine basic material classes,
- examine a method to screen-out inadequate materials, and
- investigate a method to parametrically optimize material selection.

5.2 MECHANICAL PROPERTIES

A **mechanical property** is a quantity that characterizes the behavior of a material in response to external, or applied, forces. Some of the more frequently used mechanical properties include:

Strength is a measure of the amount of tensile force per unit area that a material can withstand before it fails. If the load is small, the material elastically elongates and then relaxes without permanent plastic elongation or yielding. We can characterize the strength of materials using a tension test as shown in Figure 5.2. Fracture failure is when the material pulls apart. Brittle materials behave elastically until they fracture. Ductile materials behave elastically; then they yield plastically as the load increases, until fracture. **Stress** is a measure of force intensity per unit area. $\sigma = P/A$ and **Strain** is a measure of relative elongation, $\varepsilon = \Delta L/L$. We can graph these quantities in a stress-versus-strain diagram, as shown in Figure 5.3. A brittle material is shown as the curve OB. A ductile material is shown as curve OBCD. The curve CE shows that the ductile material elastically recovers some of its plastic deformation.

Yield strength, denoted as S_y, is the tensile stress (i.e., force/area) at which a material yields.

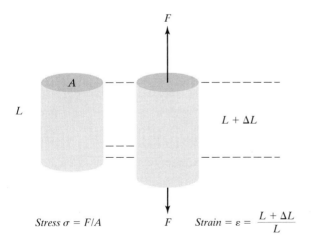

F

A

L

$L + \Delta L$

Stress $\sigma = F/A$ F *Strain $= \varepsilon = \dfrac{L + \Delta L}{L}$*

FIGURE 5.2

A tension test uses standard-sized specimens to determine the amount of stress required to produce a given strain in a material.

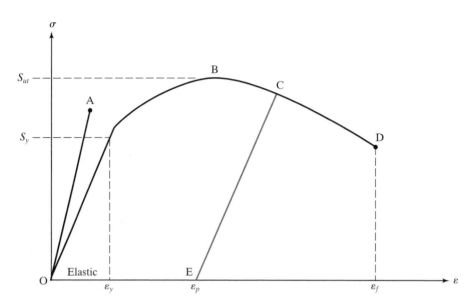

FIGURE 5.3

Stress-versus-strain diagrams for brittle and ductile materials.

Ultimate tensile strength is the largest tensile stress a material can sustain. It is also called **tensile strength** and is denoted as S_{ut}.

Shear strength is the largest stress a material can sustain under torsion before it yields or fractures.

Compressive strength is a measure of the amount of compressive force per unit area that a material can withstand before it fails. Brittle materials are weak in tension compared to ductile materials. However, when brittle materials are subjected to compressive loads, their compressive strength is often two or three time larger than their tensile strength.

Stiffness is the resistance to stretching, bending, or twisting loads. Stiffness is measured by the **modulus of elasticity** (E), which is the tangent slope of the stress-versus-strain curve.

Ductility is the ability of a material to plastically deform. It is measured by the percent elongation and or percent reduction in area.

Toughness is the ability of a material to plastically deform before fracturing. It is measured by the modulus of toughness.

Hardness is the ability of a material to resist localized surface indentation or deformation. It is measured by the Brinell Hardness Number (BHN).

Fatigue strength is the ability of a material to undergo a number of cyclic loads without fracturing. A measure of fatigue strength is the endurance limit, which is the stress at which steels fracture when given a million load cycles.

Creep resistance is the ability of a material to resist stretching while under loads over long time periods at elevated temperatures. It is measured by the amount of stress that causes rupture within 1,000 hours at the elevated temperature.

Impact strength is the ability of a material to absorb sudden dynamic shocks or impacts without fracturing. The Charpy or Izod test is used to measure impact strength (ft-lbs).

Coefficient of friction is a relative measure of the amount of friction force between two surfaces. It is equal to the ratio of the friction force divided by the force normal to the surface.

Wear coefficient is a measure of the amount of surface removal due to rubbing and sliding.

Mechanical properties are shown for a variety of materials in Table 5.1.

TABLE 5.1 Selected Materials Properties

Class	Member	Heat treat	Elastic modulus Mpsi	Tensile strength kpsi	Yield strength kpsi	Elongation %	Hardness Bhn	Density lb/in^3	Expansion coeff 10^{-6}F
Aluminum			10.5					0.1	12
	2014	annealed		27	14	18	45		
	2014	T4		62	42	20	105		
	295	T4		32	16	8.5			
	356	T6		33	24	3.5			
Copper			17.5					0.32	9.4
	C85200			38	13	35			
	C86200			95	48	20			
	C93200			35	18	20			
Magnesium			6.5					0.065	14.5
	AZ91B-F			34	23	3			
Nickel			30					0.3	7
	Hastelloy	as-cast		134	67	52			
	Inconel 600	annealed		96	41	45			
Steel									
	1020	annealed	30	57.3	42.8	36.5	111	0.28	6.7
	4340	annealed	30	108	68.5	22	217	0.28	6.3
	304A	annealed	27.5	83	40	60		0.28	8
Titanium			16.5					0.16	4.9
	Ti-35A			35	25	24			
	Ti-65A			65	55	18			
Zinc			12					0.24	15
	AG40A			41		10	82		
	ZA-12	die-cast		57	46	2	117		

TABLE 5.1 *Continued*

Class	Member	Heat treat	Elastic modulus Mpsi	Tensile strength kpsi	Yield strength kpsi	Elongation %	Hardness Bhn	Density lb/in^3	Expansion coeff 10^{-6}F
Polymers									
	ABS			6		5-20			16
	PTFE			3.4		300			40
	Nylon 6/6			12		60			
	Polypropylene			5		10-20			20
	Polycarbonate			18.5					13
	Polystyrene			13.5					19

5.3 PHYSICAL PROPERTIES

A **physical property** is a quantity that characterizes a material's response to physical phenomena, other than mechanical forces. Some of the more frequently used physical properties include (ASM, 1997):

Density is the amount of matter per unit volume. Density is directly proportional to weight. Two measures for density are mass density and weight density.

Coefficient of thermal expansion is a measure of the amount a material elongates in response to a change in its temperature.

Melting point is the temperature at which a solid changes to a liquid. It is a measure of a material's ability to tolerate elevated temperatures.

Specific heat is the amount of heat required to increase the temperature of a unit mass 1 degree.

Corrosion resistance is the ability of a material to resist oxidation, direct chemical attack, or surface degradation by galvanic currents.

Thermal conductivity is a measure of heat flow across a surface, per unit area, per unit time, per unit of thickness, per degree of temperature difference.

Electrical conductivity is a measure of the ability to conduct electricity. It is equal to the ratio of electric current to the given voltage difference.

Selected material properties and their units of measurement are summarized in Table 5.2.

5.4 MATERIAL CLASSES

The periodic table lists 103 elements. However, a large number of materials are possible when these are combined in various proportions as compounds. Much of the effort of materials scientists and engineers focuses on finding the specific recipes that produce compounds that exhibit the right kind of molecular structure, resulting in the right profile of properties.

TABLE 5.2   Fundamental Material Properties

Characteristic	Behavior	Property	Units
Strength	strong, weak	ultimate strength	MPa (ksi)
Elastic strength	elastic then plastic	yield strength	MPa (ksi)
Stiffness	flexible, rigid	modulus of elasticity	MPa (Mpsi)
Ductility	draws, forms easily	% elongation, % area reduction	dimensionless
Hardness	resists surface indentation	Brinell No.	MPa (ksi)
Corrosion resistance	resists chemicals, oxidation	galvanic series	activity number
Fatigue resistance	endures many load cycles	endurance limit	MPa (Mpsi)
Conductivity (heat, electric)	conducts, insulates	thermal conductivity electrical conductivity	(Btu/hr) / (F-ft), Mhos
Creep resistance	time dependent stretching	creep strength	MPa (ksi)
Impact resistance	shock, impact loads	Charpy energy	N-m, (ft-lbs)
Density (mass)	heavy, light	mass density	kg/m^3, $(slugs/ft^3)$
Density (weight)		weight density	N/m^3, (lbs/ft^3)
Temperature tolerance	softens, or melts easily	melting point	degrees C, F

Materials can be separated into a number of categories: metals, polymers, ceramics, composites, and other. If we define these categories as "families," similar to Ashby (1999), then each family includes a number of subfamilies that also include classes and subclasses. Let's take a look at the metals family in Figure 5.4.

Metals can be divided into ferrous and nonferrous subfamilies. Ferrous metals contain significant amounts of iron. The classes of ferrous metals are cast iron, carbon steel, alloy steel, and stainless steel. Within the carbon steel class, however, there are a couple dozen specific subclasses. The subfamily of thermoplastics, for example, can be divided into classes including: ABS, acetal, acrylic, nylon, polycarbonate, polyethylene, polypropylene, polystyrene, and vinyl. And within each of these are dozens of subclasses that represent the many different compounds or recipes.

Metals. The **metals** family of materials can be described as ductile, strong, stiff, electrically conductive, thermally conductive, fatigue-resistant, creep-resistant, impact-resistant, heavy or massive, temperature-tolerant, medium-hard, but not very corrosion-resistant.

Polymers. The **polymers** family of materials can be described as strong flexible, electrically and thermally insulating, not creep-resistant, impact-resistant, lightweight, temperature-sensitive, soft, and corrosion-resistant.

Thermoplastic polymer is material that can be repeatedly softened by heating and hardened by cooling.

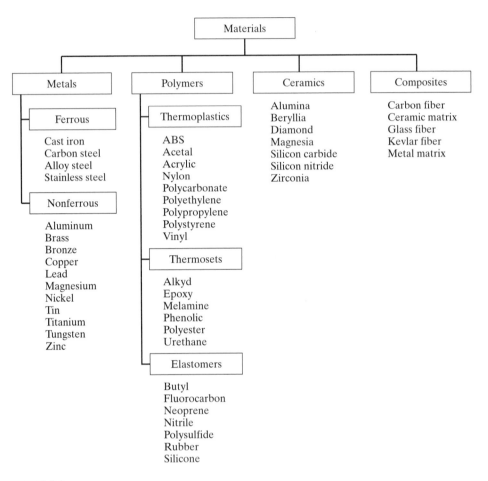

FIGURE 5.4

Material families, subfamilies, and classes.

Thermoset polymer is material that permanently sets by heating and thus curing.

Ceramics. The **ceramics** family of materials can be described as strong in compression, weak in tension, brittle, stiff, electrically and thermally insulating, not impact-resistant, medium-weight, very temperature tolerant, very hard, and corrosion-resistant.

Composites. The **composites** family of materials are heterogeneous mixtures of polyester or epoxy resins and fibers made from materials including glass, carbon, Kevlar, fibers, and metal. They can be stiff, strong, light, non-conducting, and moderately corrosion-resistant. But, they are sensitive to temperature.

Other Materials. Other materials include glasses, woods, leather, and other natural materials such as cotton, silk, cork, concrete, and hemp.

A summary of general material properties by family is presented in Table 5.3.

TABLE 5.3 Approximate Property Profile by Material Family

Characteristics	Metals	Ceramics	Polymers
Strength	Strong	strong–C weak–T	weak
Elastic strength	very	some	some
Stiffness	very	very	flexible
Ductility	ductile	brittle	—
Hardness	medium	hard	soft
Corrosion resistance	poor	good	excellent
Fatigue resistance	good	—	—
Conductivity (heat/electric)	conductor	insulator	insulator
Creep resistance	good	—	poor
Impact resistance	good	poor	good
Density	high	medium	low
Temperature tolerance	good	super	poor

5.5 MATERIAL SELECTION METHODS

Whether a part will satisfy its functional performance requirements depends a lot on its geometry. The geometry depends upon the chosen manufacturing process, which in turn depends upon the selected material. However, there are many feasible materials and manufacturing processes to choose from. Which do we determine first, the material or the process?

Screening Methods. When we use the *materials-first approach* (Dixon and Poli, 1995), we screen out materials that will not satisfy the functional requirements of the part. Namely, we compare application information from the engineering design specification to the mechanical and physical properties of material classes. We typically include criteria regarding the nature of the applied loads and the operating environment. This screening will eliminate a number of infeasible material classes. Those remaining will be compatible with some manufacturing processes.

To narrow the feasible processes further, however, we do a secondary screening, considering part information, including: the geometric complexity of the part, the production volume, and part size. As we shall discover in the next chapter, certain shapes are not compatible with some processes. Geometric complexity refers to the type and number of features, including: holes, notches, bosses, rotational symmetry, enclosed cavities, and uniformity of walls and/or cross-sections. Similarly, some processes are not feasible when producing some larger part sizes or production quantities.

When we use the *manufacturing-processes-first approach*, we first screen out manufacturing processes that will not satisfy part considerations, including: part size, production quantities, or geometric complexity. The remaining feasible processes will be compatible with some material classes. We further screen these by comparing material properties to the EDS requirements.

Either approach will lead to the same subset of material classes and compatible manufacturing process since we are doing successive eliminations or

TABLE 5.4   Application and Part Information Considered during the
Material-First or Process-First Approach to Materials Screening

Material-First Approach	Process-First Approach
Application Information	**Part Information**
1. Applied loads	1. Production volume
magnitude	2. Part size (overall)
cyclic nature (fatigue)	3. Shape capability (features)
rate (slow, impact)	boss/depression 1D
duration (creep)	boss/depression > 1D
2. Ambient conditions	holes
temperature	undercuts (internal/external)
moisture, humidity	uniform walls
chemical liquids/vapors	cross sections (uniform /regular)
sunlight (ultraviolet)	rotational symmetry
3. Conducting (elec/therm)	captured cavities
4. Safety/Legal (FDA, UL, etc)	
5. Cost	

screenings based on the same criteria. Table 5.4 lists the pertinent information considered for each approach.

Rating/Ranking Method. To refine our material choices, we can also rate or
rank their relative performance using material indices, as proposed by
Ashby (1999). Most parts perform a basic mechanical function such as a
shaft transmitting a torque or a bracket supporting a force. Functional requirements can be directly related to forces and moments causing tension,
compression, beam bending, and column buckling. The performance of the
part depends upon its geometry and its material properties. Incorporating
analytical relations from the engineering sciences, performance functions
can be obtained in the form:

$$p \leq f_1(F)f_2(G)f_3(M) \tag{5.1}$$

where the performance, p, depends upon the functional requirements, F,
the geometric parameters, G, and the material properties, M. A material
index for a column made of a material with a given modulus of elasticity
E and density ρ, subjected to a buckling load for example, would have a
material index

$$M_B = E^{1/2}/\rho \tag{5.2}$$

Materials that have higher index values perform better, relative to materials with lower M_B values. Plotting E versus ρ in a material selection
chart provides a graphic comparison of different materials, as shown in
Figure 5.5. Strength versus cost/volume is shown in Figure 5.6. A thorough development of this method, including many examples, is presented
in Ashby (1999).

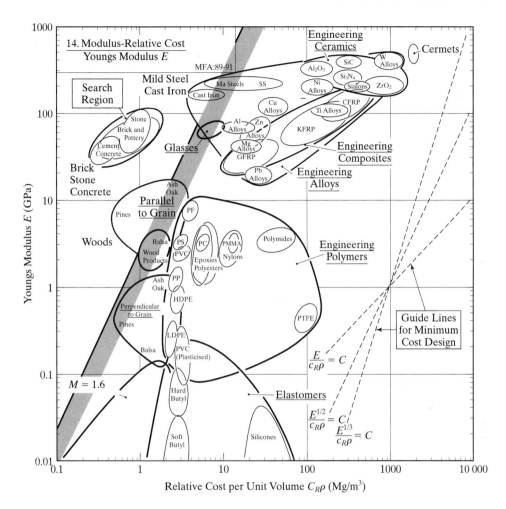

FIGURE 5.5

Material selection chart—modulus versus cost/volume.
(Reprinted from *Materials Selection in Mechanical Design*, M. F. Ashby, Copyright 1999, with permission from Elsevier Science.)

Example

Use Figure 5.5 to determine which of the engineering polymers are the least stiff, that is, the most flexible, and have a relative cost that is lower than that of nylons.

Examining the figure we find that PP, HDPE, LDPE, and PVC have relative costs lower than the nylons. However, the least stiff of the four are polyvinyl chloride (PVC) and low-density polyethylene (LDPE).

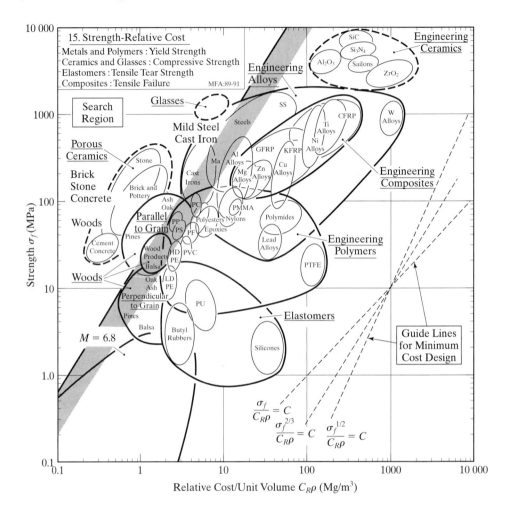

FIGURE 5.6

Material selection chart—strength versus cost/volume.
(Reprinted from *Materials Selection in Mechanical Design*, M. F. Ashby, Copyright 1999, with permission from Elsevier Science.)

Example

Use Figure 5.6 to determine which of the engineering alloys that have at least 500 MPa yield strength. Of those, which are the more expensive?

Examining the figure we find that the following alloys that are strong enough: mild steel, cast iron, steels, stainless steels (SS), and tungsten alloys (W). Further, we find that tungsten is the most expensive, then stainless steels, with mild steels and cast iron the least expensive.

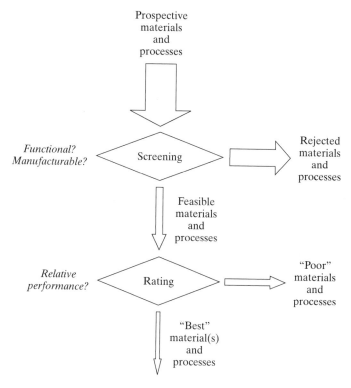

FIGURE 5.7

Screening and evaluation of materials.

We use screening and rating methods to help us narrow our material choices during the configuration design phase. The general process of screening and rating is shown in Figure 5.7. Using the materials-first approach, or the manufacturing-processes-first approach, we examine the suitability and compatibility of alternative materials and manufacturing processes. The remaining materials and manufacturing processes can then be rated, using the weighted-rating method, discussed in an earlier chapter. As we proceed from screening through evaluation, we see that the number of materials and processes decreases (shown by the decreasing width of the arrow). The resulting "best" materials and manufacturing processes are recommended for detailed analysis and evaluation that occurs in the parametric design phase discussed in Chapter 8.

5.6 SUMMARY

- Product function, geometry, material properties, and manufacturing processes are interdependent.
- Mechanical properties characterize the behavior of a material in response to external or applied forces.

- Physical properties include: electrical and thermal conductivity, corrosion resistance, heat capacity, thermal expansion, and friction.
- The major families of materials are metals, polymers, ceramics, and composites.
- Material and manufacturing process selection can be done by either the materials-first approach or the manufacturing-process-first approach.
- Screened materials can be evaluated using the weighted-rating method or performance functions.

REFERENCES

Ashby, M. F. 1999. *Materials Selection in Mechanical Design*. Oxford, UK: Pergamon Press.
ASM Handbook, vol. 20. Materials Park, OH: ASM International. 1997.
Dieter, G. E. 2000. *Engineering Design*, 3d ed. New York: McGraw-Hill.
Dixon, J. R., and C. Poli. 1995. *Engineering Design and Design for Manufacturing*. Conway, MA: Field Stone Publishers.
Groover, M. P. 1996. *Fundamentals of Modern Manufacturing*. Upper Saddle River, NJ: Prentice Hall. 1996.

KEY TERMS

Ceramics	Hardness	Strength
Coefficient of friction	Impact strength	Stress
Coefficient of thermal expansion	Mechanical property	Tensile strength
	Melting point	Thermal conductivity
Composites	Metals	Thermoplastic
Compressive strength	Modulus of elasticity	Thermoset
Corrosion resistance	Physical property	Toughness
Creep resistance	Polymers	Ultimate tensile strength
Density	Shear strength	Wear coefficient
Ductility	Specific heat	Yield strength
Electrical conductivity	Stiffness	
Fatigue strength	Strain	

EXERCISES

Self-Test. Write the letter of the choice that best answers the question.

1. _____ The quantity that characterizes the behavior of a material in response to an external, or applied, force is called:

    a. strength

    b. material property

    c. mechanical property

    d. physical property

2. _____ A measure of the amount of force per unit area that a material can withstand before it fails is called:

   a. strength

   b. weight

   c. intensity

   d. hardness

3. _____ The resistance to stretching, bending, or twisting loads is called:

   a. ductility

   b. impact strength

   c. compressive strength

   d. stiffness

4. _____ The ability of a material to undergo a number of cyclical loads without fracturing is called:

   a. impact strength

   b. compressive strength

   c. fatigue strength

   d. yield strength

5. _____ The ability of a material to plastically deform is called:

   a. ductility

   b. hardness

   c. stiffness

   d. impact

6. _____ The ability of a material to resist stretching under loads over long time periods at elevated temperatures is called:

   a. impact strength

   b. creep resistance

   c. comprehensive strength

   d. ductility

7. _____ The ability of a material to resist localized surface indentation or deformation is called:

   a. ductility

   b. stiffness

   c. yield

   d. hardness

8. _____ The amount of matter per unit volume is called:

   a. connectivity

   b. melting point

   c. conductivity

   d. density

9. _____ The measure of the rate of heat flow between two surfaces, per unit area, per unit time, per unit of thickness, per degree of temperature difference is called:

   a. corrosion resistance

   b. specific heat

   c. coefficient of thermal expansion

   d. thermal conductivity

10. _____ The temperature at which a solid changes to a liquid is called:

    a. melting point

    b. specific heat

    c. thermal expansion

    d. thermal conductivity

11. _____ All of the following are classes of ferrous metals except:

    a. cast iron

    b. alloy steel

    c. bronze

    d. stainless steel

12. _____ Polymeric materials that can be repeatedly softened by heating and hardened by cooling are called:

    a. thermoplastic

    b. acrylic

    c. thermoset

    d. polystyrene

13. _____ Materials that are heterogeneous mixtures of polyester or epoxy resins and fibers made from materials, including glass carbon and Kevlar, are called:

    a. ceramics

    b. thermoplastics

    c. polymers

    d. composites

14. _____ All the following are ceramics except:

    a. zirconia

    b. magnesia

    c. silicon nitride

    d. carbon fiber

15. _____ Product function is interdependent on:

    a. material properties

    b. product geometry

    c. manufacturing processes

    d. all of these

2: a, 4: c, 6: b, 8: d, 10: a, 12: a, 14: d

16. Use data from Table 5.1 to select the member, from each class, that is the:

Class	Strongest	Weakest	Most ductile
Aluminum			
Steel			
Polymers			

17. Use Figure 5.5 to determine which materials are "inexpensive," assuming that the required modulus of elasticity is about 60 Gpa.
18. Use Figure 5.5 to determine which materials are the least stiff, or most flexible, assuming that the "relative cost" should be about 100.
19. Assuming that the "relative cost" should be about 100, use Figure 5.6 to determine which materials are the strongest.
20. Assuming that the "relative cost" should be less than 100, use Figure 5.6 to determine which materials have a large compressive strength.

Selecting Manufacturing Processes

When you have completed this chapter you will be able to

- Characterize basic manufacturing processes
- Differentiate and select primary, secondary, and tertiary processes
- Select appropriate manufacturing processes
- Describe and use manufacturing terms and concepts
- Estimate the costs of manufacturing a product

6.1 INTRODUCTION

Regardless of the type of material, even the simplest product involves many manufacturing processes. Let's examine that Trek mountain bicycle shown in Figure 6.1.

The Trek model 4300 includes a frame, rear-wheel assembly, front-wheel assembly, steering fork with suspension, handlebar, hand controls, seat post with saddle, pedal crank subassembly, and Derailleur system.

The product is shipped as a few subassemblies. Before the product is displayed, the retailer attaches the front wheel, handlebar, seat post, and saddle. The retailer also checks for proper operation of the 24-speed shifter, and adjusts the spoke tension of each wheel. We can understand, however, that assembling components is an important manufacturing process whether the components are assembled by the retailer or at the factory.

A number of other manufacturing processes are used to make the frame subassembly. To start with, standard lengths of aluminum tubing are cut to approximate lengths for the top tube, down tube, seat tube, and rear stays. The rear stays are then formed in a bending process. The ends of each tube are shaped and cleaned. The tubes and stays are then welded together using the Gas Tungsten Arc

FIGURE 6.1

Trek mountain bicycle. (Courtesy of Trek Bicycle Corporation.)

welding process. The welded frame is then cleaned and finally painted. Quality inspections are also completed in between. Note that the type of process and its sequence is important. Part of production planning, in fact, is determining the appropriate sequence of primary, secondary, and tertiary manufacturing processes.

Other processes are used in making bicycles, including: molding (tires), chrome plating or anodizing (handlebars, wheel rims), swaging (seat posts), stamping (sprockets), and casting (hand controls).

As we plan the manufacture of any product, we need to answer the following questions:

- How do we choose the specific manufacturing processes?
- How do the materials selected influence the choice of manufacturing processes?
- Would product function or performance issues influence our choice of processes?
- What criteria should we use to select processes?
- Which criteria are most important?
- Who will make the final decisions?

In the remaining sections we will examine some of the more widely used manufacturing processes. We will investigate their: compatibility with alternative materials, ability to generate complex shapes, limitations as to the maximum or minimum part size, feasible production quantities, and overall economic viability. Then we will describe a general method that will help us select appropriate manufacturing processes.

6.2   MANUFACTURING PROCESSES

Manufacturing processes are used to transform raw materials into products. Processes are used to change a material's form as in metal casting, sheet metal bending, or machining. Processes can alter the microstructure and properties of a material such as increasing the yield strength of steel by cold rolling, or heat treatment. Processes can also change the chemical composition of a material such as in chrome plating or galvanizing steel.

Manufacturing processes can be categorized as bulk deformation, casting, sheet metalworking, polymer processing, machining, finishing, and assembly, as shown in Figure 6.2.

Primary manufacturing processes are used principally to alter the material's shape or form. We use casting, for example, to transform rectangular ingots of steel into cylindrical brake drums. Injection molding is used to convert pellets of thermoplastic polyethylene into telephone handsets. Rolling is used to flatten slabs of steel. And, sheet metalworking is used to band steel sheets into refrigerator housings.

Secondary manufacturing processes are used to add or remove geometric features from the basic forms. The brake drum casting, for example, undergoes

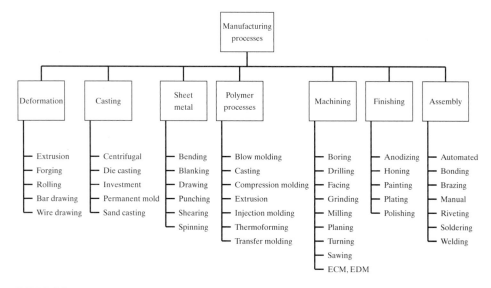

FIGURE 6.2

Types of manufacturing processes.

secondary machining operations such as turning and grinding of the friction surface. Refrigerator housings are frequently drilled or hole-punched. In injection, molded parts will have surplus flash-trimmed.

Tertiary manufacturing processes relate to surface treatments such as polishing, painting, heat treating, and joining. A part, such as the bike frame, can undergo a sequence of many processes, or operations, before it is assembled as the final product.

6.2.1 Bulk Deformation

Bulk deformation changes the shape or form of bulk raw materials caused by compressive or tensile yielding. Bulk materials include billets, blooms, and slabs, as compared to sheet materials. Ore is initially processed into solidified ingots. Ingots are subsequently rolled into small rectangular blooms and smaller rectangular billets. Ingots and blooms are rolled onto flat rectangular slabs. Some of the more common bulk deformation processes are described below.

Extrusion   As shown in Figure 6.3, in **extrusion**, heated metal plastically yields as it is pushed through a **die**, producing long pieces with a constant cross-section. Rectangular, circular, triangular, L-shaped, and U-shaped cross sections are common. Tubing and piping can also be extruded using a mandrel to form the hollow core. Custom cross sections are also produced. Typically extrusions are shipped in 40-foot lengths, the same length as a tractor trailer. Cross-sections can range from about 1/4 to 12 inches in diameter. Economical production quantities range from 1,000 to 100,000 pieces. Extrusions are made from ductile metals including aluminum, steel, zinc, copper, and magnesium.

Forging   As shown in Figure 6.4, in **forging**, material is plastically compressed between two halves of a die set by hydraulic pressure or the stroke of a hammer. In closed die forging the material is compressed on all sides. During open die forging the workpiece is hammered between two die halves that do not restrict

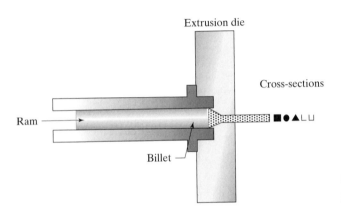

FIGURE 6.3

Extrusion of metals.

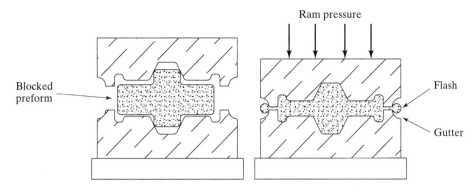

FIGURE 6.4

Closed-die forging.

the side flow of material. Often bulk material is preformed in blocker dies. Forging is typically done when the metal is hot, requiring less work. Moderately complex shapes can be forged with a maximum size limited to roughly 36 inches. Economic production quantities range from 1,000 to 100,000 pieces.

Rolling   Two or more cylindrical rollers plastically compress material, forming sheets, bars, or rods. **Rolling** is used to produce billets, blooms, and slabs from ingots. The billets are subsequently rolled into rectangular or circular bars. Slabs are rolled into plates and sheet metal. Blooms are rolled into structural shapes including I-beams, channel beams, angle, and wide-flange beams. The processing of ingots into the various shapes is shown in Figure 6.5. Hot rolling requires less work than cold rolling to deform materials but produces an oxidized surface finish. Cold rolling requires more work but increases the material's yield strength and produces a superior surface finish.

Bar Drawing   As shown in Figure 6.6, in **bar drawing**, bar stock is pulled through a die of reduced cross section. Cold drawing increases the material's strength and improves the surface finish. Straight lengths of circular or rectangular bar stock are produced. Sizes range from 1/8 to 6 inches in cross section.

Wire Drawing   In **wire drawing**, bar stock is pulled through a set of successively narrowing dies, forming a long strand of wire that is usually wound on a spool as a continuous process. Cross section sizes range from 0.001 to about 3/8 inches. Drawing is used with ductile metals such as aluminum, steel, and copper.

6.2.2 Casting

Casting is one of the oldest processes, dating back thousands of years, in which molten metal is poured into a cast to solidify. Casting can produce complex

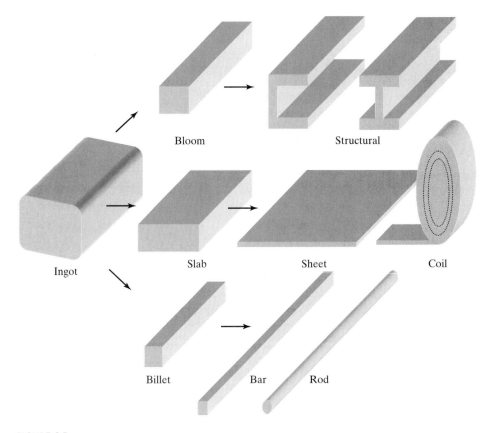

FIGURE 6.5

The rolling process deformation process initially produces blooms, slabs, and billets. These are subsequently rolled into structural, sheet, coil, bar, and rod shapes.

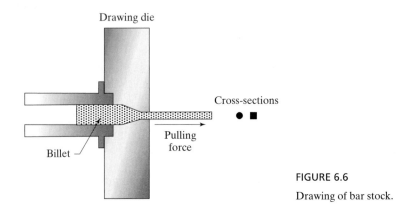

FIGURE 6.6

Drawing of bar stock.

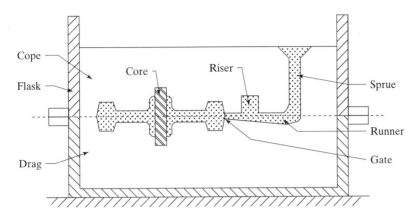

FIGURE 6.7

Closed-mold sand casting.

geometries within broad tolerances. For dimensional accuracy castings are typically machined in secondary operations. Three of the most commonly used casting processes are sand casting, die casting, and investment cast.

Sand Casting   In **sand casting**, molten metal solidifies in a sand mold. The standard mold is formed by packing sand around a pattern with the same external
shape as the part to be cast. The pattern is removed before casting and later
reused. As shown in Figure 6.7, molten metal flows down the sprue into the runner, through the opening to the part, called a gate and into the part cavity forming. The riser is a small reservoir of liquid metal that feeds material into the part
as the part cools and shrinks. Preformed cores can be placed in the mold to create voids, such as cylindrical holes. Open-mold casting is used for parts with a
flat surface on one side (e.g., manhole covers, jewelry). Closed-mold casting
uses a cope half and a drag half, enclosed in a flask. The mold is destroyed in
order to remove the part. The parting line is where the two mold halves meet.
Sizes for sand castings can range from inches to feet. Some industrial engine
blocks have been cast as large as 12 feet in cross section and 30 feet long. Economical production quantities can be very small, for example between 1 and 10.
All nonferrous and ferrous metals are used. Moderately complex shapes are
possible. Sand cores can be used to produce internal and external undercuts.
Parts made by sand casting have a granular surface finish.

Die Casting   In **die casting**, molten material is injected under high pressure
into permanent die set (i.e, mold halves) usually made of steel. Because the steel
mold is reused, die casting is faster than sand casting. But, the steel die set can
be more expensive to machine.

   The surface finish of die-cast parts is smoother than sand-cast parts. As
shown in Figure 6.8, a shot of molten metal is ladled into the opening, and then

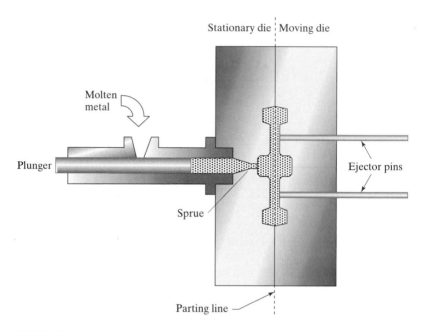

FIGURE 6.8

Cold-chamber die casting.

hydraulically rammed into the mold halves by the plunger. Hot-chamber die-casting machines, which automate the process even further, are also used. The solidified part is pushed out of the mold by **ejector pins**. Maximum part sizes are limited by the clamping capacity of the injection machines, typically around 30 inches by 30 inches. Small parts, less than an inch, can be die-cast. Moderately complex shapes are possible. Economic production volumes are generally over 10,000 pieces. In order not to melt the steel mold, common die-casting materials are low-melting-point metals, including alloys of aluminum, zinc, magnesium, and brass.

Investment Casting (Lost-Wax Process)   In **investment casting**, molten metal solidifies in a ceramic cast made by coating a wax pattern with liquid slurry, then dried. The wax is melted out prior to casting. As shown in Figure 6.9, the wax pattern is made in a metal mold. Multiple parts can be combined on the pattern tree. The wax is melted and drained from the ceramic mold. Molten metal is poured into the ceramic mold that is then destroyed after the part solidifies. Shape complexity and production volume are similar to die casting. Casting materials include alloys of aluminum, zinc, magnesium, brass, steel, and stainless steel. Investment casting is typically used for runs of less than 10,000 pieces.

6.2.3   Polymer Processing

In polymer processing, part shapes are created by solidification of thermoplastic polymers or curing of thermosetting polymers. Parts made of thermoplastic

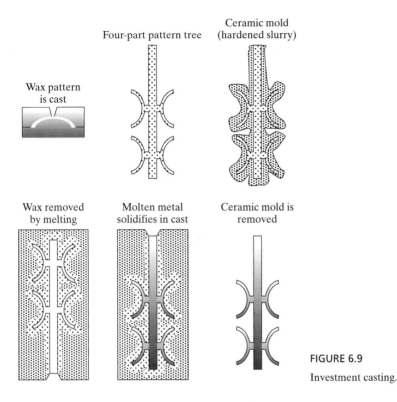

Wax pattern
is cast

Four-part pattern tree

Ceramic mold
(hardened slurry)

Wax removed
by melting

Molten metal
solidifies in cast

Ceramic mold is
removed

FIGURE 6.9

Investment casting.

and thermosetting materials can be made in many shapes and sizes, with little waste of raw material, requiring few, if any, finishing operations. Some of the more common processes include: blow molding, injection molding, compression molding, transfer molding, thermoforming, and casting.

Blow Molding   In **blow molding**, as shown in Figure 6.10, a molten parison of thermoplastic material is injected with air, then expands to the shape of the mold. Blow molding is used to produce hollow parts with thin walls. Captured cavities shapes are possible, such as a bottle. Sizes are limited to about three feet in diameter.

Compression Molding   In **compression molding**, charge of thermoset or elastomer is formed between heated mold halves under pressure while the polymer cures. As shown in Figure 6.11, compression molds are simpler than injection molds, not requiring a complex system of sprue, runners, and risers. Compression molding is commonly used for automobile tires and record albums. Transfer molding injects the polymer into the cavity permitting molded inserts. However, the thermoset or elastomer sprues and runners, as shown in Figure 6.12, are not reground and reused as with thermoplastics. Maximum part sizes are typically less than 24 inches by 24 inches. Minimum part sizes can be on the order of 1/8 to 1/4 inch in cross section. Economic production volumes start at about 10,000 pieces.

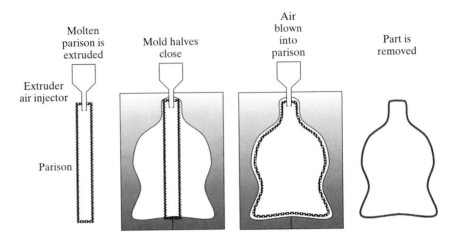

FIGURE 6.10

Extrusion blow molding.

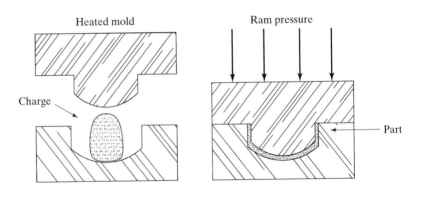

FIGURE 6.11

Compression molding.

Injection Molding In **injection molding**, thermoplastic pellets are melted and then injected under high pressure into a metal mold. As shown in Figure 6.13, pellets of thermoplastic are placed into the hopper. As the pellets auger into the barrel, the pellets begin to rub against each other, causing enough heat to ultimately melt the pellets. The auger then rams the molten plastic into the mold via the sprue, runner, and gate network. The **cavity** half of the mold forms the outside of the part. The **core** half of the mold forms the inside. As the part cools, it shrinks onto the core, requiring ejector pins to remove the part. The mold halves slide on guideposts (not shown). External and internal undercuts can be formed. **Side cores** are shaped metal parts that move in and out of the mold to form external undercuts. An **internal undercut** is a feature that restricts the removal of the part

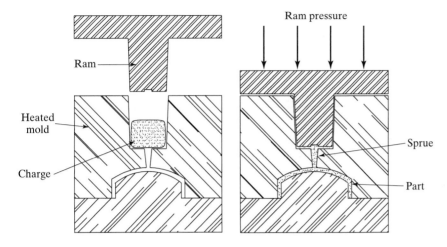

FIGURE 6.12

Transfer molding.

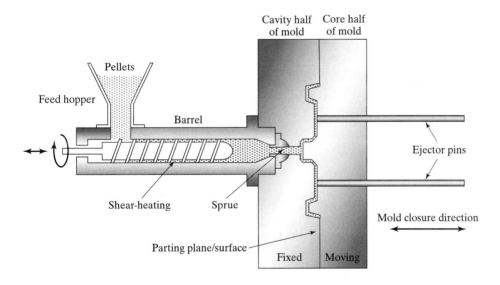

FIGURE 6.13

Injection molding of thermoplastic materials.

from the core half of the mold. An **external undercut** is a feature that restricts the removal of the part from the cavity half of the mold. A **parting line** separates the core and cavity halves. Undercuts are shown in Figure 6.14. A system of liquid channels cools the mold halves. The mold halves move along an axis called the **mold closure direction**. Very complex shapes are possible including internal and external undercuts. Part sizes are limited by the planting capacity of the injection

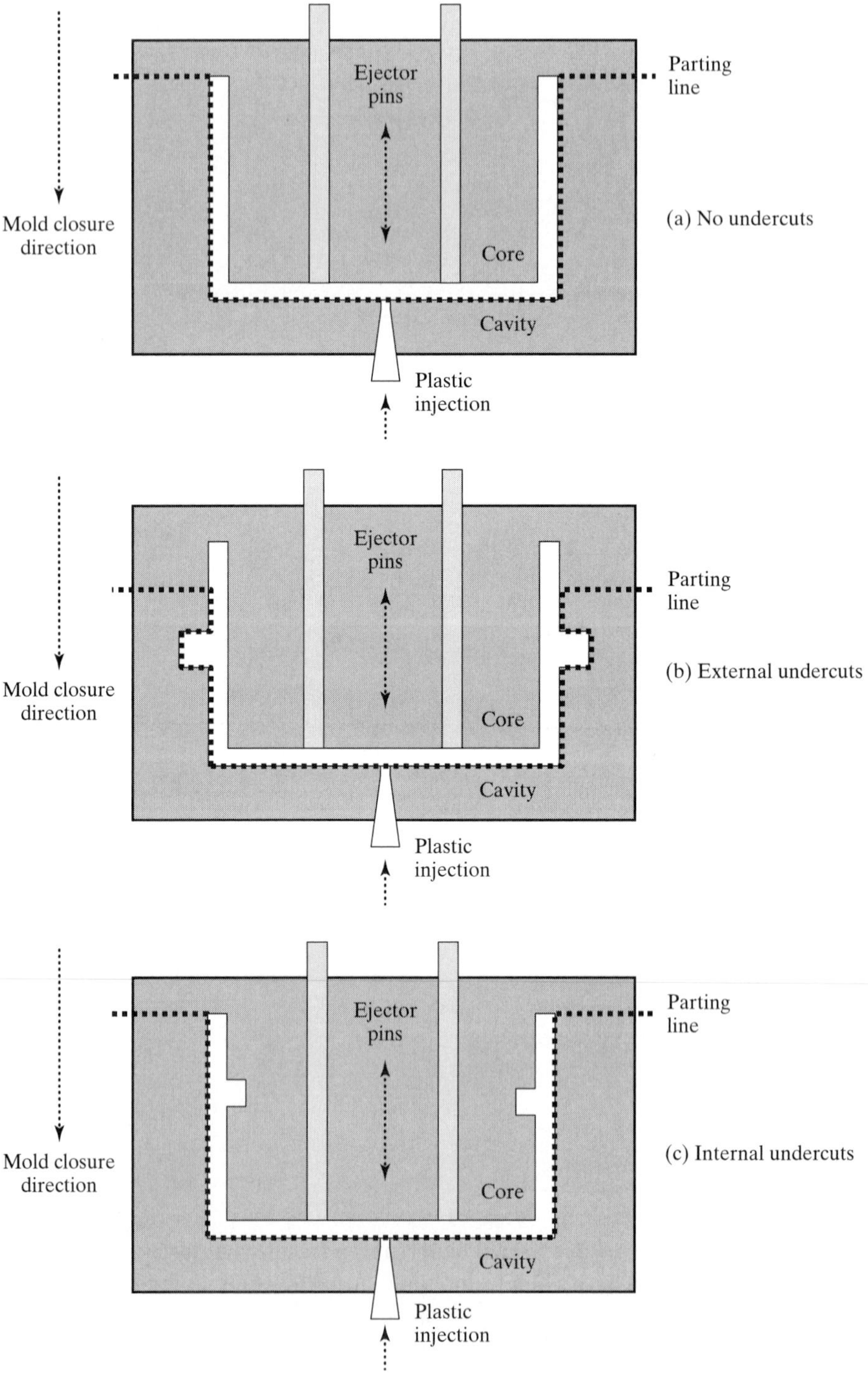

FIGURE 6.14

The boss features create external undercuts that restrict the removal of the part from the cavity half of the mold in (b), and an internal undercut that restricts removal of the part from the core half of the mold, as in (c).

molding machine. Maximum part sizes are typically less than 24 inches by 24 inches. Minimum part sizes are on the order of 1/8 to 1/4 inch in cross section. Economic production volumes start at about 10,000 pieces.

Thermal Forming   **Thermal forming** includes vacuum-forming thin sheets of thermoplastic. A vacuum is drawn sucking the molten sheet onto the pattern forming the part. Parts as large as 6 feet by 6 feet have been vacuum-formed.

Casting (Polymers)   Gravity casting of polyurethane materials is done using silicone or rubber molds. Complex shapes can be cast, including internal and external undercuts, since the mold will flex upon part removal.

6.2.4   Sheet Metalworking

In **sheet metalworking**, permanent deformation of thin metal sheets is produced by bending or shearing forces. Sheet metalworking processes are often called **stamping** processes. The forces are produced by mechanical or hydraulic presses. Hydraulic presses pressurize fluid that pushes against a piston to compress die halves. Faster, but less controllable, mechanical presses use an electric motor to energize a flywheel, connected to a ram or hammer that strokes the punch. Die sets include a collection of components, including one or more moving **punches** fastened to a punch plate and a stationary die mounted to a die holder. In addition to bending and punched features, secondary **side-action features** such as a hole or boss can be produced by machinery that acts on the sides of the stamped parts. Sheet metalworking processes produce shapes with moderate complexity. Part sizes are typically less than 24 inches by 24 inches. Economic production quantities start at about 10,000 pieces. Common materials include alloys of steel and aluminum. Common sheet-metalworking processes include **drawing**, **bending**, **punching**, **embossing**, **shearing**, and **blanking**.

Shearing. Shearing is cutting or separating sheet metal along a straight line. Shearing is used to size sheets for subsequent operations.

Blanking. Blanking is shearing of a smaller, shaped piece, called a blank, from the stock. Blanks are used in deep drawing.

Drawing (sheet). As shown in Figure 6.15, the punch plastically deforms a blank of sheet material into the die, forming cupped-, box-, or hollow-shaped parts. Drawing is used to produce products such as soda cans, ammunition cartridge casings, and pots and pans.

Punching. Producing features such as slots, notches, extruded holes, and holes using a punch and die, punching can be done on a **compound die**, which performs two or more metal deformation processes with one stroke of press, or on a **progressive die**, which is divided into sections, called stations, performing multiple operations with each stroke of the press.

Embossing. Embossing is forming plastic indentations to form ribs, beads, or lettering on the surface of metal.

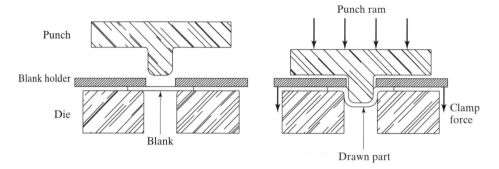

FIGURE 6.15

Sheet metal drawing.

Bending. Bending is plastic deformation of sheet metal using a matched punch-and-die set, or a descending punch to wipe-form the blank over the edge of a die.

6.2.5 Machining

Machining involves removing material from the workpiece by a sharp cutting tool that shears away chips of material to create a desired form or features. Machining is a **subtractive process** that produces **manufactured waste** and can, therefore, be expensive. Machining is often used as secondary process to true-up critical dimensions or surfaces or to smooth the surface finish. Even though machining is an expensive process, it is often used for low-volume production since an expensive mold or die-set need not be prepared. Some of the more common machining processes are listed below and shown in Figures 6.16 and 6.17.

Sawing involves removing material with a toothed blade.

Milling involves removal of material from a flat surface by a rotating cutter tool.

Planing involves removing material using a translating cutter as the workpiece feeds.

Shaping involves removing material from a translating workpiece and a stationary cutter.

Boring is increasing the diameter of an existing hole with a cutting tool by rotating the workpiece.

Drilling is removing material using a rotating bit to form a cylindrical hole.

Reaming is refining the diameter of an existing hole.

Turning is removal of material from a rotating workpiece.

Facing is material removal from a turning workpiece using a radially fed tool.

Grinding is removal of material from a surface using an abrasive spinning wheel.

Electric discharge machining is removal of material by means of a spark.

Operation	Block diagram	Most commonly used machines	Machines less frequently used	Machines seldom used
Turning	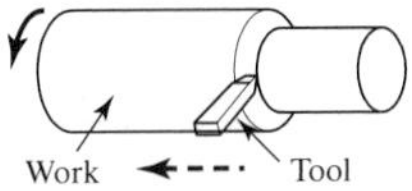	Lathe turning control	Boring mill	
Grinding	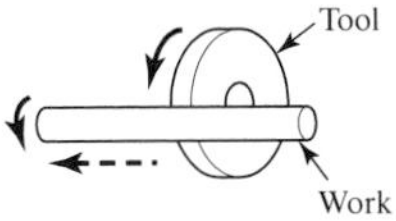	Cylindrical grinder		Lathe (with special attachment)
Sawing (of plates)	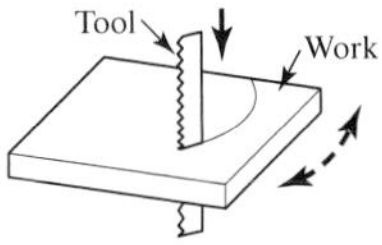	Contour or band saw	Laser Flame cutting Plasma arc	
Drilling	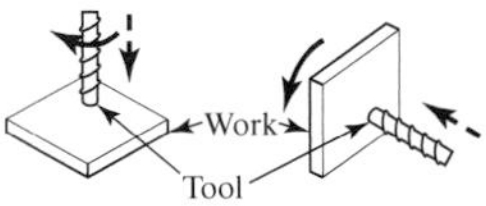	Drill press Machining center Vert. milling machine	Lathe Horizontal boring machine	Horizontal milling machine Boring mill
Boring	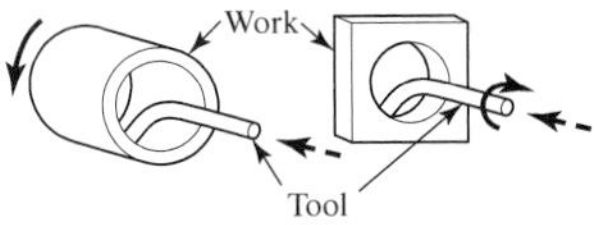	Lathe Boring mill Horizontal boring machine Machining center		Milling machine Drill press
Reaming	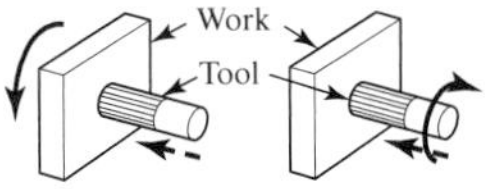	Lathe Drill press Boring mill Horizontal boring machine Machining center	Milling machine	
Grinding	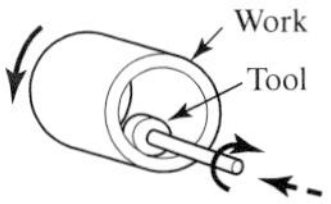	Cylindrical grinder		Lathe (with special attachment)
Sawing	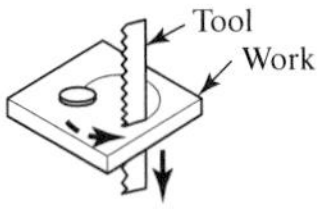	Contour or band saw		
Broaching	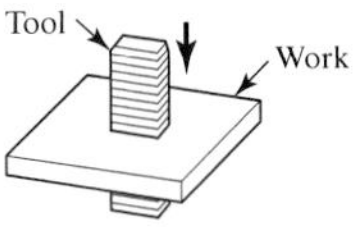	Broaching machine	Arbor press (keyway broaching)	

FIGURE 6.16

Machining cylindrical surfaces. (*Materials and Processes in Manufacturing*, 8th ed.; E. P. DeGarmo, J. T. Black, and R. Kohser; Copyright © 1997. This material is used by permission of John Wiley & Sons, Inc.)

Operation	Block diagram	Most commonly used machines	Machines less frequently used	Machines seldom used
Facing	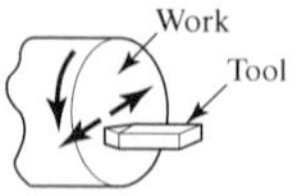 Work / Tool	Lathe	Boring mill	
Broaching	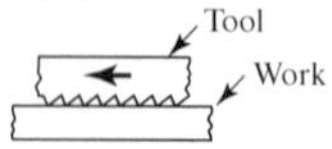 Tool / Work	Broaching machine		Turret broach
Grinding	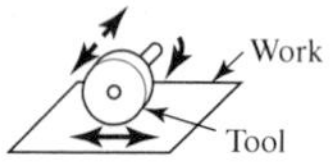 Work / Tool	Surface grinder		Lathe (with special attachment)
Sawing	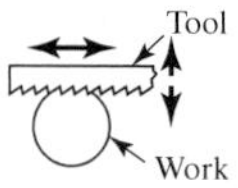 Tool / Work	Cutoff saw	Contour saw	
Shaping	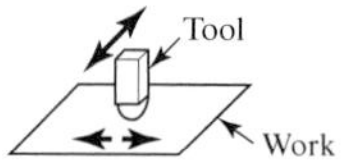 Tool / Work	Horizontal shaper	Vertical shaper	
Planing	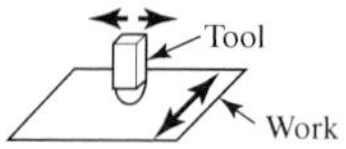 Tool / Work	Planer		
Milling	slab milling 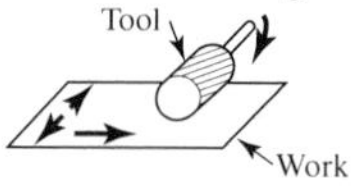 — Tool / Work	Milling machine	Lathe with special milling tools	
	face milling 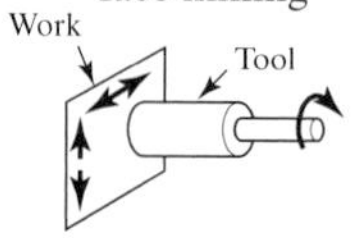 — Work / Tool	Milling machine Machining center	Lathe with special milling tools	Drill press (light cuts)

FIGURE 6.17

Machining flat surfaces. (*Materials and Processes in Manufacturing*, 8th ed.; E. P. DeGarmo, J. T. Black, and R. Kohser; Copyright © 1997. This material is used by permission of John Wiley & Sons, Inc.)

6.2.6 Finishing

Finishing is preparing the final surface for aesthetics and protection from the environment.

Surface Roughness Machining processes produce different surface roughnesses, as shown in Table 6.1. Fine grinding is sometimes used to reduce the surface roughness of parts to between 8 and 16 micro-inches (Dieter, 2002). However, for superior smoothness, surfaces are often polished between 1 and 32 micro-inches. Polishing uses abrasive powders embedded in a rotating leather or

TABLE 6.1   Surface Roughness (adapted from SME, 1992)

Process	Roughness (μ inch)
Machining	
Turning	32–250
Boring	16–250
Milling	32–250
Drilling	63–250
Reaming	16–125
Broaching	16–125
Finishing	
Precision grinding	2–16
Lapping/honing	2–16
Super finishing	1–8

felt wheel. Honing is used for interior cylindrical surfaces. Lapping is used for flat surfaces. Buffing can produce roughness approximating 0.5–16 micro-inches.

Cleaning.   Wire brushing is used to remove grit and scale, and chemical solutions, including acid baths, are used to remove oily films.

Protection.   Polymers and ceramics require little protection from the environment. Metals, however, require some surface treatment. Oil-and-water-based painting is perhaps the least expensive coating. Steels are often plated with chrome, cadmium, or zinc (galvanizing). Aluminum alloys are usually anodized, which is a chemical surface treatment.

6.2.7   Assembly

Assembly is the process of putting together all the product's components before shipping. Products that have subassemblies will have undergone some assembly operations prior to final assembly. Assembly operations include the handling, insertion, and/or attachment of parts. Handling includes grasping, moving, orienting, and placing parts, before insertion or attachment. Attachments are usually either permanent, or temporary as in the case of servicing parts.

Permanent-attachment methods include welding, brazing, soldering, and adhesive bonding. Rivets, eyelets, staples, shrink fits, and press fits are also used.

Temporary-attachment methods include threaded fasteners such as screws, nuts and bolts, and snap fits. Polymer materials sometimes use metallic threaded inserts. Metals, of course, can be machined to form threads.

6.3 COSTS OF MANUFACTURING

The cost to manufacture a product depends on the chosen materials and manufacturing processes. If we plan to produce 50,000 or more pieces of the same part, we will want to consider investing in processes with more automated tooling such as die casting and/or injection molding, to reduce labor costs and have little material waste. On the other hand, if we are making a custom, one-off part for a special customer (e.g., the U.S. Navy), the high cost of automated tooling will not likely be economical. Therefore, we will likely consider machining, even though labor costs would be higher.

The customer pays the selling price, less any discounts. The company's net income results from deducting selling costs, administrative costs, and manufacturing costs from the total sales revenues. When a company is profitable, it will have cash available to replace worn-out manufacturing equipment, in addition to paying dividends to stockholders. We can help, of course, by making economical design and manufacturing decisions that keep total manufacturing costs low.

Let's examine the total manufacturing costs (TMC) for making a production run of quantity, q, pieces and see how some manufacturing processes differ. The three major components are total material costs, M, total tooling costs, T, and total processing costs, P. Other breakdowns of costs can be made, but the one proposed will work well for illustration purposes.

$$\text{Total manufacturing cost} = \text{Material} + \text{Tooling} + \text{Processing}$$

$$TMC = M + T + P \tag{6.1}$$

Let c_m = material cost per part
   c_w = material cost per unit weight
   w_p = weight of finished part
   w_w = weight of wasted material, scrap
   α = ratio of wasted material weight/finished weight = w_w/w_p

Then the **material cost** per part is:

$$\text{Material cost per part, } c_M = M/q = c_w(w_p + w_w) = c_w(w_p + \alpha\, w_p) \tag{6.2}$$

$$c_M = c_w w_p(1 + \alpha) \tag{6.3}$$

The total **tooling cost**, T, is composed of the total material and labor costs to make the part patterns, or die-casting molds, or sheetmetal punch-and-die sets, or jigs and fixtures used in machining. Once the tooling is made, however, the tooling cost is fixed. Consequently, the more parts made with the same tooling, the more economical it becomes. The tooling cost per part is therefore:

$$\text{Tooling cost per part, } c_T = T/q \tag{6.4}$$

The total **processing cost**, P, includes the cost of: labor, energy, floor space, and machine usage. For example, a machine shop might estimate their cost per hour, c_t to run a milling machine at \$150/hr. to pay for the machine operator, electricity usage, floor space (building depreciation or rent), and machine depreciation (or rent). Then, if the **cycle time**, which is the time needed to make a part, is t (hours per part), the processing cost per part, c_P, is:

$$\text{Processing cost per part, } c_P = c_t t \qquad (6.5)$$

The total cost per part is therefore:

$$\text{Cost per part, } c = c_w w_p(1 + \alpha) + T/q + c_t t \qquad (6.6)$$

Examining equation (6.6), we see that to reduce part costs we should:

- purchase less expensive materials
- keep our finished part weight low
- produce little manufactured waste
- make many parts per production run (i.e., batch)
- design simple parts that result in less expensive tooling
- choose a manufacturing process that has a low cycle time

Unfortunately, it's not that simple. The material selection and the choice of manufacturing process are coupled. Choosing one will affect the other, as we shall see in the following example.

Example

Assume that our company is considering making a part out of low-strength metals or thermoplastics. Three processes appear compatible with the required feature shapes: sand casting, injection molding, and machining. The marketing department estimates that the company should produce about 5,000 pieces. Data gathered to select the material and manufacturing process are shown in Table 6.2. Determine the cost per part.

TABLE 6.2   Manufacturing Data for Cost Estimates

	Alternative		
	A	B	C
Mfg. process	Sand casting	Injection molding	Machining
Material	Aluminum alloy	ABS	Bronze alloy
Part weight (lb)	1	3	2
alpha	0.05	0.01	0.2
Material cost ($/lb), cw	1	0.25	0.75
Tooling cost ($), T	10000	35000	1500
Production quantity, q	5000	5000	5000
Cycle time (hrs/part), t	0.3	0.03	0.6
Machine rate ($/hr)	30	100	75

We use equation (6.3) to calculate the material cost per part as $c_M = c_w w_p (1 + \alpha)$.

Alternative A. $c_{MA} = \$1.00/\text{lb}(1 \text{ lb/part})(1 + 0.05) = \$1.050/\text{part}$

　　　　　 B. $c_{MB} = \$0.25/\text{lb}(3 \text{ lb/part})(1 + 0.01) = \$0.738/\text{part}$

　　　　　 C. $c_{MC} = \$0.75/\text{lb}(2 \text{ lb/part})(1 + 0.20) = \$1.800/\text{part}$

Using equation (6.4) we calculate the tooling cost as $c_T = T/q$.

Alternative A. $c_{TA} = \$10,000/5,000 \text{ parts} = \$2.000/\text{part}$

　　　　　 B. $c_{TB} = \$35,000/5,000 \text{ parts} = \$7.000/\text{part}$

　　　　　 C. $c_{TC} = \$1,500/5,000 \text{ parts} = \$0.300/\text{part}$

Using equation (6.5) we calculate the processing cost per part as $c_P = c_l t$.

Alternative A. $c_{PA} = \$30/\text{hr} (0.30) \text{ hrs/part} = \$9.000/\text{part}$

　　　　　 B. $c_{PB} = \$100/\text{hr} (0.03) \text{ hrs/part} = \$3.000/\text{part}$

　　　　　 C. $c_{PC} = \$75/\text{hr} (0.60) \text{ hrs/part} = \$45.000/\text{part}$

Summing the material, tooling, and processing cost for each part we obtain:

Alternative A. $c_{MA} + c_{TA} + c_{PA} = \$1.050 + 2.000 + 9.000 = \$12.050/\text{part}$

　　　　　 B. $c_{MB} + c_{TB} + c_{PB} = \$0.938 + 7.000 + 3.000 = \$10.738/\text{part}$

　　　　　 C. $c_{MC} + c_{TC} + c_{PC} = \$1.800 + 0.300 + 45.000 = \$47.100/\text{part}$

Therefore, based on the marketing department's 5,000-run estimate, injection molding would be the least expensive process.

The alternatives were plotted as a function of production volume, as shown in Figure 6.18.

Note that for very small quantities, machining would produce the lowest cost per part. And beginning at 200 per run, sand casting would be the least expensive, up to about 4,000. And, after about 4,000 parts per run, injection molding would be the least expensive.

Example

Determine the production quantity at which the manufacturing costs for alternative A and B are the same.

We can use equation (6.6) such that $c = c_w w_p (1 + \alpha) + T/q + c_l t$ and let $c_A = c_B$.

$$\$01.050/\text{part} + \$10,000/q + \$9/\text{part} = \$0.738/\text{part} + \$35,000/q + \$3/\text{part}$$
$$\$10.050/\text{part} + \$10,000/q = \$3.738/\text{part} + \$35,000/q$$

Combining terms and multiplying both sides by q we obtain

$$(\$6.232/\text{part})\, q = \$25,000$$
$$q = 4,011 \text{ parts.}$$

Therefore, less than 4,011 parts per run, process A is less costly.

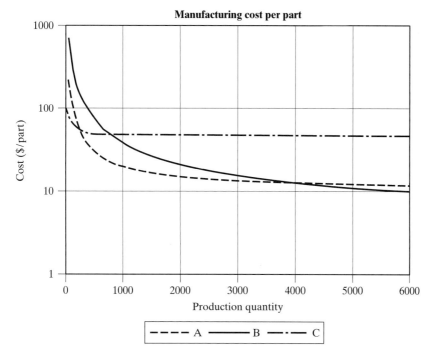

FIGURE 6.18

Manufacturing cost/part as a function of volume.

6.4 PROCESS SELECTION

Most parts undergo primary, secondary, and tertiary manufacturing processes. As a part design evolves, a variety of processes are considered. As we have learned in previous sections, some processes are not economically feasible unless significant quantities are produced. Some processes are incapable of producing large part sizes. Other processes cannot produce the desired geometric complexity, such as: the type and number of features, including: holes, notches, bosses, rotational symmetry, enclosed cavities, and uniformity of walls and/or cross sections (Boothroyd et al., 1994).

Manufacturing process and materials selection decisions occur in every phase of a part's design. As more and more information becomes available, revised cost estimates can also be made, ultimately affecting prior decisions. During conceptual design, for example, little is known about the part sizes or dimensions. Also, design is iterative. During the configuration design phase, some features may be combined to improve functionality, but demand a specialized manufacturing process or material, not previously considered.

When we use the *manufacturing-processes-first approach*, we first screen out manufacturing processes that will not satisfy part considerations, such as: production quantities, part size, or shape capability, as shown in Table 6.3. The remaining feasible processes will be compatible with some material classes. We

---

TABLE 6.3   Process-First approach first screens out manufacturing processes that are incompatible with the desired production run quantity, part size, and geometric shape capability.

---

1. Production run quantity/volume
2. Part size (overall)
3. Shape capability (geometric features)
   boss or depression in one dimension
   boss or depression in more than one dimension
   holes
   undercuts (internal/external)
   uniform wall thickness
   cross sections—uniform/regular
   rotational symmetry
   captured cavities (e.g., bottles)

---

further screen these by comparing material properties to the functional and environmental application requirements.

When we use the *materials-first approach*, we screen out materials that will not satisfy the functional requirements of the part. Namely, we compare application information from the engineering design specification to the mechanical and physical properties of material classes. We typically include criteria regarding the nature of the applied loads and the operating environment. This screening will eliminate a number of material classes. The feasible material classes remaining will be compatible with some manufacturing processes. To narrow the feasible processes further, we consider part information, including: the geometric complexity of the part, the production volume, and part size.

As mentioned in Chapter 5, either approach will lead to the same subset of material classes and compatible manufacturing process since we are doing successive eliminations or screenings based on the same criteria. Also note that during configuration design we will consider design-for-assembly and design-for-manufacture guidelines that will further help us refine our manufacturing-process decisions, such as determining secondary and tertiary processes.

6.5   SUMMARY

- Most parts undergo primary, secondary, and tertiary manufacturing processes.
- Fundamental manufacturing processes include: bulk deformation, casting, sheetmetal working, polymer processes, machining, finishing, and assembly.
- Manufacturing costs include material, tooling, and processing costs.
- Material cost per part is directly related to manufactured waste.
- Manufacturing costs are influenced by planned production quantities.
- Preliminary process selection considerations include: part size, geometric complexity, and production quantities.

REFERENCES

Boothroyd, G., P. Dewhurst, and W. Knight. 1994. *Product Design for Manufacture and Assembly*. New York: Marcel Dekker.

DeGarmo, E. P., J. T. Black, and R. Kohser. 1997. *Materials and Processes in Manufacturing*, 8th ed. Upper Saddle River, NJ: Prentice Hall.

Design for Manufacturability. 1992. *Tool and Manufacturing Engineers Handbook*, vol. 6. Dearborn, MI: Society of Manufacturing Engineers.

Dieter, G. E. 2000. *Engineering Design*, 3d ed. New York: McGraw-Hill.

Dixon, J. R., and C. Poli. 1995. *Engineering Design and Design for Manufacturing*. Conway, MA: Field Stone Publishers.

Groover, M. P. 1996. *Fundamentals of Modern Manufacturing*. Upper Saddle River, NJ: Prentice Hall.

Schey, J. A. 1999. *Introduction to Manufacturing Processes*, 3d ed. New York: McGraw-Hill.

KEY TERMS

Assembly
Bar drawing
Bending
Blanking
Blow molding
Boring
Bulk deformation
Casting
Cavity
Compound die
Compression molding
Core (mold)
Cycle time
Die
Die casting
Drawing (metal)
Drilling
Ejector pins
Electric discharge machining
Embossing

External undercut
Extrusion
Facing
Finishing
Forging
Grinding
Injection molding
Internal undercut
Investment casting
Machining
Manufactured waste
Material cost
Milling
Mold closure direction
Parting line
Planing
Primary manufacturing process
Processing cost
Progressive die
Punches

Punching
Reaming
Rolling
Sand casting
Sawing
Secondary manufacturing process
Shaping
Shearing
Sheet metalworking
Side cores
Side-action feature
Stamping
Subtractive process
Tertiary manufacturing process
Thermal forming
Tooling cost
Turning
Wire drawing

EXERCISES

Self-Test. Write the letter of the choice that best answers the question.

1. _____ Manufacturing processes that relate to surface treatments such as polishing, painting, heat trading, and joining are called:

   a. primary

   b. secondary

   c. tertiary

   d. initial

2. _____ Manufacturing processes that are used to add to or remove geometric features from the basic forms are called:

   a. primary

   b. secondary

   c. tertiary

   d. initial

3. _____ All the following are examples of bulk deformation except:

   a. forging

   b. rolling

   c. bar drawing

   d. shearing

4. _____ Changing the shape or form of bulk from materials caused by compressive or tensile yielding is called:

   a. casting

   b. machining

   c. deformation

   d. assembly

5. _____ The process in which molten metal is poured into a cast to solidify is called:

   a. polymer processing

   b. casting

   c. machining

   d. finishing

6. _____ The process of plastically compressing material between two halves of die set by hydraulic pressure or the stroke of a hammer is called:

   a. casting

   b. rolling

   c. deforming

   d. forging

7. _____ The process of solidifying molten metal in a ceramic cast made with wax patterns is called:

  a. investment casting

  b. die casting

  c. sand casting

  d. polymer casting

8. _____ Which of the following molding methods (of polymer processing) is used to produce hollow parts with thin walls?

  a. blow

  b. injection

  c. compression

  d. transfer

9. _____ Charge of thermoset or elastomeric is formed between heated mold halves under pressure while polymer cure is called:

  a. extrusion

  b. injection molding

  c. compression molding

  d. transfer molding

10. _____ Manufacturing very complex shapes, including internal and external undercuts, is possible using:

  a. extrusion

  b. injection molding

  c. compression molding

  d. transfer molding

11. _____ Cutting or separating sheet metal along a straight line is called:

  a. shearing

  b. blanking

  c. drawing

  d. bending

12. _____ Forming plastic indentations to form ribs, beads, or lettering on surface of metal is called:

  a. punching

  b. blanking

  c. shearing

  d. embossing

13. _____ Plastic deformation of sheet metal using matched punch-and-die set is called:

  a. punching

  b. blanking

  c. bending

  d. shearing

14. _____ Which other following method is used to produce products such as soda cans, cartridge casings, and pots and pans?

   a. drawing

   b. embossing

   c. blanking

   d. punching

15. _____ Which other following method is used to size sheets for subsequent operations?

   a. drawing

   b. shearing

   c. bending

   d. punching

16 _____ Removing material from the workpiece by using a sharp cutting tool that shears away chips of material to create desired form or features is called:

   a. machining

   b. casting

   c. polymer processing

   d. anodizing

17. _____ Which of the following is often used as a secondary process to true up critical dimensions or surfaces, or to smooth the surface finish?

   a. deformation

   b. casting

   c. polymer processes

   d. machining

18. _____ Removal of material from a flat surface by using a rotating cutter tool is called:

   a. planing

   b. sawing

   c. milling

   d. grinding

19. _____ Removing material using a translating cutter as the workpiece feeds is called:

   a. planing

   b. sawing

   c. milling

   d. grinding

20. _____ Removing material from a translating workpiece and a stationary cutter is called:

   a. planing

   b. shaping

   c. grinding

   d. milling

21. _____ Increasing the diameter of an existing hole by rotating the workpiece is called:
   a. drilling
   b. boring
   c. reaming
   d. facing

22. _____ The removal of material from a surface using an abrasive spinning wheel is called:
   a. electric discharge machining
   b. grinding
   c. shaping
   d. facing

23. _____ Which machining process produces the smoothest surface?
   a. drilling
   b. boring
   c. turning
   d. milling

24. _____ Handling actions, during assembly, include the following except:
   a. moving
   b. orienting
   c. placing
   d. shipping

25. _____ Manufacturing costs include the following except:
   a. materials
   b. shipping
   c. processing
   d. tooling

2: b, 4: c, 6: d, 8: a, 10: b, 12: d, 14: a, 16: a, 18: c, 20: b, 22: b, 24: d

26. Sketch how an ingot is "deformed" into billets, slabs, blooms, bars, rods, sheets, coil, and structural shapes. Label your sketch.
27. Describe how die casting differs from sand casting.
28. Describe how compression molding differs from injection molding.
29. Describe the sequence of processes used to make an aluminum bicycle frame.
30. Describe the sequence of processes used to make a stainless-steel pot.
31. Describe a possible sequence of processes used to produce a shrink-wrapped 10-pack of jewel CD cases.
32. List five permanent attachment methods used in assembly.
33. List five removable attachment methods used in assembly.
34. The following information has been gathered to analyze the costs of a proposed part to be added to an existing product. Marketing and sales have suggested making

1,000 of the new parts. The manufacturing division is considering three different processes.

Mfg. process	Alternative		
	Sand casting	Injection molding	Machining
Material	Aluminum alloy	ABS	Bronze alloy
Part weight (lb)	1	3	2
Alpha (fraction of material wasted)	0.05	0.01	0.2
Material cost ($/lb), cw	1	0.25	0.75
Tooling cost ($), T	10000	35000	1500
Cycle time (hrs/part), t	0.3	0.03	0.6
Machine rate ($/hr)	30	100	75

Use a separate sheet of paper to calculate the following (and attach).

34a. Determine the material cost per part for each of the three processes.

34b. Which is the most material-efficient?

34c. Determine the tooling cost per part for each of the three processes.

34d. Which process has the least costly tooling per part?

34e. Determine the processing cost per part for each of the three processes.

34f. Which process has the least processing cost per part?

34g. Determine the total cost per part for each of the three processes.

34h. Which manufacturing process results in the lowest total cost per part?

34i. Examine Figure 6.16. Do your numbers agree with the figure?

34j. Based on Figure 6.16, which process is "expensive" if the production quantity $> 5,000$ pieces?

35. List three examples of products made from the following processes: casting, die casting, sheet metalworking, blow molding, forging, and injection molding.

36. Describe the difference between the *manufacturing-processes-first* approach and the *materials-first* approach.

37. Visit the library and familiarize yourself with a leading periodical in the field of manufacturing, such as *Manufacturing Engineering*, published by the Society of Manufacturing Engineers.

37a. Skim through the periodical to get an overview of the content and layout.

37b. List five different categories of information presented, for example, "advertisements for equipment or services."

37c. List the types of employees who would find the information useful, for example, production supervisors.

37d. Select one advertisement that you find interesting. What does it say? What is it that strikes your interest?

37e. Look at the "Contents" page, Look under departments, and find the "Calendar." What is covered in that "department"?

37f. Select one of the "feature" articles listed on the contents page. List the title, author, month and year of magazine, and pages read. Write a one- or two-paragraph summary of the article describing the article and why is it important to design and manufacturing engineering.

38. Describe the sequence of processes used to produce the reduced vibration hammer manufactured by Stanley.

Configuration Design

When you have completed this chapter you will be able to

- Explain and apply guidelines of designing a product's architecture
- Synthesize alternative part configurations
- Describe and apply part configuration requirements sketches
- Design parts for greater ease of assembly
- Implement design for manufacture guidelines
- Evaluate alternative configurations

7.1   INTRODUCTION

As we finish the concept design phase, we determine one or more feasible concepts for the parts that go into our product. The concepts are merely abstract embodiments of physical principles, working geometries, and materials. We are eager to develop mathematical models, using our knowledge of the sciences and mathematics. We might also want to build a scale model for testing. But we can't do that just yet, because we have no specific details such as critical sizes or specific materials. Most likely, we have not even considered which of the components should be standard and therefore purchased.

For example, let's consider the design of a transmission, which converts rotational speed and torque. Assume that we selected a gear pair, shown in Figure 7.1, as the leading concept for our design, rather than a chain drive or belt-and-pulley system. We would like to proceed with parametric design and analyze gear tooth bending stresses, shaft deflections, bearing loads, and so on. But to set up the equations, or perform the computer analyses, or build a scale model we need some more information.

The abstract concept for the gear-pair has been left deliberately fuzzy. As we examine the figure, we realize that some of the remaining unknowns deal with configuration, such as:

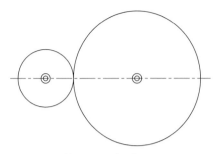

Abstract embodiment

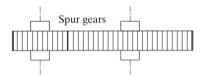

Spur gears

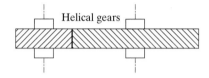

Helical gears

Alternative parts or features axis rotated 180°

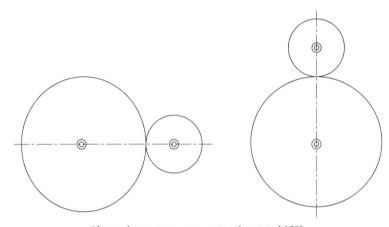

Alternative part arrangements axis rotated 180°

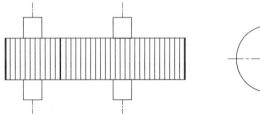

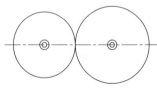

Alternative relative dimensions

FIGURE 7.1

Alternative configurations for a gear-pair transmission.

- What type of teeth should the gears have, spur or helical?
- Should the pinion be located to the left of the gear, right, above, or below?
- What size should the diameters be in relation to each other?
- What type, size, and location of shaft bearings should we consider?
- How will the drive pinion and output gear be attached to the shaft?

Configuration design deals with (1) **product architecture**, which is the selection and arrangement of components on a product, and (2) **part configuration**, which is the selection and arrangement of features on a part.

When *configuring a product*, our primary concern is architectural, that is, determining the number and type of components (standard, special, part, or subassembly), their specific function, how they are spatially arranged, and how they are interconnected. When configuring the gear-pair, we could choose different types of standard gears (helical, bevel, or spur), rearrangements of axes (horizontal, vertical, parallel, perpendicular), and different thicknesses and diameters. We could choose standard roller, tapered roller, or ball bearings to support the shaft. We could custom-design the shaft or specify a standard shaft with machined keyway. Product configuration design helps us decide what we buy versus what we design and manufacture. Product architecture also helps us to integrate separate components into multifunction components.

When *configuring a special-purpose part*, our goal is to determine the type and number of geometric features, their arrangement or **connectivity**, and the relative dimensions of selected features. For example, let's examine the custom-designed bracket shown in Figure 7.2. We can create alternative configurations by adding or removing features such as walls, holes, fillets, slots, and chamfers. In addition, we can change their arrangements and relative dimensions to generate additional configurations. We should also list unknown design variables such as hole diameter (D), wall thickness (t_w), fillet radius (R), width (W), and chamfer thickness (t_c).

When *"configuring" a standard component* (i.e., part or subassembly), our goal is to select the type of component and its rough dimensions. For example, in selecting a rolling element bearing, which is a standard subassembly, we would select one of the following types: ball bearings, roller bearings, needle bearings, and tapered roller bearings. Additionally, we would consider the bearing width relative to its height, and prepare a list of unknown design variables, such as bore diameter and bearing width.

Configuration design, therefore, is the phase of product development when we determine the number and type of parts or geometric features in our design, how they are spatially arranged or interconnected, and approximate relative dimensions of the parts or features, and develop a list of design variables, as shown in Table 7.1.

By generating a broad selection of alternative configurations, we systematically investigate different possibilities. Then, we can be selective, and choose the best configuration to develop further. In other words, to be selective we need a selection. There is no question that we will spend time and energy preparing

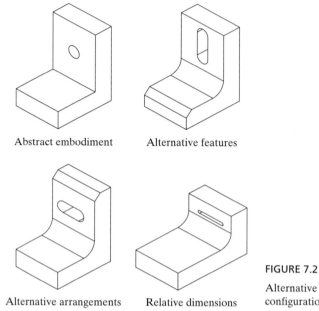

Abstract embodiment   Alternative features

Alternative arrangements   Relative dimensions

FIGURE 7.2

Alternative bracket configurations.

TABLE 7.1 Configuration Decisions for Different Product Elements

Configuration Problem	Required Decisions
Product	Type of component(s)
	Number of components
	Arrangement/connectivity
Special-purpose part	Type of geometric features
	Number of geometric features
	Arrangement/connectivity
	Relative dimensions
	Name(s) of unknown design variable(s)
Standard component	Specific type
(part or subassembly)	Names of unknown design variable(s)

different configurations. Many may not be used. But, by proceeding in a systematic way, we are likely to develop one or more configurations that meet our requirements.

> To be selective, we need a selection.

In the remaining sections we will examine methods to systematically generate, analyze, and evaluate alternative configurations. First, we will consider how the product can be configured. Then we will examine how parts can be configured. As shown in Figure 7.3, part configurations can be analyzed and refined using DFA and DFM guidelines. The refined configurations are then evaluated to determine their relative merit, and we then select one or two for parametric design.

Product development is not a simple and/or linear decision-making process. Product development is by nature fuzzy. The simplified models and diagrams presented thus far, however, give us a framework to better understand and deploy the collection of design methods used in industry. We recognize that companies

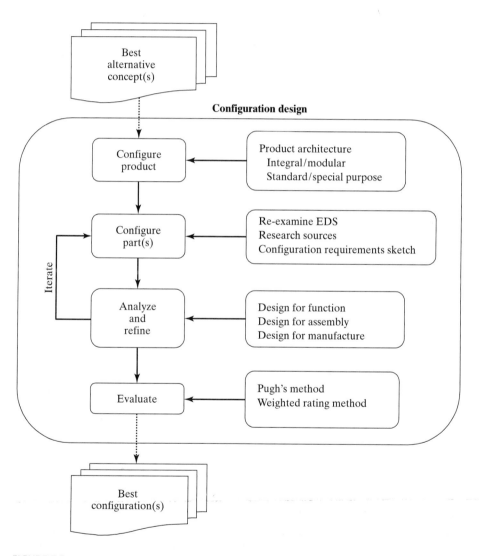

FIGURE 7.3

Configuration design, from product to part.

will also use their own methods and procedures that have been developed and fine-tuned to meet their particular purposes.

7.2 GENERATING CONFIGURATION ALTERNATIVES

At this point in the product development process we will likely have a collection of concepts, loosely conceived as a "product concept" that fulfills our customer's requirements. Of course, the abstract embodiments are, by nature, undefined. And before we can configure any of these, we need to configure the product. But, how do you configure the product when all you have is a collection of ill-defined concepts or components?

To configure a product we consider the number and type of concepts (components), their arrangement, and their relative dimensions. For example, how would we configure the transmission presented in the last section? Even though we have a gear-pair transmission concept, many product details remain fuzzy. For example:

- Will we select standard bearings or design special-purpose bearings?
- Will we purchase the pinion and gear or design our own?
- Where should we locate the shaft bearings?
- How will the bearings be removed for maintenance?
- How will the lubricant be replaced?
- Where will the transmission housing support the bearings loads?
- How will the housing seal the lubricant near the input and output shafts?

In the next two sections we will discuss ways to develop the product configuration, and then, how we might configure individual special-purpose parts.

7.2.1 Product Configuration

The fundamental goal of product configuration is to prepare a rough layout (sketch) of the product. **Layout sketches** are sketches that are drawn roughly to scale to show the geometric or spatial arrangement of the selected components and illustrate their relative shapes and sizes. Note, however, that layout sketches are not dimensioned or toleranced.

Before we draft our first layout sketch we should review the engineering design specification for the product, along with any other pertinent documentation. Of particular importance are the:

- spatial limitations with respect to overall width, height, and depth
- product interactions with other physical objects or the user
- repair and cleaning, adjusting, and lubricating of specific components
- wear of components
- desire of the customer to customize the configuration of his product

- need to include standard parts or assemblies
- need to design parts to any specific dimensional or industry standards
- need to replace consumable materials.

Next, we should review sources of good ideas, including: archives, people, the Internet, and creative methods. Or a similar product layout might be available to get us started with our own. In addition, we could examine some competitor's products.

At this point, we can develop the architecture of the product. As proposed by Ulrich and Eppinger (1995), product architecture is how the physical elements and functional elements of a product are arranged or **clustered** into physical building blocks called **chunks** and how the chunks interact. The **physical elements** are the physical concepts developed during the conceptual design phase, and/or the components such as standard or special-purpose parts and/or assemblies. The **functional elements** are the functions that a product performs.

Two types of product architecture are: modular architecture and integral architecture. In a **modular architecture**, chunks implement one or a few functions, and the interactions between chunks are well defined. Examples include office partitions, track lighting, personal computers, flashlights, and automobiles. The opposite of a modular architecture is an **integral architecture**, where a single chunk implements many functional elements or many chunks implement or share one function, and the interaction is ill-defined. An example of integral architecture is the BMW model R1100RS motorcycle since the transmission serves as a structural element as well as driveline component. By **function sharing** a component can fulfill more than one function. The BMW transmission housing shares functions and eliminates the redundant frame, saving weight and volume.

The process of configuring a product's architecture is to cluster physical elements (components) and remaining functional elements into chunks, then geometrically arrange the chunks.

Ulrich and Eppinger recommend the following procedure to develop a product architecture:

1. *Create a schematic* (of the products elements), including physical elements that the team has agreed upon and those functional elements that have not been reduced to a physical concept. Keep the number of elements under 30. Separate the others to be worked on later.

2. *Cluster elements into chunks.* Start with a single element and cluster it with another when it is beneficial to:
   - share functions
   - enable standardization (use or create standard parts or modules)
   - integrate geometry (i.e., dimensional precision between features)
   - exploit vendor capabilities (e.g., such as injection molding expertise)
   - fully utilize manufacturing process capabilities
   - accommodate consumer variety (options, customization)

- exploit common interfaces (120 VAC, air duct connectors, PC bus)
- manage changing technologies or components
- provide for wear, upgrades, maintenance, consumables

3. *Make a rough geometric layout.* Use either 2D or 3D sketches and/or CAD drawings or build a physical prototype. Arrange chunks so that the interfaces remain feasible. Recluster any element to maintain feasible interactions and interfaces. Discuss the layout with others on the team. Refine the layout as necessary.

4. *Identify the fundamental and incidental interactions.* Examine the flow of matter, energy, and signal between chunks. Are the desired functions preserved? Identify incidental interactions such as unwanted vibration, or heat generation, or EMI, and show on incidental interaction graph.

An example of product architecture is shown in Figures 7.4a–d.

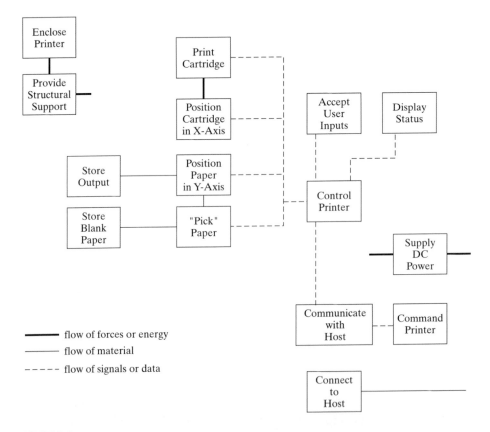

FIGURE 7.4a

Product architecture schematic of elements. (K. T. Ulrich and S. D. Eppinger, *Product Design and Development,* Copyright 1995, McGraw-Hill. Reproduced with permission of The McGraw-Hill Companies.)

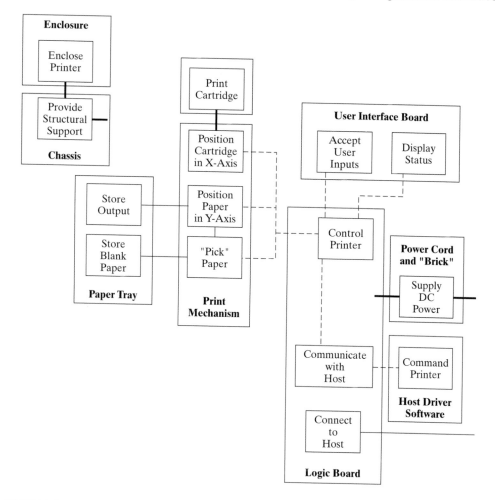

FIGURE 7.4b

Product architecture—clustered diagram. (K. T. Ulrich and S. D. Eppinger, *Product Design and Development,* Copyright 1995, McGraw-Hill. Reproduced with permission of The McGraw-Hill Companies.)

7.2.2 Part Configuration

A part is a single-piece component that is made out of one material and needs no assembly. A part has geometric **features** such as: walls, ribs, projections, fillets, bosses, rounds, cylinders, tubes, cubes, cones, spheres, holes, slots, notches, chamfers, and grooves. The **arrangement** of features includes both locations and orientations. When features are connected to each other we say they are contiguous features. Features, also, have different widths and heights with respect to each other, which we call **relative dimensions**. A "tall" feature would be high relative to its width. A needle bearing would have a small diameter relative to its length, whereas a roller bearing has similar diameter and length.

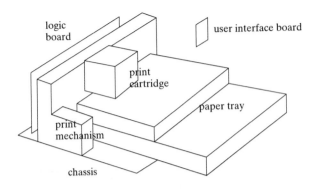

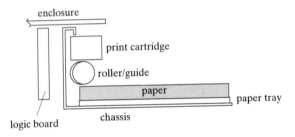

FIGURE 7.4c

Product architecture—geometric layout. (K. T. Ulrich and S. D. Eppinger, *Product Design and Development*, Copyright 1995, McGraw-Hill. Reproduced with permission of The McGraw-Hill Companies.)

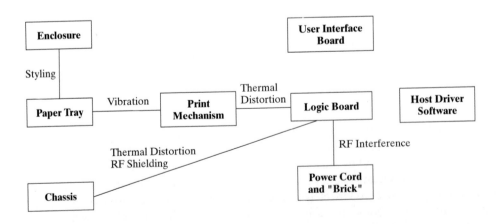

FIGURE 7.4d

Product architecture—incidental interaction diagram. (K. T. Ulrich and S. D. Eppinger, *Product Design and Development*, Copyright 1995, McGraw-Hill. Reproduced with permission of The McGraw-Hill Companies.)

When we configure a part we make decisions about the types of part features, the number of features, how the features are arranged, and what the relative dimensions are. We also establish a list of the design variable names. Later, we will find values for those variables during the parametric design phase.

As mentioned for product configuration, we should reexamine the engineering design specification along with any other pertinent product design documentation. What are the principal loads and how does the part interact with the other parts in the assembly? Are there any controlling constraints, such as weight, space, or size?

Then, too, we should consider researching archives, people, the Internet, and creative methods, to stimulate our thinking, and make sure that we are not "reinventing the wheel." Would a similar part satisfy the requirements, if modified slightly? Is a standard part available? A little time spent doing some background research can sometimes pay off handsomely.

At some point, however, the pencil has to hit the paper and alternative part configuration sketches need to be prepared. A method proposed by Dixon and Poli (1995) systematically generates a variety of part configurations. The method uses part configuration requirements sketches. It is fairly simple and focuses on essential-part requirements.

Step 1: Prepare a **configuration requirements sketch** for the part, which is a sketch, to approximate scale, of the essential surroundings of the part, including:

- forces
- flows (heat or other energy)
- features of mating parts
- support points or areas
- adjacent parts
- obstructions or forbidden areas

Note that the requirements sketch does not show any features of the part to be configured, only its surroundings. We make a number of photocopies of the configuration requirements sketch. Let's assume we make two copies.

Step 2: Prepare a few alternative **noncontiguous part configurations**. We add a partial wall or boss to the configuration requirements sketch at *mating or coupling points or surfaces*. The partial wall or boss is not connected to anything else, except the mating point or surface. It is called a "noncontiguous" feature. Set that noncontiguous configuration aside for the moment. Take a fresh configuration requirements sketch and add a partial wall or boss to a different point or area that is feasible. Set this alternative noncontiguous configuration sketch aside. And so on. Make a couple photocopies of each of the noncontiguous configurations. Let's assume we drew two noncontiguous sketches and that we make three copies of each.

Step 3: Prepare alternative **contiguous configurations**. We take one of the alternative noncontiguous sketches and connect the noncontiguous walls or bosses with material, as a single contiguous part. We do the same with the remaining five alternatives.

Step 4: Refine the configuration sketches. We improve the sketches by considering how the part will perform, how it will be manufactured, the material usage, and how it will be assembled. In this example we would refine the six sketches. As we can see, even in this simple example, a variety of alternatives can be systematically generated with this method.

Example

Let's consider the configuration design of a kitchen sink sponge holder. The user inserts the soggy sponge into the holder, letting it drain and air-dry. The holder can also be used to drain and dry pot-scrubbing pads. The concept design includes a box-like container that fits into a suction cup that is attached to the kitchen sink. A thermoplastic material and injection molding is anticipated for the first run of about 50,000 units. The force from the suction cup is expected to support the box and sponge. The wet sponge exerts the downward force.

Prepare two or more contiguous part configuration sketches.

As shown in Figure 7.5(a), the basic concept is a box supported on the side of a sink shown in Figure 7.5(b). We prepare a configuration requirements sketch, Figure 7.5(c), showing the downward sponge force and the upward suction cup force, along with a rough scale dotted line for the sink wall. We then add two chunks of material connecting the forces. This becomes our noncontiguous part configuration in Figure 7.5(d).

We then prepare four contiguous configurations, shown in Figure 7.6. The side walls of the container have not been shown to simplify the force flow illustration. Configuration (a) shows a basic connection from the suction cup pin to the bottom of the container. This exposes much of the suction cup, which is better hidden in configuration

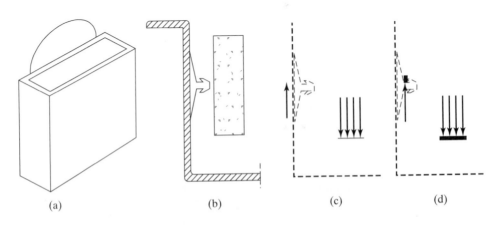

(a)   (b)   (c)   (d)

FIGURE 7.5

Configuration requirements sketch for a sponge holder.

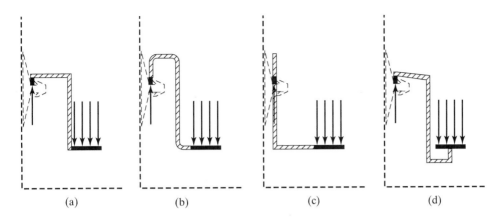

FIGURE 7.6

Contiguous configuration sketches for a sponge holder.

(b). Then in configuration (c) we see that the back wall can be pierced to eliminate material. Alternative (d) provides a drain trough. These four feasible configurations can now be analyzed, evaluated, and refined, as discussed in a later section of this chapter.

7.3   ANALYZING AND REFINING CONFIGURATION ALTERNATIVES

When analyzing configuration alternatives we consider the likelihood that the parts will function as expected, that the parts can be assembled, and that they can be manufactured. We use our general knowledge about working principles, materials, and manufacturing processes to logically reason whether the configurations will be satisfactory. We typically do not, however, prepare engineering analyses, such as comprehensive calculations, for any of the alternative configurations we have generated. We may sometimes, however, prepare some back-of-the-envelope, first-order estimates. The reason we wait is because comprehensive performance calculations are completed during parametric design. And, since *that* will take a lot of work, we would, first, like to reduce the number of alternative configurations, by screening out some of the less prospective candidates. Then, we will evaluate the remaining successful configurations using Pugh's or the weighted-rating evaluation method to select the best one or two for the parametric design phase.

To "logically reason ... the likelihood," we develop a configuration "checklist" using three criterion categories:

- Design for function,
- **Design for assembly**, and
- **Design for manufacture**

The checklist is used to systematically review each configuration to determine whether the candidate is likely to perform the required function. We use our experience and knowledge of the basic sciences to make this gut-level screening. Our goal is to screen out infeasible configurations, not to judge their relative merit. Those configurations that are screened out can be reconfigured, if time and resources permit, then reanalyzed for feasibility.

7.3.1   Design for Function

The product needs to work according to the customer's expectations, and it should be easy to maintain, and last a long time. In other words, it should not fail, and it should be easy to use and economical to use and maintain. Products fail because parts are not strong enough, or are too flexible, or they buckle, or jam due to thermal expansion. Their performance can degrade over time because of wear and/or corrosion. They also become uneconomical because they use too much fuel or energy or cost too much to repair and be disposed. These items are listed, along with a few others relating to function, in Table 7.2.

7.3.2   Design for Assembly

Design for assembly (DFA) is the name used to describe a set of practices that aim to reduce the costs of part handling, insertion, and fastening. Product components should be easy to handle, insert, or mate, and fasten together. This is true whether the product is assembled manually or automatically. When **handling** a part we typically grasp it from a storage bin or location, move it to a new location for mating with the rest of the assembly, and correctly orient the part. Then we insert or mate the part and fasten it.

TABLE 7.2   Design for Function Checklist Items

1. Strong
2. Stiff or flexible
3. Buckle-resistant
4. Thermal expansion
5. Vibrate
6. Quiet/noise
7. Heat transfer
8. Fluids transport/storage
9. Energy-efficient
10. Stable
11. Reliable
12. Human factors/ergonomics
13. Safe
14. Easy to use
15. Maintainable
16. Repairable
17. Durable (wear, corrosion)
18. Life-cycle costs
19. Styling/aesthetics

Boothroyd et al. (1994) have shown that part handling is largely influenced by part features or attributes including: part symmetry, size, thickness, weight, nesting, tangling, fragility, flexibility, temperature, slipperiness, stickiness, and the need for two hands, tools, or optical magnification. And further, these researchers have shown that insertion and fastening are influenced by factors including: accessibility, resistance (force) to insertion, visibility, ease of alignment and positioning, depth of insertion, and the type of fastener used. The researchers have also developed an excellent method to estimate the time required to assemble a product and evaluate the product's assembly design efficiency. The method can be done manually using tables and spreadsheets. However, the method is also incorporated into a computer software package that automates many of the table "lookups" and calculations.

The Society of Manufacturing Engineers (SME, 1992) recommends the following principles of design for assembly, including:

- minimizing part count
- encouraging modular assembly
- designing parts with self-fastening features (snap-fits, press-fits)
- using standard parts
- stacking subassemblies from the bottom up
- designing parts with self-locating features (e.g., chamfers, aligning recesses/dimples)
- eliminating reorientation (i.e., insertion from two or more directions)
- facilitating parts handling (grasp, orient, move)
- minimizing levels of assembly (number of assemblies)
- eliminating (electric) cables

These principles are illustrated in Figures 7.7a–d (Otto and Wood, 2001).

To screen our candidate configurations, with regard to assemble-ability, we can develop a checklist of important criteria from the handling features and insertion factors reported by Boothroyd et al., the principles proposed by the Society of Manufacturing Engineers, and Figures 7.7a–d.

As we review the design for assembly recommendations, we find many of them deal with varying product architecture and part interface configurations. For example, reviewing the SME guidelines we vary product architecture by (1) encouraging modular assembly, (2) using standard parts, (3) stacking from bottom up, and (4) minimizing the number of parts. Similarly, we vary part interface configurations by (1) self-fastening features, (2) self-locating features, and (3) eliminating reorientation.

7.3.3   Design for Manufacture

Design for manufacture (DFM) is the set of practices that aim to improve the fabrication of individual parts. For example, we may have selected materials compatible with selected manufacturing processes. However, will the manufacturing

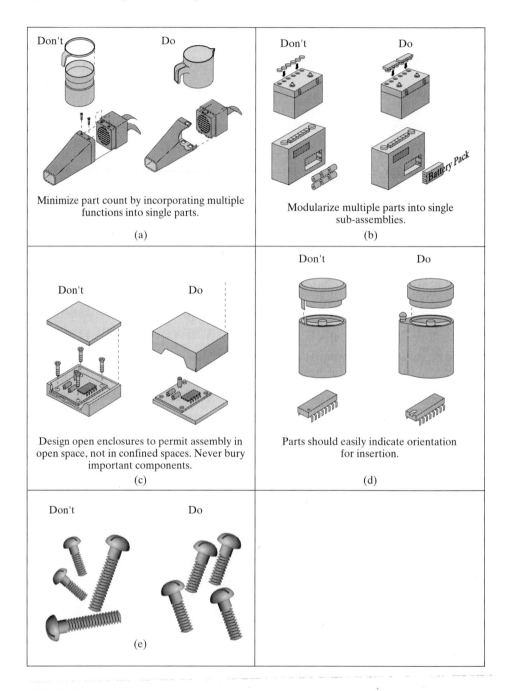

(a) Minimize part count by incorporating multiple functions into single parts.

(b) Modularize multiple parts into single sub-assemblies.

(c) Design open enclosures to permit assembly in open space, not in confined spaces. Never bury important components.

(d) Parts should easily indicate orientation for insertion.

(e)

FIGURE 7.7a

Design for assembly examples. (*Product Design: Techniques in Reverse Engineering and New Product Development*, by Otto/Wood, © 2001, Prentice-Hall. Reprinted by permission of Pearson Education, Inc., Upper Saddle River, NJ.)

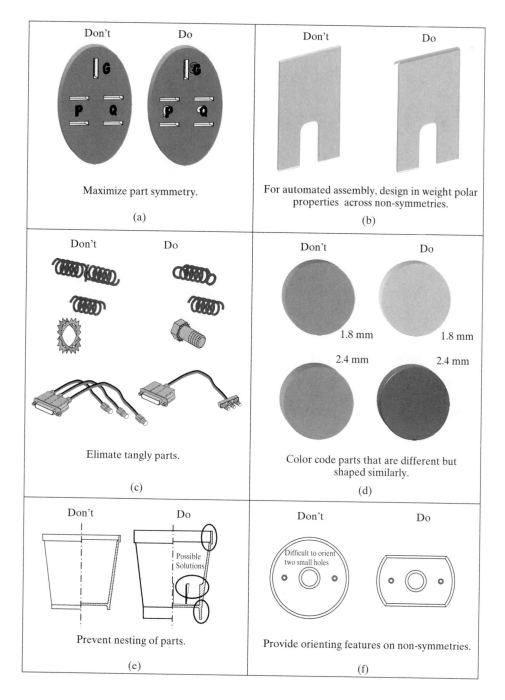

FIGURE 7.7b

Design for assembly examples continued. (*Product Design: Techniques in Reverse Engineering and New Product Development*, by Otto/Wood, © 2001, Prentice-Hall. Reprinted by permission of Pearson Education, Inc., Upper Saddle River, NJ.)

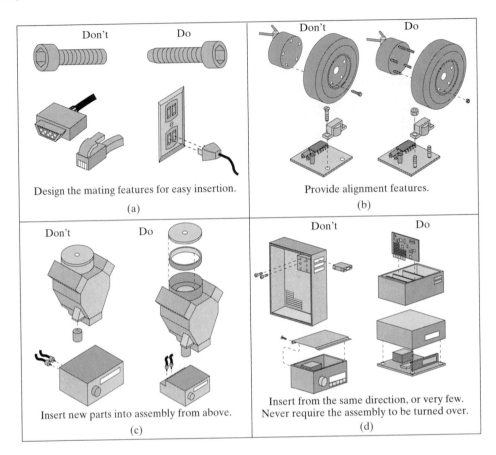

FIGURE 7.7c

Design for assembly examples continued. (*Product Design: Techniques in Reverse Engineering and New Product Development*, by Otto/Wood, © 2001, Prentice-Hall. Reprinted by permission of Pearson Education, Inc., Upper Saddle River, NJ.)

processes produce the configured part features? Further, will the processes be economical with respect to materials, processing, and tooling costs?

To screen alternative-part configurations for their manufacturability, we can develop a set of design for manufacture guidelines or checklist. Table 7.3 is a checklist for parts made by molding/casting, sheet metalworking, and machining. It was prepared by reviewing the capabilities of the manufacturing processes presented earlier in the text.

Examples of design for manufacture guidelines are shown for injection molding, sheet metal working, casting, and machining as shown in Figures 7.8–7.13.

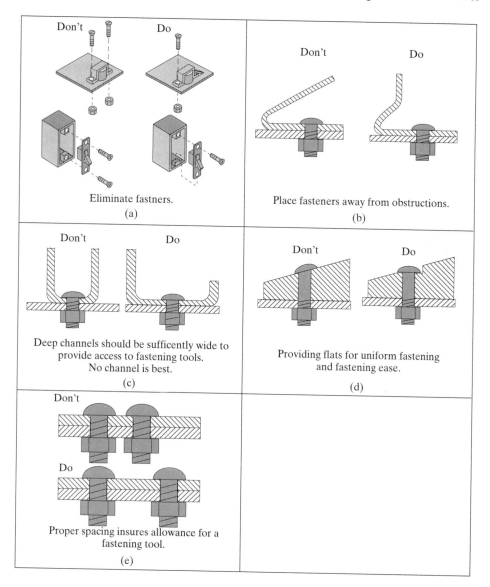

FIGURE 7.7d

Design for assembly examples continued. (*Product Design: Techniques in Reverse Engineering and New Product Development*, by Otto/Wood, © 2001, Prentice-Hall. Reprinted by permission of Pearson Education, Inc., Upper Saddle River, NJ.)

TABLE 7.3 Checklist for Parts Made by Molding/Casting, Sheet Metalworking, and Machining

Design for molding/casting
avoid designing parts with thick walls or heavy sections
design parts without undercuts
choose material for minimum tooling, processing, and material costs
design external threads to lie on parting plane/surface
add ribs for stiffening

Design for sheet metalworking
minimize manufactured scrap (cut-off versus blanking)
avoid designing parts with narrow cutouts or projections
keep side-action features to a minimum or avoid completely
reduce number of bend planes

Design for machining
use standard parts as much as possible
use raw material available in standard forms (sheet, rolls, bar, etc.)
employ standard features (holes, slots, chamfers, fillets, rounds, etc.)
specify liberal tolerances and surface finishes
avoid sharp internal corners on turned parts

As we review Table 7.3 and Figures 7.8–7.13, we find that many design for manufacture recommendations focus on part features and how they are configured. For example, eliminate or reduce undercuts, add stiffening ribs, avoid narrow projections, reduce number of bend panes, and specify standard machining features. These recommendations aim to optimize part manufacture, but note, however, they say little about how the parts assemble as a product. Therefore, we must analyze, evaluate, and refine our configurations using both DFA and DFM.

7.3.4 Refining Alternative Configurations

As we compare our alternative configuration sketches against the checklists for function, manufacturability, and assembly, we will undoubtedly find that we could make some refinements. In some cases, it will mean that we have to add or delete some feature(s), or in other cases, rearrange the feature(s), and in others, change their relative dimensions. And, in some cases, it may even mean that we have to eliminate that configuration altogether.

The process of making revisions to our configurations and recomparing them to our "checklists" is, of course, iterative in nature. And, the more experienced we become as designers, the fewer revisions we will need to make to our sketches.

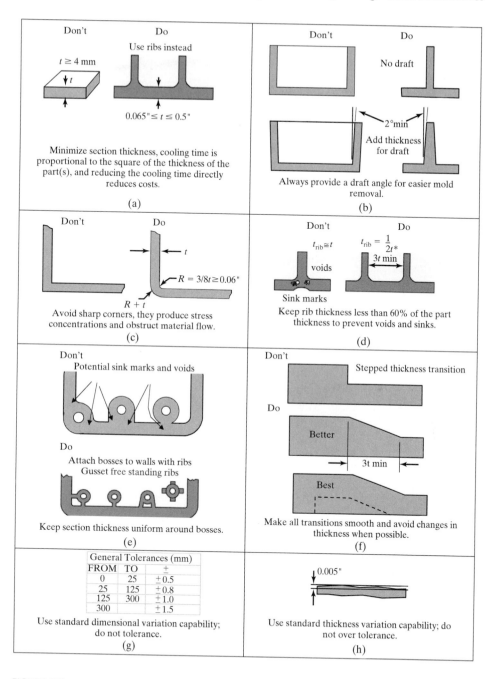

FIGURE 7.8

Design for manufacture of injection-molded parts. (*Product Design: Techniques in Reverse Engineering and New Product Development*, by Otto/Wood, © 2001, Prentice-Hall. Reprinted by permission of Pearson Education, Inc., Upper Saddle River, NJ.)

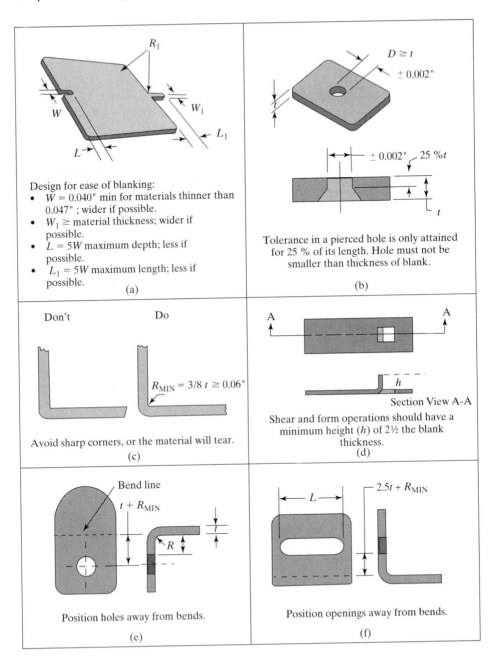

FIGURE 7.9

Design for manufacture of sheet metal parts. (*Product Design: Techniques in Reverse Engineering and New Product Development*, by Otto/Wood, © 2001, Prentice-Hall. Reprinted by permission of Pearson Education, Inc., Upper Saddle River, NJ.)

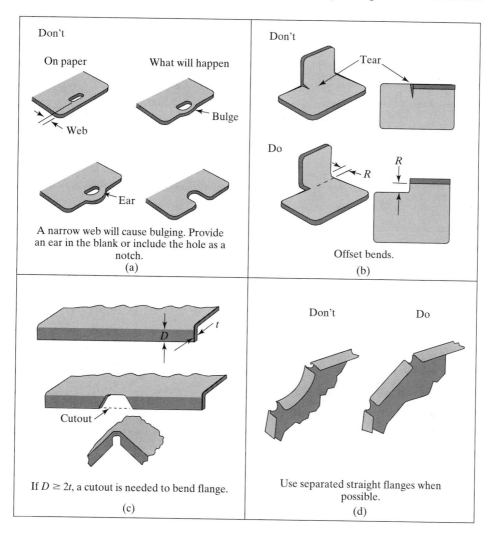

FIGURE 7.10

Design for manufacture of sheet metal parts continued. (*Product Design: Techniques in Reverse Engineering and New Product Development*, by Otto/Wood, © 2001, Prentice-Hall. Reprinted by permission of Pearson Education, Inc., Upper Saddle River, NJ.)

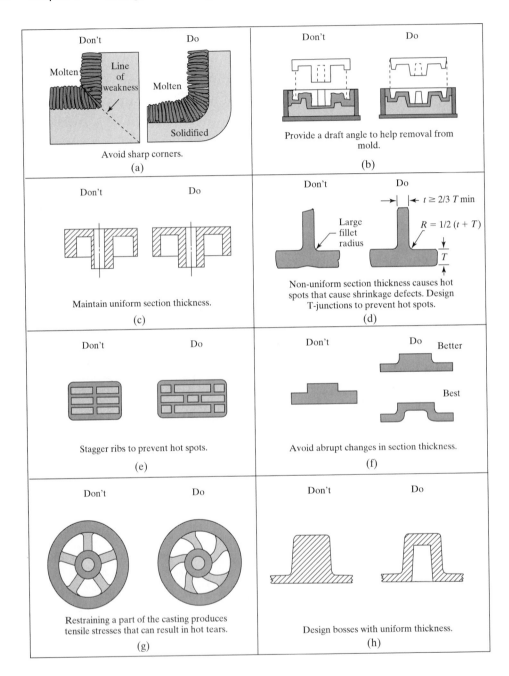

FIGURE 7.11

Design for manufacture of cast parts. (*Product Design: Techniques in Reverse Engineering and New Product Development*, by Otto/Wood, © 2001, Prentice-Hall. Reprinted by permission of Pearson Education, Inc., Upper Saddle River, NJ.)

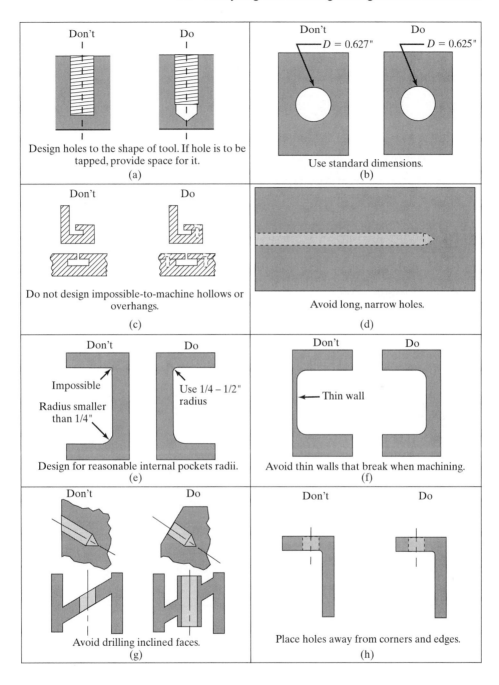

FIGURE 7.12

Design for manufacture of machined parts. (*Product Design: Techniques in Reverse Engineering and New Product Development*, by Otto/Wood, © 2001, Prentice-Hall. Reprinted by permission of Pearson Education, Inc., Upper Saddle River, NJ.)

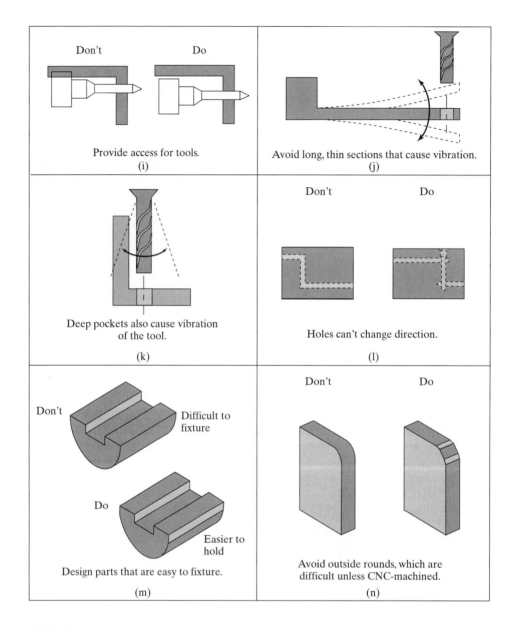

FIGURE 7.13

Design for manufacture of machined parts continued. (*Product Design: Techniques in Reverse Engineering and New Product Development*, by Otto/Wood, © 2001, Prentice-Hall. Reprinted by permission of Pearson Education, Inc., Upper Saddle River, NJ.)

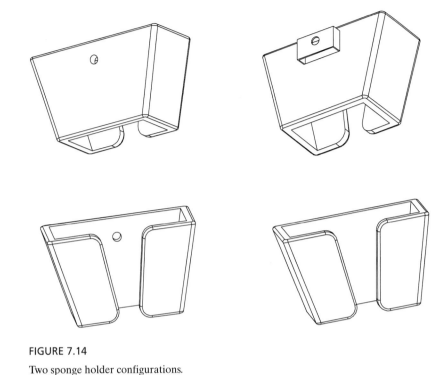

FIGURE 7.14

Two sponge holder configurations.

Example

Use the four contiguous part configuration sketches of the sponge holder and develop two refined configurations for evaluation.

The sponge holder was revised after careful consideration of the above design for function, assembly, and manufacture guidelines shown in Figure 7.14. We see that the bottom has been removed and the front has been opened for air circulation, easy cleaning, and sponge egress, thereby improving overall function. Sides have been tapered to improve packaging nesting and reduce material usage. Walls have been made uniformly thin and fillets have been added to improve molding.

7.4 EVALUATING CONFIGURATION ALTERNATIVES

As we finish making refinements to our configurations, we may need to reduce the number of alternatives. Using Pugh's method or the weighted-rating method, we select evaluation criteria for function, manufacture, and assembly. As a team, we also establish importance weights, and then rate our alternatives.

TABLE 7.4 Weighted-Rating Evaluation of Sponge Holder Configurations

| Criteria | Importance Weight | Sponge Holder Configuration Ratings | | | |
| | | With hole | | With bracket | |
		Rating	Wt. Rating	Rating	Wt. Rating
Function					
Drains well	15	3	0.45	3	0.45
Dries quickly	10	3	0.30	3	0.30
Stays clean	10	2	0.10	3	0.15
Sponge inserts easily	15	2	0.40	4	0.80
Manufacture					
Material usage	10	3	0.30	2	0.20
Tooling costs	15	3	0.45	2	0.30
Processing costs	5	3	0.15	3	0.15
Assembly					
Handling	5	3	0.15	3	0.15
Insertion	5	3	0.15	3	0.15
Number of parts	10	3	0.30	3	0.30
	100%				
Weighted rating			2.75		2.95

Example

A weighted-rating evaluation was completed for the two configurations and is shown in Table 7.4.

Selecting the holder with bracket configuration appears to be the better choice for further development. To complete the configuration design, we would prepare a list of design variables for the next phase. Some of the important design variables for the above example would include: holder width, height, depth, wall thickness, radii, bracket dimensions, hole location and size, and specific thermoplastic material.

7.5 COMPUTER-AIDED DESIGN

Configuration design activities revolve around product architecture and part configuration. We make significant use of hand-drawn sketches to work out and improve our ideas and communicate them to our team members. We use spreadsheets for checklists and weighted-rating evaluations. Similarly, we can make use of solid modeling packages. Note, however, that at this phase we are not trying to produce a set of working drawings such as detail part drawings or assembly drawings with bills of materials. Rather, our goal is to generate, analyze, and evaluate alternative configurations, *not document* our design efforts.

Computer-aided design (CAD) hardware and software can be used to assist in the creation and development of a design. The simplest packages represent parts as *two-dimensional models*. Similar to the classic drafting board, orthographic views of a three-dimensional object are created using points, lines, circles, and arcs. With this type of software three-dimensional images cannot be produced. A major disadvantage of two-dimensional systems is that if we change the length of a side we have to revise any lines, arcs, or circles attached to the side. Wireframe packages model a part using a three-dimensional database to store and manipulate the edges of the part. Once the **wireframe model** database is generated, however, generation of the orthographic views is essentially automated. In some cases, complex wireframe parts can be confusing because of all the "hidden" lines that are exposed. *Surface* modeling packages define the faces of a part in terms of flat and curved surfaces in addition to representing edges. This has the advantage of automatically determining hidden lines and surfaces.

Solid modeling packages also represent the wireframe and surface geometry necessary to define the edges and faces of a part. A solid model contains topographical information that defines the association of the geometrical entities such as faces that share a common edge makes filleting a straightforward operation. Solid models also capture interior versus exterior features so that sectioned views can be automated. Most important is that parts can be created having three-dimensional features such as ribs, slots, holes, chamfers, and fillets. The creation process is similar to a sequence of manufacturing processes. Further, these features are not stored as arcs, lines, or points, as in two-dimensional packages, but rather as solid objects. Representing part information as a virtual solid has a number of advantages:

Design Intent Is Captured. Solid modeling captures the features the designer intended. That is, as the geometry of each part is inevitably modified during the iterative steps of a product development process, each part will respond to these changes according to how the designers have planned. For example, if a cylindrical boss, having a concentric hole through it, is moved, then the hole moves with it and remains concentric.

Feature-Based Modeling. The creation of parts is simplified by automating the creation of standard manufacturing features such as bosses, cutouts, holes, ribs, fillets, chamfers, and draft. Rather than drawing an arc connected by lines, for example, the solid-modeling designer selects the "fillet" feature to automatically generate the arc to the desired radius.

Constraint-Based. Control of part geometry is maintained using constraints. For example, selected surfaces can be constrained to be parallel or perpendicular, holes can be made concentric, rounds can be made tangent, or the location of a slot can be fixed.

Parametric. Solid modeling is also parametric, in that the parameters of a model may be modified to change the geometry of that model. Instead of fixing the diameter of a hole as 5 inches, solid modeling uses a variable for the diameter that can be modified as the design evolves. Therefore, when a dimension is changed, the geometry of the part is automatically updated.

Fully Associative. Changes made to any feature of a part are automatically reflected in other views or files. For example, the four alternative configurations of the bracket shown in Figure 7.2 were produced with solid modeling starting with only one version. Then, chamfers, fillets, and slots were quickly added as well as the change in the length of the bottom (a parameter). During these changes all the other features stayed attached or associated.

Assemble-ability Check. With solid-modeling software, parts can be modeled, visualized in photorealistic renderings, revised, and improved upon within the computer before engineering drawings are ever produced. Once the individual parts of a product have been created, they can be assembled within this same CAD software and checked for proper fit and interference-free motion.

Downstream Benefits. A major advantage of solid modeling is that the database can be exported to other downstream software packages. These include **computer-aided engineering (CAE)** analysis software, used to predict product performance and finite element analysis (FEA) software, when the forces, pressures, and thermal and support conditions to which individual parts or an entire assembly are subjected can be simulated and the resultant patterns of stress, deformation, and temperature predicted. From these results part designs can be revised and improved upon to avoid stress concentrations, large deformations, and undesirable temperatures. In addition, the same database can be exported to a rapid prototyping system to automatically build physical parts, often within a matter of hours. The prototypes serve to validate fit, form, and function. Once the design is finalized, the solid-modeling software can be used to generate standard engineering drawings. Computer-aided manufacturing (CAM) software can also read the solid-model database to generate machine tool paths that can be downloaded to a computer numerical control (CNC) machine tool to automatically manufacture production parts to within specified tolerances.

7.6 SUMMARY

- Product configuration design determines the type and number of components, their arrangement, and their relative dimensions.
- Part configuration design determines the type and number of geometric features, their arrangement, and their relative dimensions.
- Standard part configuration design determines the type or class and approximate dimensions.
- Product configuration is closely linked to product architecture.
- Part features include: walls, ribs, projections, fillets, bosses, rounds, cylinders, tubes, cubes, spheres, holes, slots, notches, chamfers, and grooves.
- Configuration requirements sketches can be used to develop alternative part configurations.
- Configuration analysis includes considerations of function, manufacture, and assembly.
- Alternatives may undergo significant revisions during successive iterations.
- Solid-modeling CAD systems can be useful during configuration design as well downstream in parametric design and manufacturing.

REFERENCES

Boothroyd, G., P. Dewhurst, and W. Knight. 1994. *Product Design for Manufacture and Assembly*. New York: Marcel Dekker.
Dieter, G. E. 2000. *Engineering Design*, 3d ed. New York: McGraw-Hill.
Dixon, J. R., and C. Poli. 1995. *Engineering Design and Design for Manufacturing*. Conway, MA: Field Stone Publishers.
Otto, K. P., and K. L. Wood. 2001. *Product Design: Techniques in Reverse Engineering and New Product Development*. Upper Saddle River, NJ: Prentice Hall.
Pahl, G., and W. Beitz. 1996. *Engineering Design*, 2d ed. New York: Springer-Verlag.
Society of Manufacturing Engineers. 1992. *Design for Manufacturability*, Tool and Manufacturing Engineers Handbook, vol. 6. Dearborn, MI: SME.
Ulrich, K. T., and S. D. Eppinger. 1995. *Product Design and Development*. New York: McGraw-Hill.

KEY TERMS

Arrangement (location/orientation)
Chunk
Cluster
Computer-aided design (CAD)
Computer-aided engineering (CAE)
Configuration design
Configuration requirements sketch
Connectivity
Contiguous configurations
Design for assembly (DFA)
Design for manufacture (DFM)
Features
Function sharing

Functional element
Handling
Integral architecture
Layout sketches
Modular architecture
Noncontiguous part configurations
Part configuration
Physical element
Product architecture
Relative dimensions
Solid modeling
Wireframe model

EXERCISES

Self-Test. Write the letter of the choice that best answers the question.

1. _____ When configuring a product, we need to decide:
    a. geometric features
    b. component types
    c. design variable
    d. a, b, and c

2. _____ When configuring a special-purpose part, our goal is to determine:
    a. the type and number of geometric features
    b. their arrangement or connectivity
    c. their dimensions relative to each other
    d. a, b, and c

3. _____ When configuring a standard component we consider the:

   a. type of component

   b. cost

   c. vendor

   d. a, b, and c

4. _____ Functional and physical elements arranged as a physical building block of a product are called:

   a. clumps

   b. clusters

   c. chunks

   d. collection

5. _____ The steps to develop a product architecture include:

   a. create a schematic

   b. cluster elements into chunks

   c. make a rough geometric layout

   d. identify the fundamental and incidental interactions

   e. all of these

6. _____ Examples of part features are:

   a. bosses

   b. chamfers

   c. both a and b

   d. neither a nor b

7. _____ A configuration requirements sketch includes:

   a. forces

   b. energy flows

   c. mating parts

   d. supports

   e. all of these

8. _____ A configuration "checklist" includes all the following criteria except:

   a. DFM

   b. design for function

   c. DFA

   d. parametric design

9. _____ Handling activities include the following except:

   a. orienting

   b. fastening

   c. moving

   d. grasping

10. _____ Design for assembly includes methods to improve the following except:
   a. fastening
   b. fabricating
   c. inserting
   d. handling

11. _____ A set of practices that aim to improve the fabrication of individual parts is:
   a. DFA
   b. CAD
   c. DFM
   d. CAE

12. _____ All the following are examples of modular architecture except:
   a. office partitions
   b. coffee cup
   c. personal computers
   d. track lighting

2: d, 4: c, 6: c, 8: d, 10: b, 12: b

13. Stereo AM/FM radios/receivers are manufactured in both modular and integral architectures. Give an example of each. List the modular and integral aspects of each example. Which is better in your opinion, modular or integral? Why?

14. What items should be included in a configuration requirements sketch of a special-purpose part?

15. A manufacturer of outdoor-tree lights produces "standard" 100-light strings that use small white neon bulbs on two small-wire strings about 20 feet long. The company would like to make and sell a string "holder." The holder would be made of one-piece plastic and have a handle/or handhold. It would be configured so that one or two strings would be "wrapped" on it, something like a spool of thread. The holder and wrapped strings would then be stored until the next lighting season.

   15a. Prepare an $8\frac{1}{2} \times 11$ configuration requirements sketch of the problem. Photocopy it.
   15b. Prepare a *noncontiguous* configuration sketch on the photocopy. Make five photocopies of it.
   15c. Prepare five alternative *contiguous* configuration sketches. Identify/list the design variables.
   15d. Refine configurations and complete the sketches with labeling. Attach the five sketches.

16. A design team is configuring the frame of a custom-designed, limited-production, gasoline-engine-powered mini-bike. List three analysis criteria for:
   16a. Design for manufacture
   16b. Design for assembly
   16c. Design for function

17. The configuration of standard parts or subassemblies includes identifying the type of component and the design variables.

    17a. List five different types of fasteners.

    17b. Select one of the types and list five design variables.

18. List 10 types of geometric features that can be configured on a special-purpose part.

19. The configuration of standard components includes identifying the type of component and the identification of design variables.

    19a. List five different types of rolling element bearings.

    19b. Select one of the types and list five design variables for it.

20. List five things that can impede handling during assembly.

21. List five things that can impede insertion and fastening during assembly.

22. Sketch a pictorial view of a custom-designed switchplate for a teenager's bedroom. A switchplate is the polymer or metal plate that covers the standard, wall-mounted light switch. Be creative in your design (i.e., nonrectangular). Identify and label three or more geometric features that have been configured. Select a material and list some manufacturing processes that could be used to make 25,000 of them. Attach a sketch of the plate with the required information.

23. Review the DFA figures and list three guidelines that could be adopted in your design project.

24. Review the DFM figures and list three guidelines that could be adopted in your design project.

Parametric Design

L E A R N I N G
O B J E C T I V E S

When you have completed this chapter you will be able to

- Describe information flow through design phases
- Establish engineering characteristics, evaluation parameters, and constraints
- Identify and characterize design variables
- Select and use analytical or experimental methods to analyze alternative designs
- Develop parametric spreadsheets to refine designs
- Characterize and assess overall customer satisfaction
- Explain design for robustness
- Describe and use failure modes and effects analysis

8.1 INTRODUCTION

Parametric design includes a number of decision-making processes, much like the other phases of design. The processes use information (input) from prior phases to arrive at logical decisions (output). What makes parametric design special and particularly challenging is that we will employ analytical and experimental methods to predict and evaluate the behavior of each of the design candidates to make these decisions.

Let's examine the types of input and output information that we will process as we make our decisions in the various design phases.

8.1.1 Information Flow in the Design Phases

As we recall the phases of design as shown in Figure 8.1, information about the customers' needs is processed in the formulation phase usually resulting in a list of customer requirements/product functions, the importance of each, a list of engineering characteristics that quantitatively describe how well the functions

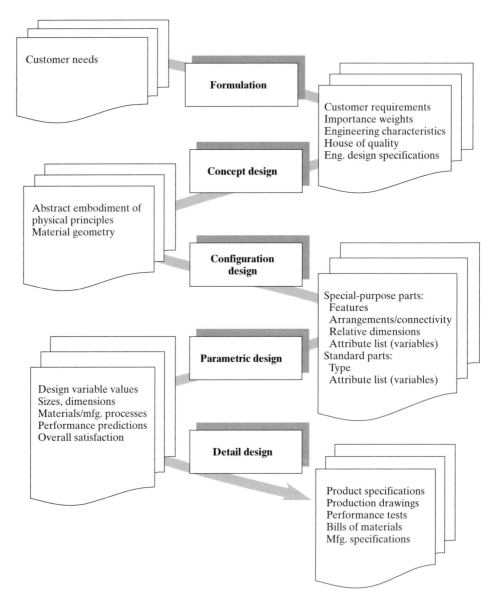

FIGURE 8.1

Design phase information input/output.

are performed, a detailed engineering design specification, and ideally, the house of quality diagram.

We use that information as input to the concept design phase, wherein we make decisions about the physical principles, abstract embodiment, primary manufacturing processes, and material classes.

Similarly, we use that information as input to the configuration design phase, wherein we make decisions for special-purpose parts relating to the geometric features on the part, their arrangement and/or connectivity, and their relative dimensions. And we establish a list of attributes or variables that we will need values for. And for standard parts, we select the specific type and attributes that we will need values for.

In the **parametric design** phase, we determine the values of those attributes, typically called **design variables**. These usually relate to specific sizes, lengths, radii, diameters, material types, and manufacturing process requirements. We do this by predicting the performance of alternative candidate designs using analytical methods and/or experimental methods. If the performance of the candidates satisfies all the design constraints, we call them **feasible designs**. Then we evaluate the feasible designs to determine which candidate is the best.

Finally, during the detail design phase, we complete the remaining decisions, resulting in comprehensive product specifications, drawings, manufacturing specifications, performance tests, and bills of materials.

Next, we examine the parametric design of a simple flange bolt.

8.1.2 Parametric Design: Pipe Flange Bolt Example

At a petrochemical plant, eight bolts will be used to fasten a cooling water pipe to the inlet flange of a high-pressure chemical reactor vessel. The piping system is subjected to an operating pressure of 1,000 psi. The flanges will be subjected to temperatures up to 250°F. Each bolt will be required to withstand a 4,000-pound design overload, which is four times the typical operating load of 1,000 pounds. The company has suggested a hex-head bolt, shown in Figure 8.2. Some of the design variables for bolts include: bolt diameter d, overall length L, number of threads per inch, and type of material. Assume that for this example, we have been asked to determine the bolt diameter only.

The principal function of a bolt is to fasten parts together so that they can be disassembled later for maintenance or repair. As the bolt is tightened, a tensile force develops over its length. The greater the force, the better the clamping action. Also, since the head of the bolt consequently develops higher friction forces, there is an improved resistance to loosening.

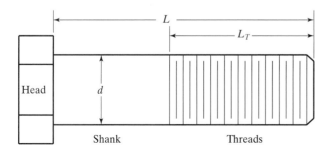

FIGURE 8.2

Hex-head bolt.

However, a bolt can fail to function in a number of ways, including: (1) the bolt head can twist off during tightening, (2) the bolt threads can strip off, and (3) the bolt can experience excessive tightening or excessive operating loads that it permanently elongates or even fractures.

Let's consider only the last mode of failure for this example. The ability of a bolt to withstand the tensile force without a permanent measurable elongation or "set" is called the proof load F_p, and is directly proportional to cross-section area A of the bolt and the material's proof strength S_p, by the relation:

$$F_p = AS_p \tag{8.1}$$

The Society of Automotive Engineers (SAE) has categorized proof strengths of bolt materials as Grades 1–9, with strengths ranging from 33,000 psi to 120,000 psi, respectively. Assume that we select a Grade 5 material, whose proof strength is 85,000 psi.

Therefore, for this design problem, we need to design the bolt strong enough to have a proof load that is greater than the 4,000 (pounds) design load. This is a constraint for this problem. Consequently, we need to find an area and, therefore, diameter such that:

$$F_p \geq 4{,}000 \text{ lbs.} \tag{8.2}$$

By substituting equation (8.1) into the constraint equation (8.2), we obtain:

$$AS_p \geq 4{,}000 \text{ lbs.} \tag{8.3}$$

We find that that the proof load constraint can be satisfied if the area:

$$A \geq \frac{4{,}000}{S_p} \tag{8.4}$$

$$A \geq \frac{4{,}000(\text{lbs.})}{85{,}000(\text{lbs./in.}^2)} \tag{8.5}$$

$$A \geq 0.047 \text{ in.}^2 \tag{8.6}$$

and since cross-sectional area is related to diameter d as:

$$A = \frac{\pi d^2}{4} \tag{8.7}$$

substituting, we find that

$$\frac{\pi d^2}{4} \geq 0.047 \text{ in.}^2 \tag{8.8}$$

and further, that

$$d^2 \geq \frac{0.047(4)}{\pi} = 0.0598 \text{ in.}^2 \tag{8.9}$$

$$d \geq 0.245 \text{ in.} \tag{8.10}$$

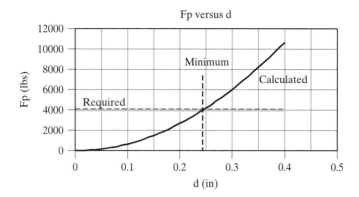

FIGURE 8.3

Flange bolt proof load versus diameter.

The calculated diameter is then rounded up to the nominal size 0.25 in., so that a standard size could be purchased.

  As shown in Figure 8.3, as we increase the diameter, the calculated proof load increases exponentially. The required proof load is constrained to be equal to or greater than 4,000 (lbs.) and is shown as a horizontal line. The minimum feasible diameter is shown as the dashed line at 0.245 (in.) and, therefore, satisfies the constraint. Selecting a diameter larger than 0.245 (in.) would improve the safety and reliability of the design and could be considered in the final design.

  This parametric design problem is rather simple, in that we have only: (1) one governing constraint (designed proof load must be greater than 4,000 [lbs.]) and (2) one algebraic relation to substitute for the area of a circular cross section. Therefore, for this design problem, we can directly solve for the diameter that satisfies the constraint. There is no need to iterate. Some engineers call this type of problem **inverse analysis**, because we can rearrange an analysis equation(s) to find the unknown value.

  The parametric design of most mechanical parts is much more difficult. Let's examine the overall process to find out how we can be systematic to help us complete better designs with less effort.

8.2   SYSTEMATIC STEPS IN PARAMETRIC DESIGN

Systematic parametric design has five major steps, as shown in Figure 8.4:

  Step 1: Formulate the parametric design problem.
  Step 2: Generate alternative designs.
  Step 3: Analyze/predict the performance of the alternatives.
  Step 4: Evaluate the performance of each alternative.
  Step 5: Refine/optimize.

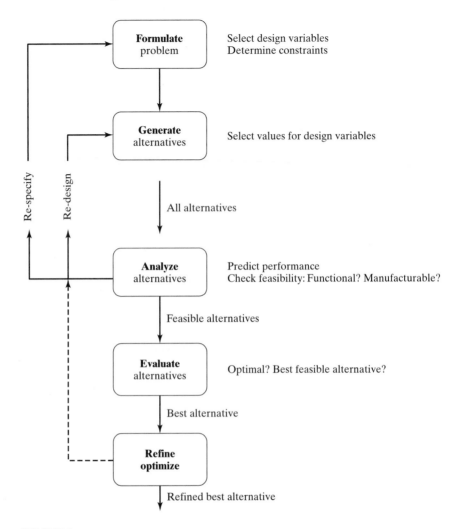

FIGURE 8.4

Parametric design decision-making processes.

Step 1. Formulate the problem. Practicing engineers often comment that they spend 40–50 percent of their time gathering appropriate data, clarifying important details, considering different analytical and experimental analyses, and then planning how they will complete a project. All these activities relate to problem formulation. As we start a parametric design problem, we need to familiarize ourselves with the problem parameters and also plan ways complete the design.

Performance Parameters/Solution Evaluation Parameters. By reviewing the house of quality chart and the engineering design specifications we reaffirm what functions the customer wants the product to perform. We then select engineering characteristics to measure the predicted performance of the functions. **Solution evaluation parameters** are the engineering characteristics that are selected to evaluate how well a candidate design "solves" the problem (Dixon and Poli, 1995).

Solution evaluation parameters depend upon the product and part being designed, but often include: cost, weight, speed, efficiency, safety, and reliability. We also denote parameter symbols, units (of measurement), and any lower and or upper limits of the parameter. As mentioned in Chapter 3, we can also describe the customer's satisfaction with respect to each solution evaluation parameter, as a satisfaction curve. Finally, we agree on analytical or experimental methods to determine their values.

> In the flange bolt example, the customer wants the bolt to clamp the pipe flange to the pressure vessel flange. The clamping force can be measured by the proof load and is, therefore, an example of a solution evaluation parameter. The higher it is, the better the clamping force, and the more satisfied the customer. We selected F_p as its symbol and pounds-force as the units. A lower limit, 4,000 (lbs.), is also established. The bolt proofload is constrained to be larger than 4,000 (lbs.). An analytical formula will be used to predict the proofload, although experimental tension tests could have been used.

Design Variables (DVs). Parameters under the control of the designer, which influence the candidate's performance, are called design variables. Design variables usually relate to part dimensions, tolerances, and/or material properties. In addition to the design variable names, we establish appropriate symbols, units, and upper and/or lower limits or bounds. For discrete variables we determine permissible values (e.g., $2 \times 4, 2 \times 6$ lumber, $1/2$ diameter bolts).

> The diameter is selected as the design variable in the flange bolt example. Larger diameters result in larger areas and improved performance. Permissible values are identifiable using handbook data.

Problem Definition Parameters (PDPs). Parameters that describe specific conditions of use, such as operating conditions are called **problem definition parameters**. In addition to their names, we establish appropriate symbols, units, and values.

> The operating pressure of 1,000 psi and the limiting temperature of 250°F are identified as PDPs for the flange bolt example.

Preliminary Plan for Solving the Problem. For small design problems, we often jump right in and start calculating things. For larger, more involved problems, we need to make a preliminary plan based on considerations including:

1. Do we have analytical models/formulas for our problem?
2. Are the assumptions used in our models the same as our problem?
3. Will we need to perform pilot scale or bench-top experiments to validate our analytical formulas?

4. How much time and money do we have to solve the problem?
5. Are "ballpark" computations required, or do we need more thorough and precise calculations?
6. Do we have knowledge about acceptable industry standards?
7. Will manufacturing specifications need to be generated from our analyses?
8. Do we understand the customers' function requirements versus satisfaction well enough?
9. Are we sufficiently qualified or competent to complete the required design?

For large design problems, design teams will consider the questions above and will prepare a design project proposal for upper management to review and approve. Design project proposals include items such as a background section, goals or mission of the project, a scope of work (of the work tasks to be performed), schedule, and budget. Regardless of the size of your project, however, it is always a good idea to prepare an outline of what you are going to do and when and how you are going to do it.

> Recognizing the simple nature of the flange bolt calculations, we went straight to making a few calculations. As mentioned previously, this was not the best way to "plan" the solution of the problem.

Step 2. Generate alternative designs. We select different values for the design variables to generate different candidate designs. These values can come from our own experience, from our company's experience, or from industry standards. Sometimes we need to make educated guesses.

Step 3. Analyze the alternative designs. We predict the performance of each candidate design using analytical and/or experimental methods:

Analytical Methods. Formulas from physics, mathematics, and the engineering sciences are most often used. Sometimes, advanced computer-aided-design packages, such as finite element analysis, computational fluid dynamics, and motion simulation, are used.

Experimental Methods. Oftentimes, the complexity of the design is beyond the accuracy or assumptions of our analytical models. In these cases, we can build scale models and/or full-scale models of the product or critical parts of the product, and test their performance. We can use wind tunnels, for example, to analyze the performance of complicated wing geometries, or check control surfaces.

> The performance of each design candidate is checked so that every performance constraint is satisfied. These designs are called feasible designs. If the constraints are violated, we reiterate back to generating another alternative and then analyzing it.

> The formula for the area of a cylindrical cross section (8.7) and the formula relating the proofload of a bolt to its area and strength (8.1) were used in the bolt example. The performance constraint considered in the bolt example was that the proofload be larger than 4,000 (lbs.). Since the formulas were

straightforward, we were able to juggle the equations into a sequence that did not require iteration. More complicated problems will not be "solvable" by equation juggling.

Step 4. Evaluate the results of the analyses. All the feasible designs are evaluated to determine which is the best design. Usually one or more criteria are identified in the formulation phase and used to determine the "best" feasible design alternative.

> The proofload was established during formulation as a rough measure of customer satisfaction. We assume that the higher the proofload, the more satisfied our customer would be. That would mean that a superlarge bolt should be chosen as the "best" design.
>
> This, however, ignores other aspects of real design problems such as weight and cost limitations. Unfortunately, since we do not know how the customer feels about these issues, we can only surmise that he/she might be satisfied with the smallest bolt that meets the force constraint. As we shall see in the next section, we should always try to ascertain customer satisfaction (with respect to each solution evaluation parameter).

Step 5. Refine/optimize. If no feasible design candidates exist, we select new values for the design variables, and thereby generate new design candidates. These are subsequently analyzed and evaluated for feasibility and optimality. This is the solid "redesign" iteration loop shown in Figure 8.4. If, after considerable effort has been expended, we cannot find *any* feasible candidates, we might ask whether we have set our design specifications too restrictively. Perhaps one or more constraints could be relaxed, for example. If so, we are "respecify"-ing the problem, also shown in Figure 8.4.

As we substitute new values for each of the design variables into the system of analysis equations we become familiar with how each design variable affects part or product performance. Using this familiarity, we can sometimes generate new values that satisfy the constraints. Dixon and Poli (1995) call this "physical reasoning." In other words, by becoming familiar with the causes and effects, we can logically reason the "physics" of the problem.

Optimal design methods can automatically regenerate new values of the design variables to improve expected performance and satisfaction (Arora, 1989; Papalambros and Wilde, 1989; Rao, 1984; Reklaitis et al., 1983; Siddall, 1982; Vanderplaats, 1984). For single-attribute optimization, a single criterion is chosen, such as minimizing the weight of a part. The criterion is a function of the design variables, and is called the **objective function**. Values of the design variables are usually constrained to some upper and lower limits. The predicted performance of a part is also related to the design variables. For example, the strength of a bolt is a function of its diameter. Design variable functions and limits are called **constraints**. Excel's Solver feature can be used to optimize many typical design problems found in mechanical engineering. This is shown as the dashed regenerate loop in Figure 8.4.

A more detailed example using this systematic procedure is presented in the next section.

8.3 SYSTEMATIC PARAMETRIC DESIGN: BELT-AND-PULLEY EXAMPLE

Let's consider the systematic procedure presented in the last section using a more detailed example.

8.3.1 Design Problem Formulation

A $\frac{1}{2}$-hp electric motor, running at 1,800 rpm, will be used to drive a grinding wheel operating at 600 rpm. A flat belt-and-pulley drive system configuration has been selected, as shown in Figure 8.5. The design team has also determined that:

■ the drive motor will have a 2-inch-diameter pulley mounted to its $\frac{1}{2}$-inch output shaft,

■ candidate designs should be able to utilize the full horsepower available,

■ the customer desires a compact system design,

■ the drive pulley will slip first, before the driven pulley,

■ the purchasing department has located a vendor that can provide a flat belt that can withstand a maximum 30-pound tensile load,

■ the coefficient of friction between the belt and pulley is 0.3,

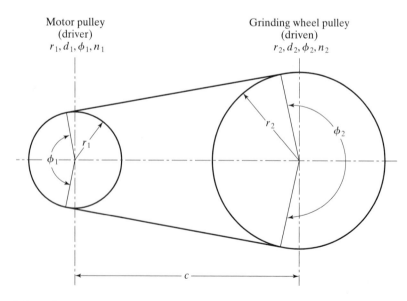

FIGURE 8.5

Belt-and-pulley-drive system for motor and grinding wheel.

- other design engineers in your group will design the mountings, bearings, and protective equipment, therefore,
- parametric design efforts should focus on distance between centers and driven-pulley diameter.

Solution Evaluation Parameters (SEPs)   The principal function of the pulley is to transform the power of the motor from a high speed to low speed. In the transformation, the smaller motor torque is converted to a larger torque, according to the conservation of energy law. Also, the belt-and-pulley system would fail to perform its principal function if the belt slipped or if the belt broke owing to excessive tension. Finally, the customer would be more satisfied with a compact design.

Since we know that the tension forces in the belt are limited by the amount of friction between the belt on the driver pulley, up to the point of impending slip, we could determine the torque that the belt can deliver to the pulley, T_b, and compare it with the maximum torque, T_m, that the motor can supply. Also, we should calculate the maximum belt tension, F_1, to make sure that it does not exceed the 35 (lbs.) limit.

We can therefore summarize the solution evaluation parameters in Table 8.1.

TABLE 8.1   Solution Evaluation Parameters for Belt-Pulley System

	Parameter	Symbol	Units	Lower Limit	Upper Limit
1	belt torque	T_b	lb-in.	T_m	—
2	belt tension	F_1	lbs	—	35
3	center distance	c	in.	small	—

Design Variables (DVs)   The value of the center distance, c, directly affects the compactness of the design and is to be determined by the designer. Also, as the center distance is increased, more of the belt wraps around the pulley (i.e., is in contact with the surface of the pulley), increasing the ability of the belt to grip the pulley and thereby satisfy the torque requirements of the motor. The design variables are summarized in Table 8.2.

TABLE 8.2   Design Variables for Belt-Pulley System

	Design Variable	Symbol	Units	Lower Limit	Upper Limit
1	center distance	c	in.	small	—
2	driven-pulley diameter	d_1	in.	—	—

Problem Definition Parameters   Studying the design problem data, we find a number of "givens" that define design problem conditions such as the friction coefficient, belt strength, motor power, and motor pulley diameter. Therefore, we identify these as the problem definition parameters in Table 8.3.

TABLE 8.3   Problem Definition Parameters for Belt-Pulley System

	Parameter	Symbol	Units	Lower Limit	Upper Limit
1	friction coefficient	f	none	0.3	0.3
2	belt strength	F_{max}	lbs.	—	30
3	motor power	$\dot{W}$	hp	$1/2$	$1/2$
4	motor pulley diameter	d_1	in.	2	2

Plan for Solving the Design Problem   Using analytical relations from physics and mathematics we can use a hand calculator or build a spreadsheet to analyze a variety of engineering characteristics, including:

1. grinding wheel pulley speed, n_2,
2. angle of wrap as a function of the center distance, c,
3. belt torque, T_b,
4. maximum belt tension, F_1,
5. slack-side belt tension, F_2, and
6. initial tension (before torque is applied), F_i.

Then we will check that the constraints are not violated. Specifically, we will make sure that the belt will deliver the full motor torque to the grinding-wheel pulley and that the belt tension does not exceed the belt strength limit.

8.3.2   Generating and Analyzing

Following our plan, we need to develop an analytical means to predict the behavior of the system. We can model the behavior of the system using relations from physics and mathematics and develop a system of equations to analyze the performance of each candidate design as we substitute different values for the design variables.

For example, we know that a motor will deliver power $\dot{W}$ (hp), to a pulley rotating at n rpm when producing a torque T_m lb.·ft. according to equation 8.11:

$$\dot{W} = \frac{T_m n}{5252} \text{ (hp)} \tag{8.11}$$

If the belt is not permitted to slip on the pulleys, the pulley speeds are related to the ratio of the pulley diameters.

$$n_2/n_1 = d_1/d_2 = r_1/r_2 \tag{8.12}$$

Then, we can determine the angle of wrap ϕ_1 using basic geometric relations.

$$\phi_1 = \pi - 2 \sin^{-1}\left(\frac{d_2 - d_1}{2c}\right) \tag{8.13}$$

Similarly, we know that a belt, with coefficient of friction f that is in contact with the pulley for an angle of wrap ϕ, has a maximum belt tension F_1, on the taut side of the pulley, and is related to the belt tension F_2 on the slack side of the pulley according to:

$$\frac{F_1}{F_2} \le e^{f\phi} \qquad (8.14)$$

Then, by examining the free-body diagram, shown in Figure 8.6, we know that for static equilibrium, we can sum the moments about the bearing B to obtain the torque T_m, delivered by the belt to the driver pulley of radius r_1, as:

$$T_b = (F_1 - F_2)r_1 \qquad (8.15)$$

As a point of interest, we can show how the angle of wrap affects the torque capacity of the belt at impending slip, by substituting 8.14 into 8.15, and find:

$$T_b = (F_1 - F_2)r_1 = \left(F_1 - \frac{F_1}{e^{f\phi}} \right)r_1 = F_1\left(1 - \frac{1}{e^{f\phi}} \right)r_1 \qquad (8.16)$$

Examining equation 8.16, we see that the right-hand term gets smaller as the angle of wrap increases and consequently the torque increases.

The maximum torque delivered by the motor T_m can be determined by rearranging 8.10 as:

$$T_m = \frac{\dot{W}\,5252}{n} = \frac{0.5(5252)}{1800} = 1.46 \text{ (lb.}\cdot\text{ft.)} = 17.52 \text{ (lb.}\cdot\text{in.)} \qquad (8.17)$$

The belt must be able to provide a torque equal to or greater than 17.52 (lb. · in.); otherwise it will slip.

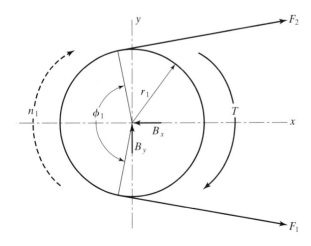

FIGURE 8.6

Free-body diagram of motor pulley.

Using equation 8.12 we can predict the driven-pulley speed, n_2, as a function of the design variable, diameter d_2.

$$n_2 = n_1 \frac{d_1}{d_2} = 1{,}800 \frac{2}{d_2} \tag{8.18}$$

Note that equation 8.18 is an equality constraint, having one unknown design variable, d_2. A feasible candidate design must satisfy this constraint. Since the grinding-wheel speed has been specified as 600 rpm, we can obtain a feasible value of the driven-pulley diameter by rearranging equation 8.18.

$$d_2 = d_1 \frac{n_1}{n_2} = 2 \left(\frac{1{,}800}{600} \right) = 6 \text{ (in.)} \tag{8.19}$$

Therefore, the only feasible value of the design variable d_2 is 2 in.

Generating an Initial Value of the Design Variable c   Since the customer will be more satisfied with a compact design, we would like the distance between the pulley centers c to be small. The closest that the two pulleys can be is when their radii are almost touching, theoretically speaking, or

$$c_{min} = r_1 + r_2 = \frac{d_1 + d_2}{2} = \frac{6 + 2}{2} = 4 \text{ in.} \tag{8.20}$$

Now we can use equation 8.13 to find the angle of wrap on the driving pulley.

$$\phi_1 = \pi - 2 \sin^{-1} \left(\frac{6 - 2}{2(4)} \right) = 2.09 \text{ rad} = 120 \text{ deg} \tag{8.21}$$

To find the tensile force F_1 that satisfies the motor torque constraint we use 8.16 to obtain

$$F_1 = \frac{T}{\left(1 - \dfrac{1}{e^{f\phi}} \right)} = \frac{17.52}{\left(1 - \dfrac{1}{e^{0.3(2.09)}} \right)} = 37.5 \text{ lbs.} \tag{8.22}$$

We find the tension on the slack side of the belt as

$$F_2 = \frac{F_1}{e^{f\phi}} = \frac{37.5}{e^{0.3(2.09)}} = 20.0 \text{ lbs.} \tag{8.23}$$

The initial tension in the belt before the torque is applied is obtained as

$$F_i = \frac{F_1 + F_2}{2} = \frac{37.5 + 20.0}{2} = 28.8 \text{ lbs.} \tag{8.24}$$

We check the belt torque using equation 8.15.

$$T_b = (F_1 - F_2)r_1 = (37.5 - 20.0)1 = 17.5 \text{ lb.} \cdot \text{in.} \tag{8.25}$$

We note that to satisfy the torque constraint, a 37.5-lb. tension will be necessary for F_1. But that level of tension exceeds the belt strength constraint of 35 lbs. Therefore, a center distance of 4 in. is an infeasible value. It causes a violation of a constraint and will need to be increased (to increase the angle of wrap, and consequently reduce the tension in the belt).

To reduce the effort in computing the expected performance of the system as a function of center distance, c, we can develop a spreadsheet as shown in Table 8.4.

Redesigning: Finding Feasible Values Since the belt tension is constrained to a maximum of 35 lbs., the initial value chosen for c is infeasible. We therefore

TABLE 8.4 Spreadsheet for Analyzing Belt-Pulley Performance

Flat Belt-and-Pulley Design

Problem Definition Parameters

Parameter	Sym.	Units	Value			
friction coefficient	f	none	0.3			
belt strength	F_{max}	lbs.	35			
motor power	$\dot{W}$	hp	0.5			
motor speed	n_1	rpm	1800			
motor pulley diameter	D_1	inches	2			

Design Variables

Variable	Sym.	Units	Lower	Value	Upper	
driven-pulley diameter	d_2	inches	—	6	—	
center distance	c	inches	4.0	4	12	

Performance Calculations

Eng. Characteristic	Sym.	Units	Value	Constraint Type	Constraint Value	Condition
motor torque	T_m	lb.-in.	17.51			
grinding-wheel speed	n_2	rpm	600	=	600	Satisfied
angle of wrap	ϕ_1	degrees	120.0			
belt tension—taut	F_1	lbs.	37.5	$\leq$	35	Unsatisfied
belt tension—slack	F_2	lbs.	20.0			
initial belt tension	F_i	lbs.	28.8			
belt torque	T_b	lb.-in.	17.51	$\geq$	17.51	Satisfied

TABLE 8.5 Belt Tension for Alternative Center Distances

Center distance c (in.)	Belt Tension F_1 (lbs.)
4	37.5
4.96 (*Goalseek*)	35.0
6	33.5
8	32.0
10	31.2

increase the center distance and recalculate the belt tensions. Using the spread-sheet, we obtain the following belt tensions, F_1 in Table 8.5.

Using Excel's Goalseek feature we find when $c \geq 4.96$ in., the belt-strength constraint is satisfied. However, to provide some extra capacity or compensate for some belt wear, or a slight decrease in friction coefficient, we usually select a larger value, perhaps 8, or even 10 in. Selecting $c = 8$ in., for example, we would obtain a 1.09 factor of safety n, from:

$$n = \frac{F_{allowable}}{F_{design}} = 35/32 = 1.09 \tag{8.26}$$

Factor of safety is a term used to express ratio of loads or stresses as exemplified in equation 8.26. Factors of safety less than one mean that the design will not support the load without "failing." Factors of safety greater than one indicate that the design is "safe." Factors of safety are also discussed later in the chapter.

Finding feasible values for the design variables is what generating and analyzing is all about **infeasible designs** are those whose design variable values do not satisfy the constraints. We have found that when $d_2 = 2$ in. and $c = 8$ in. all the constraints are satisfied (speed, torque, and tension). But are these feasible values the best values?

Trade-offs   As shown in Table 8.5, as we increase the center distance, we obtain lower values of belt tension, and consequently higher factors of safety. And that's good. But, we are also increasing the size of the system. Unfortunately this design problem, like most design problems, exhibits a **trade-off**, wherein one attribute improves as the other degrades. Trade-offs are caused by the interdependency of variables, typically referred to as **coupling**. Both compactness and belt tension are coupled to the center distance. We trade-off higher belt tensions as we satisfy the customer's desire for compactness. We could ask, "What is the best design?" Is more safety (lower belt tension) and a less compact system better than less safety and more compactness?

This question is difficult to answer. However, as it has been said, "beauty is in the eye of the beholder." Why not consider these trade-offs from the customer's perspective. That is, how does the customer feel about the importance of each of these attributes? Also, is there a way to assess overall satisfaction of the customer? This is the process of evaluating, which we consider in the next section.

8.3.3   Evaluating

During evaluation, we choose the best design candidate from among those feasible designs that we generated and analyzed earlier. Often our company will want the least expensive design. Other times we may want to maximize an aspect of system performance to obtain the fastest, or the lightest, or the most fuel-efficient design. Like many design teams, we could, therefore, select one of these criteria as the measure of "best," to determine our final recommendation.

Most design problems, however, have multiple attributes or criteria that usually involve trade-off decisions. In the belt-pulley system, for example, we need to make compromises between center distance and belt tension (i.e., compactness versus safety).

Therefore, we need a rational method that can help us make these necessary compromises among multiple criteria. As in concept design and configuration design, the weighted-rating method can be used and is recommended whenever possible.

Step 1: Establish a set of evaluation criteria.

Step 2: Rate the feasible designs for each criterion.

Step 3: Weight the ratings according to importance.

Step 4: Sum the weighted ratings to calculate an overall weighted rating.

Step 1. Establish a set of evaluation criteria, importance weights, and satisfaction levels.   If we wish to maintain high standards of quality we need to incorporate the "voice of the customer." Therefore, evaluation criteria are often developed from the list of solution evaluation parameters, or from engineering characteristics. Recall, that we use the house of quality, and/or other marketing research tools, to identify the customer's: (1) functional requirements for the product, (2) key engineering characteristics (that measure how well the functions are performed), and (3) the importance of each functional requirement.

> For our belt-pulley system, let's assume that the product development team determined that as long as the motor horsepower was fully utilized, the two most important engineering characteristics were belt tension and center distance (compactness). We identified these earlier during the parametric design formulation step as solution evaluation parameters.
>
> The team was also able to determine that the customer considers belt tension as "very important" and center distance as "important." Assume that we interpret "very important" with a weight of 0.6 and "important" with a weight of 0.4. Note that we can certainly use other numerical values for importance levels. The very act of choosing numerical values often leads to lots of discussion among design team members. And while it may be time-consuming and frustrating, it is very constructive in that it leads to a better understanding of the customer's desires.

Step 2. Rate the feasible designs for each criterion.   We can use the same rating scale of 0–5 (unsatisfactory to very good), as we used in Table 4.2 when evaluating alternative-concept designs. It should be noted, however, that the people doing the rating may or may not reflect the voice of the customer. An estimate of customer satisfaction is preferred, however. Recall that in Chapter 3, we considered satisfaction curves as a means to assess the customer's satisfaction. We arbitrarily let minimum satisfaction be equal to the numerical value of 0 and maximum satisfaction be equal to the number 1. These are similar to the ratings of 0 and 5, discussed before. The main difference is that in satisfaction curves, we try to link the value of each criterion, or SEP, to a value of satisfaction. This linking of satisfaction versus SEP values is facilitated using graphs and curve-fits. Let's see how we can use these to evaluate our belt-pulley designs.

Let's assume that the product development team also prepared estimates of customer satisfaction with respect to belt tension and center distance (graphed in Figures 8.7 and Figure 8.8).

When the satisfaction curves are sketched, we can rate each feasible design by reading values off the curves and entering them into the weighted-rating table.

To facilitate the weighted-rating calculations we can find analytical formulas that fit the curves (i.e., curve-fits) and then automate the computations using a spreadsheet. Basic formulas from geometry can be used as in our example. Special computer programs can also be used. And Excel's Trendline feature is quite capable in curve-fitting the following relationships: linear, polynomial, logarithmic, exponential, and power.

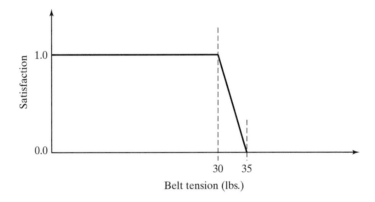

FIGURE 8.7

Customer satisfaction versus belt tension.

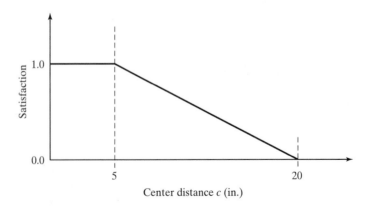

FIGURE 8.8

Customer satisfaction versus center distance.

Since the curves are straight lines in this example, we can fit a simple, straight-line equation using the two-point formula

$$(y - y_1) = \frac{(y_2 - y_1)}{(x_2 - x_1)}(x - x_1) \tag{8.27}$$

We substitute satisfaction S for y, and solution evaluation parameter for x, noting that the maximum value of S is 1, and the minimum value of S is 0. Therefore, for an SEP with a decreasing satisfaction curve we find the satisfaction S as:

$$S_{x-} = \frac{(x_2 - x)}{(x_2 - x_1)} \tag{8.28}$$

And for an SEP with an increasing satisfaction curve, we find the satisfaction S as:

$$S_{x+} = \frac{(x - x_1)}{(x_2 - x_1)} \tag{8.29}$$

We set up the belt-pulley satisfactions as:
for belt tension F_1:

$$S_{F_1} = \frac{(35 - F_1)}{(35 - 30)} \tag{8.30}$$

and for compactness c:

$$S_c = \frac{(20 - c)}{(20 - 5)} \tag{8.31}$$

Step 3. Weight the rating according to importance.   In this step we multiply the satisfaction rating by the importance weight.

Step 4. Sum the weighted ratings to calculate an overall weighted rating.   We can estimate the customer's overall satisfaction Q as the sum of the importance-weighted satisfaction levels. Specifically, let S_i be the level of the customer's satisfaction for i-th solution evaluation parameter and w_i be equal to the importance weight for that parameter. Then the overall customer satisfaction can be estimated by

$$Q = \sum w_i S_i = w_1 S_1 + w_2 S_2 + \cdots \tag{8.32}$$

For a center distance equal to 6 in. we can calculate the overall satisfaction as

$$Q = \sum w_i S_i = w_{F_1} S_{F_1} + w_c S_c \tag{8.33}$$

$$Q = 0.6\frac{(35 - F_1)}{(35 - 30)} + 0.4\frac{(20 - c)}{(20 - 5)} \tag{8.34}$$

$$Q = 0.6\frac{(35 - 33.5)}{(35 - 30)} + 0.4\frac{(20 - 6)}{(20 - 5)} \tag{8.35}$$

$$Q = 0.6(0.3) + 0.4(0.93) = 0.18 + 0.37 = 0.55 \tag{8.36}$$

Other values of overall satisfaction are calculated and summarized in Table 8.6. Note how values of the design variable, c, are coupled to the values of belt tension and compactness.

The satisfaction values are also graphed in Figure 8.9. The satisfaction curve for belt tension illustrates how as c increases, the customer is more satisfied. This occurs because the belt tension decreases, resulting in higher factors of safety. And similarly, as c increases, the customer is less satisfied because the design is less compact.

The overall satisfaction curve, however, illustrates that the customer's satisfaction is somewhat maximized when the center distance is about 10 in. Also, the figure indicates that the overall satisfaction is fairly insensitive from about $c = 9$ in. to about $c = 14$ in. That appears to indicate that the customer may be fairly indifferent to values of c between 9 and 14 in.

The overall satisfaction formula can be used as an aggregate objective function to optimize the belt-pulley design problem. An aggregate objective function is used to optimize problems when more than one criterion exists, in other words for **multiattribute optimization**. **Aggregate objective functions** combine a number of separate objective functions into one scalar function. Using Excel's Solver feature and the overall satisfaction formula as an aggregate objective function, an optimal value of $c = 11.38$ in. resulted in an overall satisfaction of 0.723, whereas $c = 10$ in. results in an overall satisfaction of 0.717.

The "best" feasible design candidate is therefore, $d_2 = 2$ in. and $c = 11.38$ in. Note that this set of values for the design variables satisfies all

TABLE 8.6 Weighted Satisfactions Based on Different Center Distances c

Center distance (in.)	F_1 (lbs.)	Tension S_{F_1}	$w_{F_1}S_{F_1}$	c (in.)	Compactness S_c	$w_c S_c$	Overall satisfaction Q
4.96	35.00	0.00	0.00	4.96	1.00	0.40	0.40
6	33.50	0.30	0.18	6	0.93	0.37	0.55
8	32.00	0.60	0.36	8	0.80	0.32	0.68
10	31.25	0.76	0.46	10	0.67	0.27	0.72
14	30.40	0.92	0.55	14	0.40	0.16	0.71
18	30.00	1.00	0.60	18	0.13	0.05	0.65
20	29.90	1.00	0.60	20	0.00	0.00	0.60

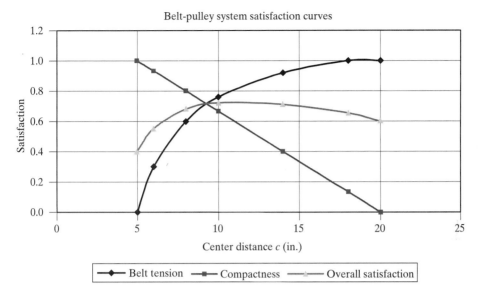

FIGURE 8.9

Satisfaction curves for belt-pulley system.

the constraints (speed, torque, and tension) and maximizes the customer's satisfaction.

8.4   DESIGN FOR ROBUSTNESS

A **robust product** is one whose performance is insensitive to variations. Variations come from a variety of sources including: (1) manufacturing, (2) wear, and (3) operating environment. When a product is manufactured, minor variations in material composition can occur. Sometimes an alternative, less costly, material with similar mechanical properties is purchased. In other cases, we might heat-treat or surface-coat the material in manufacturing processes that have resulting variations. And, as important, most of the manufacturing processes, such as machining, casting, and sheet metal stamping, will result in variations in final dimensions.

As the product is used, its moving parts undergo abrasive and adhesive wear. Also, some materials have properties that degrade over time. Steels become brittle with exposure to hydrogen gases and some polymers become opaque, etc. Then, we might also consider situations when the customer abuses the product, by subjecting it to higher-than-expected loads.

The operating environment can also cause variations in performance. Excessive humidity, dust, dirt, and extremely hot or cold environments can alter product performance. Steel, for example, can be brittle at temperatures under 25°F. Corrosive gases or vapors, such as in chemical plants or food-processing plants, can also cause performance degradation.

Design for robustness is a term used to describe a number of methods aimed at reducing the sensitivity of product performance to variations (Fowlkes and Creveling, 1995; Ullman, 1997). During parametric design, these variations in material properties, dimensions, and operating conditions can be simulated. Alternative designs can be generated and analyzed for their sensitivity to these variations.

Two methods used quite frequently are probabilistic optimal design and the Taguchi method. Probabilistic optimal design uses methods such as Monte Carlo simulation combined with nonlinear optimization algorithms to design parts with high levels of reliability (Eggert and Mayne, 1993; Haugen, 1980; Rao, 1984; Siddall, 1983). The Taguchi method exploits concepts from design of experiments to reduce the sensitivity of products or processes to "noise" (i.e., variations) (Ross, 1996).

8.5 COMPUTER-AIDED ENGINEERING

Computer-aided engineering (CAE) refers to computer software and hardware systems used in the analysis of engineering designs to validate functional performance. Computer-aided design (CAD) is sometimes used interchangeably with CAE.

CAE systems can be separated into four main categories: dynamics analysis, finite element analysis (FEA), general purpose, and other.

Dynamics analyses include the kinematics and kinetics of bodies, that is, the motion and the forces and moment that cause motion. We find that solid-modeling packages, for example, will mate parts into an assembly, and that we can manipulate various parts to animate their relative motion. However, in a dynamics analysis package such as Working Model, ADAMS, or DADS, we can apply specified loads to the parts and the package will calculate their resultant motion using fundamental equations of physics and a variety of numerical methods. These packages simulate behavior based on physics, whereas CAD solid-modeling packages illustrate motion. Since kinematics and kinetics-based packages are based on the physics of the problem, they can analyze a number of behaviors including:

- positions, velocities, accelerations,
- contacts and collisions,
- joint forces, shaking forces, and
- relative motions.

Finite element analysis (FEA) is a method that essentially divides a part into smaller discrete elements to analyze the functional performance of a part. Used in many engineering fields, the process begins with the creation of a geometric model using CAD software. Solid-modeling packages can export files to FEA and some can communicate directly with FEA packages. The part is subdivided into a mesh of smaller finite elements connected at nodes. Loads are then virtually applied to the part. Using fundamental equations from physics and engineering, an FEA package can analayze a number of behaviors including:

- stresses and strains throughout the part,
- factors of safety,
- buckling of parts in compression,
- temperature gradients and resulting heat transfer,
- fluid pressures and flows,
- natural frequencies of vibration, and
- displacements of nodes.

General-purpose software refers to computer applications covering word processing, spreadsheets, mathematics, oral presentations, and project management. We spend a significant amount of time communicating our work to others within and without our department and/or company. Having applications such as MSWord®, Excel®, MathCAD®, PowerPoint®, and MSProject® can help us complete these daily tasks more efficiently.

Other computer-aided engineering applications include:

- Quality function deployment—applications that facilitate the development of houses of quality such as QFDCapture.
- Material selection—specialized software that aids in screening materials such as GRANTA design, which implements Ashby's methodology.
- DFMA—applications that facilitate design for assembly and/or design for manufacture analyses such as the Boothroyd-Dewhurst package.
- Systems simulation—such as MATLAB Simulink, which analyzes the transient and steady-state behavior of multicomponent electrical and mechanical systems.
- Variance analysis—applications that compute tolerance stacks in assemblies.

8.6 SUMMARY

- The parametric design phase includes decision-making processes to determine the values of the design variables that satisfy the constraints and maximize the customer's satisfaction.
- The five steps in parametric design are: (1) formulate, (2) generate, (3) analyze, (4) evaluate, and (5) refine/optimize.
- During parametric design analysis we predict the performance of each alternative, reiterating when necessary to assure that all the candidates are feasible.
- During parametric design evaluation we select the best alternative.
- Many design problems typically exhibit trade-off behavior, necessitating compromises among the design variable values.

- The weighted-rating method, using customer satisfaction curves, can be used to determine the best candidate from among the feasible design candidates.

- Design for robustness methods can be used to decrease the sensitivity of candidate designs to variations due to materials, manufacturing, or operating conditions.

- Computer-aided engineering applications are useful in analyzing the functional performance of candidate designs.

REFERENCES

Arora, J. S. 1989. *Introduction to Optimum Design*. New York: McGraw-Hill.

Dixon, J. R., and C. Poli. 1995. *Engineering Design and Design for Manufacturing*. Conway, MA: Field Stone Publishers.

Eggert, R. J., and R. W. Mayne. 1993. "Probabilistic Optimal Design Using Successive Surrogate Probability Density Functions." *ASME Journal of Mechanical Design*, Vol. 115 (September): pp. 385–391.

Fowlkes, W. Y., and C. M. Creveling. 1995. *Engineering Methods for Robust Product Design*. Reading, MA: Addison-Wesley.

Haugen, E. B. 1980. *Probabilistic Approaches to Design*. New York: Wiley.

Papalambros, P. Y., and D. J. Wilde. 1989. *Principles of Optimal Design*. New York: Cambridge University Press.

Rao, S. S. 1984. *Optimization: Theory and Applications*, 2d ed. New York: Wiley.

Reklaitis, G. V., A. Ravindran, and K. M. Ragsdell. 1983. *Engineering Optimization: Method and Application*. New York: Wiley.

Ross, P. J. 1996. *Taguchi Techniques for Quality Engineering*, 2d ed. New York: McGraw-Hill.

Siddall, J. N. 1982. *Optimal Engineering Design: Principles and Application*. New York: Marcel Dekker.

Siddall, J. N. 1983. *Probablistic Engineering Design: Principles and Applications*. New York: Marcel Dekker.

Ullman, D. G. 1997. *The Mechanical Design Process*, 2d ed. New York: McGraw-Hill.

Vanderplaats, G. N. 1984. *Numerical Optimization Techniques for Engineering Design*. New York: McGraw-Hill.

KEY TERMS

Aggregate objective function

Computer-aided engineering

Constraint

Coupling

Design variable

Factor of safety

Feasible design

Finite element Analysis (FEA)

Infeasible design

Inverse analysis

Multiattribute optimization

Objective function

Optimize

Parametric design

Problem definition parameter

Robust product

Solution evaluation parameter

Trade-off

EXERCISES

Self-Test. Write the letter of the choice that best answers the question.

1. _____ We complete comprehensive product specifications, drawings, performance tests, and bills of materials in the:
   a. detail design phase
   b. configuration design phase
   c. parametric design phase
   d. concept design phase

2. _____ We make decisions about the physical principles, abstract embodiment, geometry, and material in the:
   a. detail design phase
   b. configuration design phase
   c. parametric design phase
   d. concept design phase

3. _____ We make decisions for special-purpose parts relating to the geometric features on that part and their arrangement and their relative dimensions in the:
   a. detail design phase
   b. parametric design phase
   c. configuration design phase
   d. concept design phase

4. _____ In which of the five steps of parametric design do we assess the feasible designs to determine which is the best design:
   a. analyze
   b. refine
   c. evaluate
   d. generate

5. _____ In which of the five steps of parametric design do we gather appropriate data, clarify important details, consider different analytical and experimental analysis, and then plan and decide how to complete the project:
   a. generate
   b. analyze
   c. evaluate
   d. formulate

6. _____ In which of the five steps of parametric design do we select different values for the design variables to synthesize different candidate designs:
   a. formulate
   b. generate
   c. evaluate
   d. refine

7. _____ Values of the design variables that do not satisfy the constraints are said to be:
   a. feasible
   b. nonoptimal
   c. inaccurate
   d. infeasible

8. _____ Design variables are parameters that:
   a. describe specific conditions of use
   b. are under the control of the design engineer
   c. measure product performance
   d. measure customer satisfaction

9. _____ The term used to describe a number of methods aimed at reducing the sensitivity of product performance to variations is called:
   a. weighted rating
   b. design evaluation
   c. design for robustness
   d. optimal performance

10. _____ Methods to automate the regeneration of design variable values are called:
    a. design for robustness
    b. safe design
    c. optimal design
    d. synthesis design

2: d, 4: c, 6: b, 8: b, 10: c

11. As shown in the bolt design example in the text, the equation for area can be "juggled" or rearranged to explicitly solve for diameter d. Many, if not most, engineering problems involve implicit relations such as the Secant buckling formula below (implicit in Pcr):

$$Pcr = \frac{ASy}{1 + (ec/\rho^2)sec[(Le/\rho)\sqrt{P_{cr}/4AE}]}$$

Explain how you might solve for area A, if given other variable values.

12. A customer would like to support a 2,000-pound art sculpture on a column 10 feet high. You have been hired to complete the parametric design. The customer would be unhappy if the column buckled or cost too much. Your column, he further suggests, should be able to support a design load of 6,000 pounds. Begin the parametric design by formulating the problem and complete the following tables. List pertinent constraints. Do we know if cost or safety is more important to the customer? Why? Should cost or safety be more important to the design engineer?

Other Problem Information

- The force causing the first sign of buckling is the critical load P_{cr} according to Euler's buckling formula below, E = modulus of elasticity, I = moment of area (inertia), L = column length.

$$P_{cr} = \frac{\pi^2 E I}{L^2}$$

- $E_{aluminum}$ = 10 Mpsi, E_{steel} = 30 Mpsi.

- Column available in different cross sections: circular, rectangular, box, and structural "H."

- Column cost can be approximated by the following relation: C = 35.6 I [\$].

Solution Evaluation Parameter(s)

	Parameter	Symbol	Units	Lower Limit	Upper Limit
1					
2					
3					

Design Variable(s)

	Design Variable	Symbol	Units	Lower Limit	Upper Limit
1					
2					
3					
4					

Problem Definition Parameter(s)

	Parameter	Symbol	Units	Lower Limit	Upper Limit
1					
2					
3					

13. Parametric design miniproject.
A cylindrical chemical storage tank is closed at both ends. It has diameter D and height H. The tank must store at least 150 m^3 of liquid. It is made of rolled and welded sheet steel whose material and fabrication costs are \$550/m^2 of surface area. The height is restricted to 10 m, and the diameter must be less than 5 m. The customer will be satisfied if the tank cost is small and the volume is large. He feels that cost is very important (weight = 0.7) and that volume is important (weight = 0.3). Your design team has estimated that his satisfaction for cost is 1.0 when cost is equal to, or less than, \$50,000, and it is 0.0 when cost is more than \$100,000. His satisfaction for volume is 0.0 for volumes equal to or less than 130 m^3, and 1.0 for volumes greater than 180 m^3. Straight-line equations can be used to estimate his satisfactions for values of cost or volume in between.

a. Prepare an Excel spreadsheet to analyze and evaluate your candidate designs. Label design variables, solution evaluation parameters(s), and problem definition parameter(s), as well as any other items. Label the variable names, symbol, units, and upper and lower limits, if any.

b. Write all the formulas you developed on a separate sheet of engineering paper, with your name and date, to be attached to your spreadsheet. Use a straight-line equation for each satisfaction function (i.e., for cost and volume). A convenient way of automating the satisfaction function in Excel is to use a nested "IF" statement, such as:

$$= \text{IF}([\text{cost} < 50000), 1, \text{IF}(\text{cost} > 100000, 0, 2 - 1/50000*\text{cost}])$$

c. Use your spreadsheet to analyze and evaluate points 1–6 given below. Write in your spreadsheet results on the attached summary table.

d. Examine your results. Is there any pattern? Generate three more candidate designs, by guessing and/or logical reasoning to guide your redesign. Print out a copy of the spreadsheet for the last design point tried.

e. Circle the design alternative/point on the summary table that has the "best" weighted satisfaction. (Note: a co-worker says that he can find values for D and H to optimize satisfaction to about 0.334.)

					Constraints	Weighted Satisfaction		
Design Alt.	D (m)	H (m)	Cost ($)	Volume (cu. m)	Ok? Y or N	Cost	Volume	Total
1	0	0						
2	5	10						
3	5	6						
4	6	6						
5	6	7						
6	6	5.5						
7								
8								
9								

Building and Testing Prototypes

When you have completed this chapter you will be able to

- Describe why companies build and test parts and products
- Describe various product and part tests to validate form, fit, and function
- Characterize traditional prototyping methods
- Describe various prototyping methods
- Explain the overall rapid prototyping process
- Describe the components of a test plan

9.1   INTRODUCTION

Many technical and economic issues arise during the development of a new product. For example, will the:

1. new product work at extreme temperatures and humidity?
2. housing appear old-fashioned to the customer?
3. newly added part actually work as predicted by the equations.
4. product have a good "feel" in the hands of the user?
5. customer be satisfied with the overall size?
6. fabricated parts assemble without interference?
7. finished product meet industry-mandated safety tests?
8. new housing material be compatible with the intended lubricants?
9. customer be able to easily replace worn parts?
10. new part be stiff (flexible) enough?
11. new part have an acceptable wear rate?

12. new product present an electric shock hazard?
13. part be more attractive with a mirror finish?
14. finished product meet minimum performance requirements?
15. manufacturing group be able to fabricate the new part to acceptable tolerances?
16. assembly group be able to easily insert the new part into the mating part?

The economic viability of the company hinges upon the intelligent expenditure of time and money, especially as it relates to product development. Companies can spend millions, for example, during the design phase and especially when tooling-up to launch a new, mass-produced product. The product development team, therefore, must establish a sound plan of tests, to be conducted at various stages, to ensure that the final product will be economically and technically successful.

In the next sections we will examine the types of tests that companies carry out using models and/or prototypes and how they are fabricated by traditional and rapid prototyping methods.

9.2 PRODUCT AND PART TESTING

Product and part tests can be separated into three major categories that focus on validating form, fit, and/or function. Concerns about the form of a part or product relate to its overall appearance, including shape and relative size. Concerns about fit relate to how precisely the parts are fabricated and how well they fit together in the assembly, or how well they "fit" the user. Finally, parts and products must function, or perform as expected, last a long time, and be easy to maintain. Let's regroup the list of concerns from the previous section into form, fit, and function categories.

- **Form test**—Will the part and/or product have an acceptable appearance?
    2. housing appear old-fashioned to the customer?
    5. customer be satisfied with the overall size?
    13. part be more attractive with a mirror finish?
    16. assembly group be able to easily insert the new part into the mating part?
- **Fit test**—Will the parts fit together and also fit the user, with an acceptable precision?
    4. product have a good "feel" in the hands of the user?
    6. fabricated parts assemble without interference?
    15. manufacturing group be able to fabricate the new part to acceptable tolerances?
- **Function test**—Will the part and/or product perform as required?
    1. new product work at extreme temperatures and humidity?
    3. newly added part actually work as predicted by the equations?

7. finished product meet industry mandated safety tests?

8. new housing material be compatible with the intended lubricants?

9. customer be able to easily replace worn parts?

10. new part be stiff enough?

11. new part have an acceptable wear rate?

12. product present an electric shock hazard?

14. finished product meet minimum performance requirements?

9.2.1   Product Development Tests

Most of the questions regarding form, fit, and function can be tested at earlier phases in the development cycle, before the final product rolls off the production line. For example, we can test the appearance of a product by showing customers a reduced-size scale model, made of clay or balsa wood. Another typical situation arises for a product that has many standard parts, but only a few special-purpose parts. Samples of those parts could be made and tested for function by using planned materials and manufacturing processes.

A **physical prototype** is a tangible replica of the part showing principal geometric features, such as length, width, height, holes, fillets, rounds, projections, slots, cavities, and surface texture. Prototypes may be reduced-scale, expanded-scale, or full-scale representations of the part. Prototypes may be made of materials similar to or exactly like the intended product. They can be fabricated with manufacturing processes that are similar to, different from, or exactly the same as those to be used on the intended product. As a consequence, prototypes will vary in their ability to be tested for form, fit, and function. Effective development teams will use the right type of prototype, to test critical aspects at the right time, thereby saving the company more time and money.

Let's examine the different types of tests and prototypes that are typically used during the various development phases of a product as shown in Figure 9.1.

1. **Product concept tests**. A reduced-scale or full-scale model of a new product or "product concept" is fabricated to look like the "finished" product. Usually only the exterior of the product is fabricated and shown, illustrating existing or newly improved features that compose the product concept. This appearance prototype is then shown to potential customers to obtain their reactions and their willingness to ultimately purchase the product. Product concept tests are usually done early in the development cycle, to make sure that the product will have the right look, or appearance, and have the right combination of features. An automobile manufacturer would use a clay scale model of a new sports car to examine the acceptability of a two-seat, rear-engine convertible. A large appliance manufacturer would "mock-up" a new laundry washing machine, to examine the customer appeal of enhanced "digital" controls, or the acceptance of top loading versus side loading. An electronics firm would prototype the case of a new personal digital assistant to examine the effects of various colors and overall shape.

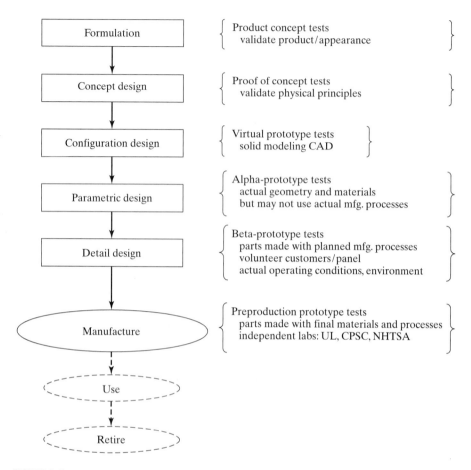

FIGURE 9.1

Prototype testing throughout the product development process.

2. **Proof-of-concept tests**. The model or prototype is built to prove that it will function or perform according to the function concept or the physical principles and abstract embodiment that were selected. Benchtop, pilot-plant, and/or laboratory experiments are performed to prove that the working principle will work in the final product. An office printer manufacturer could machine parts for a benchtop "working-model" of a new electrostatic paper feeder to replace the mechanical roller-feeder in its new line of printers. An electronics manufacturer would test the concept of a newly improved circuit using a prototype board.

3. **Virtual prototype tests**. A part or product can be prototyped or modeled inside the memory of a computer as a virtual prototype using computer-aided design packages, which can create and analyze three-dimensional, solid models. Solid modeling is often used to develop the form of a part

and examine the mobility and/or potential interference of parts in an assembly. The solid-model data file is also used to analyze forces, moments, motions, stresses, and strains, among other things. A construction equipment manufacturer would use solid modeling and finite element analysis to develop a CAD model or virtual prototype of its new gearbox and analyze its strength and/or rigidity. A small-appliance manufacturer would build a solid model of its new high-power hand-held hair blower, then generate realistic pictorial renderings, to assess customer acceptance of the new shape and size.

4. **Alpha-prototype tests**. A reduced-scale or full-scale part or product is prototyped using the same geometric features, materials, and layout as the intended final assembly (Otto and Wood, 2001). Alpha prototypes are not usually prototyped using the same manufacturing processes that will be used in the final production. Alpha-prototype tests are conducted in a company's laboratories and focus on a part or product's appearance and/or its functional performance. A screwdriver manufacturer would fabricate an alpha prototype by machining plain carbon steel to examine the "feel" of its newly shaped handle, even though in final production stainless steel and stamping would be used. The parts for a new desktop printer case could be machined from a production polymer to prove its functionality, even though injection molding would be used in final production.

5. **Beta-prototype tests**. A full-scale, functional part or product is prototyped using materials and manufacturing processes that will be used in production. Beta prototypes can be tested in the company's laboratories but are often tested by volunteers or potential customers in their home or work environment (Crawford and DiBenedetto, 2003). The tests confirm intended mass-production and assembly processes and demonstrate the approximate performance of the product. Results from the beta-prototype tests are used to make the last remaining changes to the product, and to complete the production planning and initiate the production tooling. Groups of potential customers, called "customer panels," are often used to evaluate beta prototypes of new vacuum cleaners, kitchen appliances, hand tools, and sports equipment. Potential customers are also used to test and evaluate Beta version software on their home or work computers.

6. **Preproduction prototype tests**. A full-scale part or product made and assembled with final materials and production-like processes is tested. Therefore, the tests are made on "finished" products similar to ones that customers purchase. The tests confirm and document production and assembly processes and demonstrate the actual performance of the "final" product. Test results are used to make last-minute revisions to the production tooling and assembly processes and to make minor design revisions to comply with state and federal codes. The tests are also used to verify and document that the products have no defects in design or manufacture, which could be tested in a product liability lawsuit. An electric toaster manufacturer has the Underwriters Laboratories (UL), a nonprofit independent organization, conduct

electrical shock and fire safety tests. A baby stroller manufacturer has its new product tested in an independent laboratory before sending production-run samples to the Consumer Product Safety Commission (CPSC). An automobile manufacturer has production prototypes tested for crashworthiness by the National Highway Transportation Safety Administration.

9.3 BUILDING TRADITIONAL PROTOTYPES

Craftsmen have been making models for thousands of years using simple hand tools and a variety of materials. Traditional materials include woods, clay or ceramics, construction papers, and more recently with the availability of machine tools, metals and polymers (Otto and Wood, 2001). Traditional prototype examples include:

- clay models of new auto-body styles,
- wood models of heavy-equipment patterns for metal castings,
- machined metal airplane wings for function testing in a wind tunnel,
- reduced-scale balsa wood models of large facilities, to examine equipment layout.

As mentioned in Chapter 5, materials and the processes used to shape them are interdependent. Some materials are incompatible with certain fabrication methods. For example, metals in general cannot be carved with a sculpting knife, as wood can be. Nor can construction papers be welded; rather they must be glued or adhesively taped together. Also, some materials are poorly suited for function testing, such as a paper wing that could deform under load in a wind tunnel. Therefore, the choice of material and fabrication method will largely depend on:

- *Shape-generating compatibility*. Can the material be formed into the needed geometric features to adequately represent the part?
- *Function-testing validity*. Are the material properties representative or scalable such that the part when reduced (or expanded) in size can be validly tested?
- *Fabrication costs*. Will the prototype costs for materials and labor be acceptable?
- *Fabrication time*. How long will it take to fabricate the original and one or more duplicates?

9.4 BUILDING RAPID PROTOTYPES

Traditional prototypes use tools and fabrication methods that are labor-intensive, often require significant mechanical or artistic skills, and take a long time to fabricate an original or its duplicates. **Rapid prototyping** is an alternative technology that uses computers and computer-controlled equipment to automatically and rapidly fabricate prototypes (Jacobs, 1996; Wood, 1993). Rapid prototyping processes include the following.

Numerical Control Machining (NC/CNC). Perhaps the earliest of these is numerically controlled machining of prototypes, otherwise called **NC/CNC machining**. Beginning around the early 1960s or so, if we wanted to make a prototype using an NC lathe or milling machine, we would give our drawing to an operator, who would interpret the drawing and create a program of machine control codes, which were punched as holes on a paper tape. The tape would then be read by the NC machine to control the precise feeds and speeds of the workpiece and tool, thereby generating the various part features. An example part is shown in Figure 9.2. Since the prototype is made of steel, it can be used for function testing under simulated loads.

More recently, computer numerically controlled machines (CNC) now have on-board computers, which generate and store their own programs. Software packages make efficient use of files generated by CAD solid-modeling software. The CNC software generates the appropriate tool paths and tool changes from the solid model file with a minimum of operator attention. It can be argued that NC/CNC is not fully "automatic," because a skilled operator must still be involved in running the machine and making fixture changes. However, multiple copies of a part can be made quickly and with a high level of precision.

FIGURE 9.2

Example CNC part, a wheel made for the Mars rover.
(Courtesy of Haas Automation, Inc.)

NC/CNC machines remove material from circular or rectangular base parts. Mills, for example, are used to cut out slots and trim parts to their final dimensions. NC/CNC lathes can be used to cut various shoulders or steps onto shafts. A principal advantage of machining (by hand or NC/CNC) is, of course, being able to make prototypes out of strong, hard, and/or stiff metals. Accuracies and tolerances are the best of all the prototyping methods. As such, the resulting prototypes can be subjected to operating loads approaching the actual in-use conditions.

Stereolithographic apparatus (SLA) is a material **additive process** in nature and uses a high-power laser to selectively solidify a liquid photopolymer, layer by layer, into the shape of the finished prototype. The CAD model file is first "sliced" into virtual lamina, in the computer's memory, which is then "printed" by the laser. SLA produces prototypes with superior finishes using a number of different polymers. Parts are postprocessed to dry and cure the nonsolidified resin, and supporting structures are mechanically removed. The polymeric prototypes are also weaker than metal prototypes produced with NC/CNC. The stereolithographic process is shown in Figure 9.3. An SLA prototype is shown in Figure 9.4. The prototype has an excellent surface finish, which approaches a ground, polished, and/or lapped and painted/coated part.

Fused-deposition modeling (FDM) is a process that deposits a thin filament of melted (fused) material in precise locations on a horizontal layer, using numerically controlled positioners. As the material solidifies, a prototype model is built, layer by layer, typically up to 10 inches by 10 inches in cross section and up to 16 inches high. Parts can be made from a number of materials, including: high-strength ABS plastic in multiple colors, impact-resistant ABS, polycarbonate, and polyphenylsulfone. No special room air vents or exhausts are required, and the process is fully automated.

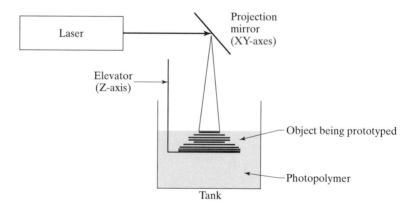

FIGURE 9.3

The SLA process uses a powerful laser to cure a photosensitive polymer, layer by layer, as the elevator lowers into the tank.

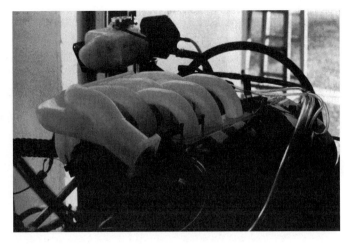

FIGURE 9.4

Example of an SLA part for rapid tooling. (Courtesy of 3D Systems.)

The materials are nonhazardous and nontoxic. Parts are postprocessed by mechanical agitation in a water solution to remove support structures using the "waterworks" option. Prototype parts are used by companies to verify form, fit, and function. An example FDM product is shown in Figure 9.5.

Laminated-object manufacturing (LOM) is a process that builds each prototype by laminating together thin layers of paper, polymer, or sheet steel, which have been cut using a numerically controlled laser. LOM prototypes can be sanded to reduce jagged edges, but are not able to be function-tested, such as for stress or strain, owing to the allotropic material properties of the laminate.

Selective laser sintering (SLS) is a process that uses a high-power laser to sinter together fusible materials, such as powdered metals, layer by layer. Sintering is the heating and fusing of small particles resulting in a hard bonded material block. The unsintered powder supports the part as the layers are sintered.

3-D inkjet prototyping is a process that selectively deposits a glue-like binder onto a layer of dry powder, layer by layer, which dries into a solid prototype. A similar process uses a print head to deposit a thermoplastic material, layer by layer. The processes work well as concept modelers (Smith, 2001). Each prototype has limited dimensional tolerances and is somewhat fragile unless coated with a hardener to help to maintain part integrity. Prototypes made with this process are typically not function-tested. An example shown in Figure 9.6 shows how prototyping can be used for rapid tooling.

Overall Rapid Prototyping Process begins with the creation of a virtual part created in the memory of a computer, as shown in Figure 9.7. Using

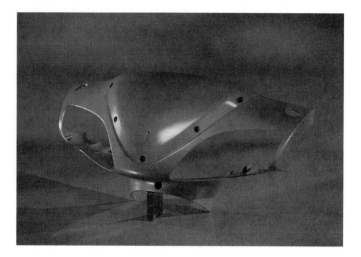

FIGURE 9.5

Example FDM product/part. (Courtesy of Stratasys Inc.)

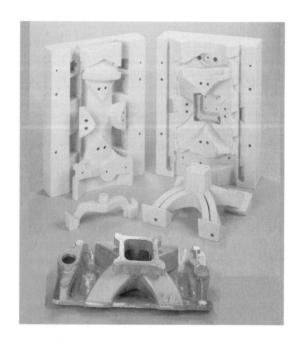

FIGURE 9.6

Example of 3-D Inkjet prototyping used to create the patterns for a set of mold halves and cores for the manifold casting. (Courtesy of Z Corporation.)

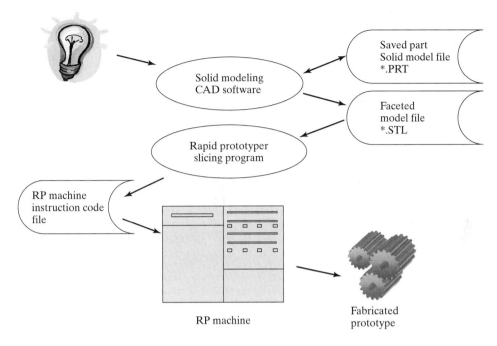

FIGURE 9.7

Overall rapid-prototyping process.

3-D solid modeling software, we can design the virtual part to include features such as: walls, ribs, holes, bosses, fillets, slots, chamfers, and rounds. Sizes are entered as parameters, that is variables, which can be modified at any time during or after the creation of the part. The resulting image can be viewed in any of the orthographic views along with a variety of oblique and isometric views. The image can also be rendered with various colors and surface textures, for marketing purposes, etc. Bills of materials and dimensioned views are also produced with similar ease. The part shape is stored as numerical data and is archived on computer disks as **solid-model part files**.

Next, all the surfaces of the part are converted to very small triangular facets. Each facet is represented by geometric data, including the x, y, and z coordinates of the three vertices and an outward-facing normal vector. The data are stored with the as *.**STL file** extension.

The *.STL file is read by a computer program for the specific rapid prototyping machine, such as a numerical controlled milling machine. The program slices the virtual solid into layers and a set of codes that instruct the prototyping machine to fabricate the part layer by layer.

Advantages/Disadvantages of Rapid Prototyping. The advantages of rapid prototyping often outweigh the disadvantages. For example:

- **Time**. The overall time from hand sketches to physical prototype can be on the order of hours or days, compared to traditional methods, which can require days, weeks, or even months.
- **Duplication ease and costs**. Duplicate prototypes can be fabricated quickly, less expensively, and with greater precision than with traditional methods.
- **Flexibility**. Changes to the virtual model and subsequent generation of revised prototypes are facilitated with rapid prototyping, because the data are stored parametrically.
- **Rapid tooling**. Prototype models and data can be conveyed electronically, and easily, to downstream departments, such as production and facilities planning, to communicate tooling concepts to mold makers, or generate master patterns for tooling, or generate tooling inserts.

The principal disadvantages are the costs of purchasing and setting up the rapid prototyping equipment and the fact that some prototyping materials are not suitable for function testing. Rather than investing in prototyping equipment, however, product manufacturers can have parts made for a nominal fee by a rapid prototyping **service bureau**. The product manufacturer e-mails the solid model part file to the service bureau, typically as an *.STL file. The bureau uses its software to convert the *.STL file to a "sliced" file format specific to the selected prototyping hardware (i.e., FDM, SLA, SLS, LOM), then fabricates the prototype and any duplicates. The part(s) is then overnight-mailed to the product manufacturer. Some parts can be processed and received the next day, others may take two or three days. Nominal fees for a rapid prototype typically run on the hundreds of dollars compared to traditional prototyping methods, which often run on the order of thousands or tens of thousands.

9.5   TESTING PROTOTYPES

Successful companies plan and conduct tests to validate form, fit, and function, thereby ensuring that any new design is economically and technically successful. Companies can progressively refine their products by using an iterative design-build-test approach on their parts and subassemblies.

Specific Tests. Prototype evaluations often include specific tests for mechanical modes of failure, manufacturability, operation/maintenance, safety (Hales, 1993), and environmental protection, as shown in Table 9.1.

Test Plans. A product development **test plan** is a description of the type of tests to be performed, the timing when they are to be completed, and the resources to be expended. A test plan provides a structure for organizing, scheduling, and managing the testing program, and a means to communicate the test program details to all the stakeholders, eliciting feedback and delegating responsibilities.

TABLE 9.1 Specific Tests for Part and Product Prototypes

1. Mechanical modes of failure:
   static strength
   fatigue
   deflection/stiffness
   creep, impact
   vibration
   thermal/heat transfer/fluid
   energy consumption / production
   friction (i.e., too much, too little)
   wear
   lubrication
   corrosion
   life, reliability

2. Manufacturability concerns:
   process compatibility/precision
   process technology readiness
   raw-material quality
   assembly

3. Operation/maintenance concerns:
   styling/aesthetics
   ergonomics
   maintenance
   repairs

4. Safety concerns:
   risk to user, products liability
   risk to consumer/society
   safety codes, standards (UL, NHTSA)
   risk to production worker (e.g., OSHA)

5. Environmental protection concerns
   air quality, noise
   water—quality, quantity
   solid waste—hazardous materials
   radioactivity—fallout

In some cases, test plans are required. For example, in the food and drug industry, test plan requirements are mandated by the Food and Drug Administration. And, products for the military are typically subject to "Mil-Spec" test plan requirements.

The basic components in a test plan document are:

Objectives. A list of items to be tested and the purposes for which the tests are being conducted. The objectives section basically covers what is being done for whom and why. If possible, objectives should be written with specificity. They should also be measurable, realistic, and timely.

Workscope. A narrative description of the work tasks to be performed, including the type of tests, test descriptions, experimental setup, experimental controls, design of experiments test matrix, and list of deliverables.

Budget. A tally of the resources to be expended, including items such as number of hours by task, cost per hour, materials to be consumed, equipment rental costs, and total costs.

Schedule. A chart that shows when each major task begins and ends, with major milestones indicated.

9.6 SUMMARY

- Companies build and test prototypes to ensure form, fit, and function.
- Product development tests include: product-concept, proof-of-concept, virtual, alpha, beta, and preproduction.

- Prototypes can be built using traditional and rapid-prototyping methods and materials.
- Rapid-prototyping methods include NC machining, stereolithographic apparatus, fused deposition modeling, laminated object manufacturing, selective laser sintering, and 3-D inkjet printing.
- The overall rapid-prototyping process takes advantage of computer-aided design automation.
- Part and product testing can include tests for: mechanical modes of failure, manufacturability, user operation and maintenance, safety, and environmental protection.
- Comprehensive product development requires the preparation and completion of a detailed test plan.

REFERENCES

Crawford, C. M., and C. A. DiBenedetto. 2003. *New Products Management*, 7th ed. New York: McGraw-Hill.

Hales, C. 1993. *Managing Engineering Design*. Essex, England: Longman Scientific.

Jacobs, B. 1996. *Stereolithography and Other RP&M Technologies*. Dearborn, MI: Society of Mechanical Engineers.

Otto, K. P., and K. L. Wood. 2001. *Product Design: Techniques in Reverse Engineering and New Product Development*. Upper Saddle River, NJ: Prentice Hall.

Smith, P. "Using Conceptual Modelers for Business Advantage." *Time-Compression Technologies*, April 2001.

Wood, L. 1993. *Rapid Automated Prototyping: An Introduction*. New York: Industrial Press.

KEY TERMS

Additive process

Alpha prototype test

Beta prototype test

Fit test

Form test

Function test

Fused-deposition modeling (FDM)

Laminated-object manufacturing (LOM)

NC/CNC machining

Physical prototype

Preproduction prototype test

Product concept test

Proof-of-concept test

Rapid prototyping

Rapid tooling

Selective laser sintering (SLS)

Service bureau

Solid-model part file

Stereolithographic apparatus (SLA)

*.STL file

Test plan

3-D inkjet prototyping

Virtual prototype test

EXERCISES

Self-Test. Write the letter of the choice that best answers the question.

1. _____ Which of the following tests are usually done early in the developmental cycle to make sure that the product will have the right appearance and have the right combination of features:

   a. beta-prototype tests

   b. preproduction-prototype tests

   c. product concept tests

   d. alpha-prototype tests

2. _____ Tests in which a reduced-scale part or full-scale part or product is prototyped using the same geometric features, materials, and layout as the intended final assembly are called:

   a. alpha-prototype tests

   b. virtual prototype tests

   c. beta-prototype tests

   d. preproduction-prototype tests

3. _____ Tests in which a full-scale, functional part or product is prototyped using materials and manufacturing processes that will be used in production are called:

   a. alpha-prototype tests

   b. virtual-prototype tests

   c. beta-prototype tests

   d. preproduction-prototype tests

4. _____ Test in which a full-scale part or product is made and assembled with the final materials and production-like processes are called:

   a. visual-prototype tests

   b. proof-of-concept tests

   c. alpha-prototype tests

   d. preproduction-prototype tests

5. _____ An alternate technology that uses computers and computer-controlled equipment to automatically and rapidly fabricate prototypes is called:

   a. rapid prototyping

   b. alpha prototyping

   c. visual prototyping

   d. beta prototyping

6. _____ The rapid-prototyping process that uses a high-power laser to selectively solidify a liquid photopolymer, layer by layer, into the shape of the finished prototype is called:

   a. fused-deposition modeling

   b. laminated-object manufacturing

   c. selective laser sintering

   d. stereolithography

7. _____ The rapid-prototyping process that deposits a thin filament of melted material in precise locations on a horizontal layer, using numerically controlled positioners, is called:

    a. laminated-object manufacturing

    b. fused-deposition modeling

    c. overall rapid-prototyping process

    d. stereolithography

8. _____ The rapid-prototyping process that builds each prototype by laminating together thin layers of paper, or sheet steel, which have been cut using a numerically controlled laser, is called:

    a. laminated-object manufacturing

    b. overall rapid-prototyping process

    c. 3-D-Inkjet prototyping

    d. fused-deposition modeling

9. _____ The rapid-prototyping process that uses a high-power laser to sinter together
fusible materials, such as powdered metals, layer by layer, is called:

    a. laminated-object manufacturing

    b. selective laser sintering

    c. fused-deposition modeling

    d. stereolithography

10. _____ The rapid-prototyping process that selectively deposits a glue-like binder onto
a layer of dry powder, layer by layer, which dries into a solid prototype is called:

    a. overall rapid-prototyping process

    b. fused-deposition modeling

    c. laminated-object manufacturing

    d. 3-D-Inkjet prototyping

11. _____ The rapid-prototyping process that begins with the creation of a virtual part
created in the memory of a computer is called:

    a. overall rapid-prototyping process

    b. fused-deposition modeling

    c. stereolithography

    d. Nc machining

12. _____ All of the following are advantages of rapid prototyping except:

    a. rapid tooling

    b. low equipment cost

    c. duplication ease and costs

    d. flexibility

13. _____ All of the following are basic components in a product development test plan except:

    a. production

    b. work scope

    c. objectives

    d. schedule

14. _____ The heating and fusing of small particles resulting in a hard bonded material block is called:

    a. deposing

    b. laminating

    c. sintering

    d. modeling

2: a, 4: d, 6: d, 8: a, 10: d, 12: b, 14 : c

15. Do an Internet search for "rapid prototyping." Browse some of the sites. List three sites found (url address).

16. Connect to my Web site at: http://coen.boisestate.edu/reggert/RJE/rapid_prototyping. htm. Browse the 3D Systems, Stratasys, and Z Corporation sites. Examine how the RP machines are used. Then select one of the sites and look for a case study, or application. Print out an image of the prototype made. Attach the printout annotated with the following items listed below:

    a. site it came from

    b. what it is

    c. how it was used

    d. why it was beneficial

17. Briefly describe the difference between engineering tests and scientific experiments.

18. Consider the engineering tests that can occur during the development of one of the following products: mini-bike, log-splitter, or wind turbine water pump. In each of the phases, select a specific component and list a specific test, type of test (form, fit, or function), and prototype in the table provided.

Ex: Mini-bike	Component	Test	Type	Prototype
Parametric	handlebar	verify hand brake position	fit	Alpha using FDM

Phase	Component	Specific Test	Type	Prototype
Formulation				
Concept				
Configuration				
Parametric				
Detail				

CHAPTER 10

Design for *X*: Failure, Safety, Tolerances, Environment

When you have completed this chapter you will be able to

- Identify product failure modes, their effects and severity
- Establish failure mode causes, occurrence likelihood, and detectability
- Determine risk priority numbers to focus remedial actions
- Identify hazards associated with the product manufacture, use, and retirement
- Describe and apply the safety hierarchy fundamentals
- Explain the difference between dimensions and tolerances
- Understand and implement worst-case and statistical tolerance design methods
- Describe and apply alternative design for environment methods

10.1   INTRODUCTION

Design for *X* is a term used to describe any of the various design methods that focus on specific product development concerns. For example, *X* can stand for safety, reliability, assembly, manufacture, tolerances, robustness, or the environment.

Since design for *X* methods address specific issues, they implement customized procedures that take advantage of specific information available at different phases of design. Design for assembly and design for manufacture, for example, were introduced in Chapter 7, "Configuration Design," because those methods utilize changes in product form, such as shape and configuration, to improve assembly and manufacture. Design for robustness was introduced in Chapter 8 "Parametric Design," since most methods depend on quantitative information available at that phase.

Tolerance design is especially useful during and after parametric design, as the final manufacturing specifications are being prepared. **Design for safety**, the failure modes and effects analysis method, and **design for the environment** are additional multiphase methodologies that can be used at the beginning, end, or throughout the product realization process. We examine these four multiphase methods next in this chapter.

10.2   FAILURE MODES AND EFFECTS ANALYSIS

Failure modes and effects analysis (FMEA) is a systematic method used to identify and correct potential product or process failures before they occur (McDermott et al., 1996). It is a stepwise procedure in which we first examine the modes or ways that each part in the product might fail to perform its intended function. A failure mode may arise from a variety of causes, and result in a number of adverse effects. Second, we analyze the severity of each effect. Then, we try to identify the potential defects in design or manufacture that could cause each failure mode, along with the likelihood that the cause would occur. Next, we examine how we might have detected the defects by design review controls, testing, and/or inspection. Finally, we consider what corrective actions we should take, including eliminating the causes of the potential failure modes and/or reducing the severity of the failure mode.

10.2.1   Failure Modes, Causes, Effects, Severity, and Detection

A **failure mode** is the way in which a part could fail to perform its desired function. For example, a helical compression spring might fail in three modes: buckling, long-term set, or coil clash. Other examples are given in Table 10.1. A list of some prevalent failure modes is given in Table 10.2.

The cause of a potential failure mode is the reason why the part may fail owing to a defect in design or manufacture. The shaft, for example, could fracture because the engineering design team might have underestimated the applied static and/or dynamic loads, or an inferior raw material was purchased for its manufacture. A hydraulic hose might leak because it was cut during assembly, the hose connector was improperly attached, or an inadequate material was selected. It should be noted that not every cause has the same chance of occurring. Some causes may be "expected" while others are "improbable."

TABLE 10.1   Possible Failure Modes of Some Example Parts

Part or Component	Failure modes
Fluid valve	Open, closed, partial open/close
Screw	Loose, stripped, corroded
Battery (dry cell)	Discharged, partially discharged
Suspension cable	Frayed, stretched, kinked
Switch	Open, closed
Shaft	Fractured, bent, seized
Hydraulic hose	Leaks

TABLE 10.2   Prevalent Failure Modes

fracture	bind
buckle	delaminate
bend	erode
crush	corrode
seize	leak
deform	open circuit
vibrate	short circuit

The effects of a failure mode are the adverse consequences that the user might experience. An open switch (failure mode) might cause a commercial meat refrigeration unit to stop functioning, resulting in spoiled inventory (effect). A fractured automobile drive shaft (failure mode) might result in a car accident (effect). Also, the definition of the "customer" is herein broadened to include the end-user as well as others involved in the fabrication and assembly of the product. A cut hose may cause a waste of hydraulic fluid during assembly, for example. The severity of an effect is an assessment of the extent of damage or injury to the product, the user, nearby people, or the environment. A small hydraulic leak on a portable log splitter, for example, might result in "very little degradation of function," whereas a hydraulic leak in the rudder control of a passenger aircraft might be "catastrophic."

Hopefully, design and manufacturing personnel will have instituted controls or inspection procedures to prevent the cause of the failure mode to occur or to detect the failure before it reaches the customer. **Detection** is an assessment of the probability that the controls will detect the cause of the failure mode. In the design of a drive shaft, for example, we may be "certain" that the calculation checking procedure (control) would detect that the static and dynamic loads were estimated incorrectly (cause), which might result in fracture (failure mode). In the case of a leaky hydraulic hose (failure mode), we might be able to "possibly" detect a manufacturing cut (cause) by inspecting 50 percent of the assemblies (control).

10.2.2 Calculating the Risk Priority Number (RPN)

The **risk priority number** is a metric used to assess the risk of a failure mode. The RPN for a part failure mode is the product of three factors: a **severity rating**, an **occurrence rating**, and a **detection rating**, RPN $= (S) \cdot (O) \cdot (D)$. Each factor is a numerical rating that ranges between 1 and 10, as shown in Tables 10.3–10.5. Note that lower rating numbers are better. Consequently, the smaller the RPN, the smaller the risk. In other words, a part with almost no risk would have a severity rating of 1, an occurrence rating of 1, and a detection rating of 1, resulting in an RPN equal to 1. Similarly, a part having the most risk of failure would be have a RPN $= 10 \cdot 10 \cdot 10 = 1000$.

The 10-step FMEA method uses a tabular layout to organize the information and facilitate the calculations, as shown in Table 10.6.

TABLE 10.3 Severity Rating for the Effects of a Failure Mode, Based on the Injury or Damage to People, Property, or the Environment

Severity (S) Rating	Type of effects	Description
10	**Catastrophic**	Causes injury to people, property, and/or the environment
9	Extremely harmful	Causes damage to product, property, or environment
8	Very harmful	Causes damage to product
7	**Harmful**	Major degradation of function
6	Moderate	Causes partial malfunction of product
5	Significant	Performance loss causes customer complaints
4	**Annoying**	Loss of function is annoying, cannot be overcome
3	Minor	Some loss of performance, but can be overcome
2	Insignificant	Very little function degradation
1	**None**	No noticeable effects in function or harm to others

TABLE 10.4 Occurrence Rating Based on the Likelihood That a Cause Will Occur

Occurrence (O) Rating	Likelihood of cause	Description	
10	**Expected**	>30%	>One per *day*
9	Very likely	30% (3 per 10)	
8	Probable	5% (5 per 100)	One per week
7	**Occasional**	1% (1 per 100)	One per *month*
6	More plausible	0.3% (3 per 1,000)	One per three months
5	Plausible		
4	**Remote**	0.006% (6 per 10^5)	One per *year*
3	Unlikely	0.00006% (6 per 10^7)	One per three years
2	Very unlikely		
1	**Improbable**	<2 per 10^9 events	>5 *years* per failure

TABLE 10.5 Detection Rating Based on the Probability That Design and Manufacturing Controls Will Detect the Cause

Detection (D) Rating	Detectability of cause	Description
10	**Impossible**	Impossible to detect, or no inspection
9	Very rare	
8	Rare	
7	**Possible**	Some chance of detecting, or 50% inspection
6	Quite possible	
5	Somewhat likely	
4	**Likely**	Quite likely to detect, or 75% inspection
3	Quite likely	
2	Almost certain	
1	**Certain**	Will be detected, or 100% inspection

TABLE 10.6   Failure Modes and Effects Analysis Template Used to Calculate Risk Priority Number Based on Ratings for Severity, Occurrence, and Detection

Failure mode	Severity (S)		Occurrence (O)		Detection (D)		RPN	Recommended Action
	Effects	S Rating	Causes	O Rating	Controls	D Rating		
(step 1)	(step 2)	(step 3)	(step 4)	(step 5)	(step 6)	(step 7)	(step 8)	(step 9)

Step 1: Determine the failure modes.

Identify how the component might fail to perform its desired function(s). Consider how a function transforms matter, energy, and signal. Review information sources such as the house of quality, function decomposition diagrams, free-body diagrams, the product engineering design specification, process flow diagrams and configuration sketches, part-detail drawings, and assembly drawings. And in the case of an existing product, obtain service and/or test records.

Step 2: Determine potential effects of each failure mode.

Establish the effects of each failure mode in terms of what the user might be subjected to.

Step 3: Select a severity (S) rating for each effect.

Delineate possible injury to users, damage to product, personal, or real property, and harm to the environment. Select an appropriate rating number from Table 10.3.

Step 4: Determine the possible causes for each effect.

Determine possible causes for each effect. Consider prevalent failure modes, effects, and causes for similar products. Brainstorm mechanical, electrical, chemical, and other physical mechanisms that could cause each effect.

Step 5: Select an occurrence (O) rating for each effect.

Estimate the probability each cause will occur and select an occurrence rating from Table 10.4.

Step 6: Determine design and manufacturing controls.

List design activities such as calculation checks and engineering tests used to validate the design. Also, identify and list inspection or product testing controls that may detect possible manufacturing defects that could cause a failure mode.

Step 7: Select a detection (D) rating for each cause.

Estimate the likelihood that the controls will detect each cause and select an appropriate detection rating value from Table 10.5.

Step 8: Calculate the risk priority number for each failure mode.

Compute the product of the severity, occurrence, and detection ratings, $RPN = (S) \cdot (O) \cdot (D)$.

Step 9: Recommend and take action.

Rank the failure modes and determine which failure mode has the greatest risk. Generate, analyze, and evaluate alternative action plans that eliminate the potential cause(s) or reduce the severity of their effects. Recommend and implement the best alternatives.

Step 10: Recalculate the risk priority numbers.

After remedial action has been implemented, revise the RPN's for the "corrected" failure mode. Reiterate steps 9 and 10.

Example

A portable log splitter has been designed for home use. A 10-horsepower gasoline engine has been selected to drive a hydraulic pump that connects to the piston and ram by a wire-reinforced rubber hose. The product development team has identified that a fluid leak is a possible failure mode for the hose. Fluid leaks have been caused in the past by poor hose material supplied by vendors. Also, during manufacture, an automated assembly machine has been known to occasionally cut the hose. The team estimates that the likelihood of a "poor material" is about 5 percent and a "cut" is about 0.5 percent. When a poor-material hose leaks it reduces the pressure to the piston/ram resulting in poor splitting, in other words, a partial malfunction of the log splitter. When a cut hose leaks it damages the pump as the oil empties onto the ground. It also causes an oil spill harming the ground. The company will inspect 50 percent of the rubber hose as it is delivered from the vendor. Also, it is quite likely that a cut hose will be detected because the assembly machine usually jams when it cuts.

Use the FMEA method to determine the risks associated with a leaky hose and recommend corrective actions.

Based on the given information we select a "moderate" severity rating = 7 for the poor material cause, and an "extremely harmful" severity rating = 9 for the cut-hose. We select a "probable" occurrence rating = 8 for the poor material, and a "probable" occurrence rating = 7 for the cut hose. Finally, we select detection ratings of 7 and 4, respectively. We enter the data into the template and calculate the RPN's as shown in Table 10.7.

TABLE 10.7 FMEA Method Template to Estimate the Relative Risks of a Failure Mode and Its Causes

| Failure mode | Severity (S) | | Occurrence (O) | | Detection (D) | | RPN | Recommended Action |
	Effects	S Rating	Causes	O Rating	Controls Tests	D Rating		
hose leaks	poor log splitting	7	poor material	8	none	7	392	establish new pressure test
	oil mess pump damage	9	cut hose	7		4	252	modify assembly machine

TABLE 10.8   Revised Risk Priority Number for Hose Leak Failure Mode

| Failure mode | Severity (S) | | Occurrence (O) | | Detection (D) | | | |
	Effects	S Rating	Causes	O Rating	Controls Tests	D Rating	RPN	Recommended Action
hose leaks	poor log splitting	7	poor material	8	none	1	56	establish new pressure test
	some oil spill no pump damage	7	cut hose	7	pressure-test hose	4	196	modify assembly machine

In discussions with the development team we establish a new inspection procedure to test 100 percent of the incoming hose material, and we will modify the assembly machine to reduce the cut size, thereby producing effects leaks with a similar severity to that of poor hose material. Revised RPN calculations result in Table 10.8.

The new pressure test procedure will make a significant reduction in the risk of a poor material causing a leak. Similarly, the risk for a cut hose will be reduced. In the first case, we changed the detectability, and in the second case we changed the severity.

A failure mode-and-effects analysis can be completed during design beginning as early as the configuration design phase. Existing products can also be evaluated. The method has also been used to reduce the risk of failure within processes.

10.3 DESIGN FOR SAFETY

An unsafe product may cause injury to people. It may also cause damage to personal or real property. A defective automobile brake system, for example, could result in a collision that might injure the driver, passengers, other drivers, and pedestrians. The collision could cause damage to vehicles, buildings, and/or other facilities. Similarly, an unsafe industrial product or system could injure employees or manufacturing facilities. For example, an unguarded machine tool could entrap an operator's hand, crushing it severely; or improperly vented fumes could explode or catch on fire, injuring workers and damaging valuable equipment and the building.

Designing and manufacturing an unsafe product can lead to customer and employee lawsuits and criminal penalties. It is also unethical. The American Society of Mechanical Engineers Code of Ethics places the safety, health, and welfare of the public as the paramount responsibility of an engineer.

10.3.1 Safety Hazards

Injuries may occur to employees or customers at anytime in the life of a product, including during:

- manufacture—when components are fabricated, assembled, and distributed,
- use—when the user sets up, operates, repairs, and maintains the product,
- retirement—when the user takes down, disassembles, recycles, and disposes of the product.

Therefore, as we design a product it is our paramount responsibility to consider the hazards that may be dangerous for anyone coming into contact with the product or its components during its lifetime. Wherever possible, we must eliminate or reduce the risk of injury or property damage.

A **hazard** is a source of danger. Hazards have the potential to injure people or damage property or the environment. Let's consider the types of injuries to people, including cuts, burns, broken bones, poisoning, and radiation. Some hazards act through direct contact. Others act indirectly via inhalation (of gases or vapors) or radiation. We also recognize that injury and damage can result from the dissipation of the kinetic energy of moving machine components or the potential energy stored in chemicals and mechanical equipment. Lindbeck (1995), for example, describes six basic types of hazards:

1. *Entrapment.* **Entrapment hazards** cause injuries when machine parts move toward each other to create pinch points where fingers, hands, or other body parts can get pinched or crushed. Examples include roller chains and sprockets, belts and pulleys, pivoting levers, meshing gears, and lifting platforms.

2. *Contact.* Contact hazards cause injuries merely by coming into contact with body parts. These hazards include hot surfaces, sharp edges, or electrically charged parts, either moving or stationary. Examples include gasoline engine mufflers, welding equipment, stamped metal parts with burs, rotating or reciprocating saw blades, and exposed electrical wires or contacts.

3. *Impact.* **Impact hazards** cause injuries as translating, rotating, or reciprocating machine parts collide with body parts. Examples include hammer blows to the hand and gear hub collars with projecting bolts.

4. *Ejection.* **Ejection hazards** cause injuries as debris particles are flung from moving machine parts. Examples include grinding-wheel disintegration chunks and circular-saw wood particles.

5. *Entanglement.* **Entanglement hazards** include rotating shafts or workpieces that can catch hair or loose clothing. Examples include entwining hair on a rotating shaft such as a lathe and catching loose clothing in a wood chipper.

6. *Noise and vibration.* Loud noises can result in temporary or permanent hearing loss. Examples include unprotected exposure to noise from engine exhausts and stamping machines. Extended exposure to vibrating hand tools can cause numbness in the limbs, referred to as hand-arm vibration syndrome (HAVS).

Hazards also include fire, explosion, electromagnetic radiation, and exposure to toxic, hazardous, or radioactive substances. Finally, we can include environmental hazards such as extremes in temperature, humidity, wind, ice, snow, and lightning.

Conditional Circumstances We should also distinguish the conditional circumstances associated with hazards. For example, a grinder has a rotating abrasive wheel. If a chunk of the wheel breaks off during use, it may have sufficient
kinetic energy to penetrate unprotected human skin tissues and is also very
likely to cause eye injury. This is a circumstance when the abrasive failed to hold
the wheel together. Sparks caused by normal grinding, on the other hand, are a
type of normal hazard that is inherent in the operation of the machine. Misuse
by the user is another conditional circumstance. For example, consider how a
user might pick up a bench grinder to grind an I-beam that is elevated 10 feet
off the floor, rather than using a portable grinder.

Finally, we have the circumstance due to regular maintenance. For example, a service person may be exposed to electrical shock when removing a protective cover from a machine. Improper maintenance and failure to properly
maintain equipment are additional circumstances that may create hazards. Lack
of exhaust muffler maintenance, for example, might lead to carbon monoxide
poisoning. We can summarize these hazard circumstances as follows:

1. Hazard is inherent during normal use.
2. Hazard originates from a component failure.
3. Hazard is caused by user misuse.
4. Hazard exists during normal maintenance.
5. Hazard is created by improper maintenance.
6. Hazard stems from lack of maintenance.

We are responsible for design and manufacture of safe products. Therefore, we
must consider not only those inherent hazards that occur during normal use, but
also those hazards that may occur under conditional circumstances.

10.3.2 Legal Responsibilities

As professionals we are ethically bound to hold paramount the safety, health,
and welfare of the public. However, under local, state, and federal laws, we are
also legally responsible for the safety of our customers and employees. Products
liability deals with customer product safety while the Occupational Safety and
Health Act(s) (OSHA) deals with employee safety.

Products Liability **Products liability** is a phrase used to describe our legal responsibility to our customers under civil action or criminal actions (Blinn,
1989). **Civil actions** are taken by the injured party, called the plaintiff, against
the manufacturer or seller, called the defendant, to recover damages for personal injury or loss to property caused by defect(s) in design or manufacture.
Criminal actions, resulting in prison sentences, are directed at individuals who
were negligent in the performance of their duties.

The attorney(s) will try to prove that the product had a defect in design or
defect in manufacture (Himmelfarb, 1985) and that the defect occurred because of the negligence of the manufacturing company or the designer(s) in
that they failed to:

Recall notices are used to alert consumers of potential product defects and their modes of failure before they happen. Then, after repair or replacement, the product can be relied upon to perform its required function.

- perform appropriate analyses,
- comply with published standards,
- make use of state-of-the-art technology, owing to ignorance,
- include reasonable safety features or devices,
- take into account how the user might misuse the product,
- consider hidden dangers that might surprise the user,
- consider variations in materials, manufacturing processes, or effects of wear,
- carry out appropriate testing, or interpret results correctly,
- provide adequate warnings.

It is important to note that in addition to diligently designing and manufacturing safe products we must also document our efforts in order to prove that we were not negligent in our duties.

Occupational Safety   Employees are protected with regard to **occupational safety** under local, state, and federal laws, the most notable being the federally mandated Occupational Health and Safety Act of 1970. Employers are therefore legally responsible for ensuring that their employees are protected from a variety of workplace hazards, including: hazardous chemicals, noise, vibration, heat, cold, dust, radiation, inhaled fibers, bacteria, molds, fungi, insects, bites, inadequate lighting, repetitive movements, and lifting.

10.3.4 Guidelines for Safe Products

A number of guidelines result directly from the discussion above:

1. Perform appropriate analyses. All analysis calculations should be completed thoroughly, checked, signed, and dated. Analyses should be as comprehensive as those typically performed in the practice.

2. Comply with published standards. Obtain, review, and implement relevant industry and government standards.

3. Use state-of-the-art technology. We must keep our knowledge up to date. Ignorance of newer, safer technology is not an excuse.

4. Include reasonable safety features or devices. Design-in safety features and devices to eliminate or protect against the hazard.

5. Take into account how the user might misuse the product. Consider how the user could misuse or abuse the product. Try to eliminate opportunities for the user to misuse the product.

6. Consider hidden dangers that might surprise the user. Consider sudden loud noises or vibrations or unusual modes of operation that might startle the operator, causing loss of control and injury.

7. Consider variations in materials or manufacturing processes, or effects of wear. Analyze the robustness of the product. Design-in allowances for wear and environmental deterioration.

8. Carry out appropriate testing and interpret results correctly. Conduct a thorough test plan of individual parts and completed assemblies. Consider all modes of failure and the hazards originated by each mode. Thoroughly document your findings and judge your results conservatively.

9. Provide adequate warnings. Include appropriate warning labels on the product and in the user/owner manual. Also consider warning devices such as audible or visible alarms.

10. Implement superior quality control. Implement a program of part and assembly inspection and function testing that is commensurate with the nature of the hazards and the probabilities of occurrence.

11. Document everything. Check, sign, date, and archive relevant documents.

Safety Hierarchy Lets consider different alternatives that can make our products safer. For example, take the case of a machine tool housing that requires removal of a cover panel during servicing, exposing moving gears and links, which could cause pinching injuries. Some of the remedial actions include:

A. have the user turn off the machine, and "red-tag" the on/off switch to alert other personnel,

B. redesign the cover/access such that the moving parts are not exposed,

C. print a warning label on the access cover and in the user manual, and

D. retrofit the machine with an "open-access-cover" alarm.

Depending on the electrical connection, the "red-tag" alternative might be the easiest and least expensive. But, an untrained, informed repairman might not "follow standard procedures," and ultimately end up with an injury. Using an alarm will most likely alert the repairman, but still expose him to the hazard. If, on the other hand, the panel cover could be redesigned such that only the stationary parts needing service be exposed, the hazard would be eliminated completely.

Over the last 30 years or so, a method has evolved that prioritizes design efforts. It is called the safety hierarchy (Pahl and Beitz, 1996; MIL-STD-1472F, 1999). The **safety hierarchy** establishes the priorities for risk reduction as follows:

1. *Eliminate the hazard.* This is a proactive approach that recognizes that it is often more effective to just "design-out" the hazard. Redesigning the cover to eliminate access to the moving parts is an example.

2. *Protect against the hazard.* This is a passive or indirect approach, which does not eliminate the hazard, but eliminates the opportunity for the user to come into contact with the hazard. Examples include machine guards, safety interlock switches, light curtains, seatbelts, and locked equipment rooms.

3. *Warn against the hazard.* This does not eliminate the hazard or the opportunity. It is a weak remedy, but in some cases the only effective alternative. Examples include warning labels, passive barriers, and alarm systems.

4. *Provide training.* Provide and require operating training.

5. *Provide personal protection.* Provide the operator or user with protective clothing or protective devices such as safety glasses, nonflammable welding aprons, or rubber gloves to handle chemicals.

Safe Design Principles   Three fundamental design principles are used to design safe products and systems: **safe-life design, fail-safe design**, and **redundant design**.

1. *Safe-life design principle.* Product components are designed to operate for their entire predicted useful life without breakdown or malfunction. Examples include hammers, screwdrivers, bridges, buildings, and electric hand drills. Product design, based on this principle, requires the designers to identify and specify: all operating conditions, misuses and abuses of the product, and appropriate maintenance and repair schedules.

2. *Fail-safe design principle.* Product is designed such that upon failure of a component, some critical functions are still performed. Failure of the component should be discernible such that the product can be shut down safely. Examples include using a castle nut and a cotter pin to prevent a nut from falling off a bolt and separating the connection, or a boiler feedwater valve failing in the open position, a "failed" roller bearing generating vibration and noise while providing some function.

3. *Redundant design principle.* Additional components or systems, configured in parallel or series, take over the principal function of the failed component or system. Examples include multiengine airplanes, standby electric generators, dual in-line oil filters, and emergency brakes.

Safety is no accident.

Anonymous

10.4 TOLERANCE DESIGN

As part of our parametric design activities, we determine appropriate values for the design variables such as lengths, widths, heights, hole diameters, thicknesses, fillet radii, chamfers, and slot widths. These values are shown on detail drawings as dimensions. A **dimension** is any quantity that sizes, locates, or orients geometric features on a part.

Example

The part shown on Figure 10.1 has dimensions A–F. Which of these are used to size a feature, locate a feature, or orient a feature?

Dimension A, D, and E are size dimensions that specify the width and height of the part and the diameter of the hole, respectively. Dimensions B and C are location dimensions that determine the center of the hole. Dimension F is an orientation dimension that specifies the angle of the inclined surface.

Many products are composed of multiple components. Some of the components may be standard parts, which we purchase from a variety of vendors. Others may be special-purpose parts that we manufacture, or outsource for manufacture by another company.

The success of our product ultimately depends on how well the components fit and function together. Rotating or translating parts, for example, need clearances to allow for motion, or fastened parts need interferences such as press fits and snap fits. Assuming that we are mass-producing the product, we can benefit by making interchangeable parts. Interchangeable parts fit and

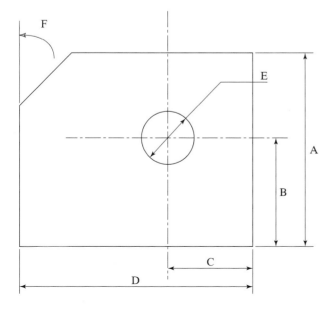

FIGURE 10.1

Example part showing size, location, and orientation dimensions.

function together during original assembly without trimming or patching operations. They also provide for convenient replacement during routine servicing. Therefore, we can tolerate just so much variation before the fit or function becomes unacceptable.

To maintain control over dimensional variation, we specify upper and lower limits on part dimensions, that is, maximum and minimum sizes. The specified difference between maximum and minimum size limits of a part feature is called a **tolerance**. We might be tempted to specify smaller tolerances to achieve greater interchangeability. The realities of manufacturing are that interchangeable parts are not easy to make. Sand-casting processes, for example, can achieve only rough dimensional precision. Investment casting is better, and machining processes are better yet.

The reasons for dimensional variations in machining include how fast we remove material, the mechanical properties of the material, the quality of the tool, and the rigidity of the clamping fixtures. All of the manufacturing processes, in fact, have similar limitations. Of course, if we handcraft each part, we can achieve fine tolerances. For example, we can slow down the feed rate during machining, use the most rigid fixture, or use the best tools. In general, however, we find that the cost of producing a part increases rapidly as we specify finer tolerances that are beyond the typical ability of commercial equipment, as shown in Figure 10.2.

Our goal as a product development team, therefore, is to design a part that balances the costs of finer tolerances with the benefits of improved interchangeability, fit, and function.

10.4.1 Worst-Case Tolerance Design

Worst-case tolerance design is a relatively simple approach that assures that almost all of the parts fit together. It assumes that each manufacturing process will produce parts with the "worst" precision within its capability. If, for example, a machine produces parts to within ±0.005 inches and we need the part to

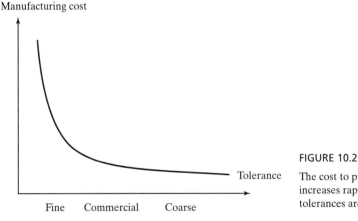

FIGURE 10.2

The cost to produce a part increases rapidly as finer tolerances are specified.

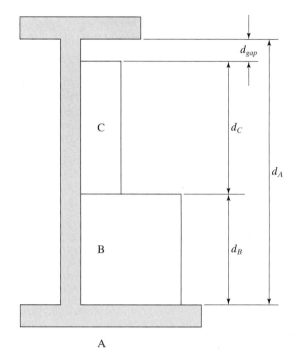

FIGURE 10.3

Assembly consisting of parts B and C mating in part A.

be big, the method will assume that the part will be short by −0.0005 inches. It is considered to be a conservative method because it ignores the many parts that are made with little or no deviation from the nominal.

Consider the three-part assembly having a clearance fit shown in Figure 10.3. Parts designed for a **clearance fit** are specified with an intended space between mating parts. A clearance fit is need for a bolt to fit into a hole, for example. An **interference fit** is specified with an intended lack of space between mating parts, such as a gear press-fitted onto a shaft. We note that parts B and C are assembled into part A. Based on fit and function requirements, assume that a 0.015 clearance fit is required. Assume that the nominal dimensions for A, B, and C are 3.000, 1.650, and 1.300 inches, respectively. Further, assume that part A can be made to within ±0.015 inch, B within ±0.003 inch, and C within ±0.020 inch. Will the assembly have the desired clearance fit? If not, how do we specify the tolerances?

We can define the manufactured lengths of A, B, and C as $d_A = 3.00 \pm 0.015$, $d_B = 1.650 \pm 0.003$, and $d_C = 1.300 \pm 0.020$. The clearance gap d_g can be obtained from a simple equation, as shown in equation (10.1).

$$d_{gap} = d_A \pm t_A - (d_B \pm t_B + d_C \pm t_C) \qquad (10.1)$$

We can rearrange the equation to separate the nominal dimensions from the tolerances to obtain equation (10.2). We see that the clearance is equal to the nominal gap minus the tolerance stack within the summation sign.

$$d_{gap} = (d_A - d_B - d_C) - (\pm t_A \pm t_B \pm t_C)$$

$$= d_{gap\,Nominal} - \sum_i \pm t_i \qquad (10.2)$$

We find that the worst case of clearance gap is found by substituting the maximum material condition (MMC) dimension of each part into equation (10.1). The worst clearance, or smallest gap, results from having the smallest part A space to mate with the largest parts B and C. This results in equation (10.3), as follows.

$$d_{gap} = d_A - t_A - (d_B + t_B + d_C + t_C) \qquad (10.3)$$

Substituting the numerical values, we determine the clearance gap as:

$$d_{gap} = 3.000 - 0.015 - (1.650 + 0.003 + 1.300 + 0.020) \qquad (10.4)$$

$$d_{gap} = 2.985 - (1.653 + 1.320) = 2.985 - 2.973 \qquad (10.5)$$

$$d_{gap} = 0.012 \text{ in.} \qquad (10.6)$$

Since the minimum clearance required is 0.015, this is not an acceptable situation. We see that even under the worst circumstance for part A and the worst circumstance for parts B and C, we still cannot guarantee that the assembly will fit together. In all likelihood, the worst-case assembly does not occur that often. In fact, the major drawback of this method is that it may be too conservative.

We can remedy this situation a couple of ways:

Tighten some tolerances. We can ask manufacturing engineering to estimate the impact of tightening the tolerance on one of the parts. If, for example, we squeeze out 0.003 inch by improving the manufacturing processes for part A or part C, we would have the desired clearance. This may mean spending more time and labor to produce either part, which adds cost to the assembly. We might be able to afford a higher-precision machine. But, until we ask, we won't know what the facts are.

Redesign the nominal values. We can reexamine the nominal sizes for the parts. Can A be 0.003 inch larger? Can either B or C be 0.003 inch smaller? This may not be possible for standard parts. So we have to dig in to the circumstances and find out.

Redesign some of the parts. If the top wall of part A, for example, could be eliminated, we would not have a need for a clearance gap at all.

Consider selective assembly. Assume that during assembly, larger B and C parts could be separated from the smaller ones. Using **selective assembly** they could be selected and assembled onto larger A parts (Stoll, 1999). This, of course, would require inspection and inventory costs. It also may not be suitable for parts that need replacement in the field.

10.4.2 Statistical Tolerance Design

The **statistical tolerance design method** uses stochastic measures of each manufacturing process to estimate the probability that parts will fit and/or function based on the actual statistical capability of the processes. It is not as conservative as the worst-case method and is considered to be the most widely used method in industry (Creveling, 1997).

The statistical method assumes that dimensions and their tolerances are random variables, x_i and s_i. Therefore, we can approximate the mean value of a function of those variables $\phi(\mathbf{x})$, such as clearance gap, by the function evaluated at the mean values of the random variables, shown in equation (10.7). The standard deviation can be similarly estimated by taking the root of the sum of the squared deviations, shown in equation (10.8).

$$\overline{\phi}(\mathbf{x}) = \phi(\overline{x}_1, \overline{x}_2, \overline{x}_3 \ldots \overline{x}_n) \tag{10.7}$$

$$s = (s_1^2 + s_2^2 + \cdots s_n^2)^{1/2} = \left(\sum_1^n s_i^2 \right)^{1/2} \tag{10.8}$$

Therefore, we can estimate the mean value of the clearance gap by substituting the mean values of the part dimensions, that is, the nominal values, as shown in equations (10.9) and (10.10), resulting in a clearance gap mean value of 0.050 inch.

$$\overline{d}_{gap} = \overline{d}_A - (\overline{d}_B + \overline{d}_C) \tag{10.9}$$

$$\overline{d}_{gap} = 3.000 - (1.650 + 1.300) = 0.050 \text{ in.} \tag{10.10}$$

The upper limit for a manufacturing process is typically assumed as three standard deviations above the mean value, that is, $t^+ = +3s$, and the lower limit is assumed to be $t^- = -3s$. Rearranging the variables, we find that $s = t/3$.

The equation for the standard deviation of the gap can be obtained as:

$$s_{gap} = (s_A^2 + s_B^2 + s_C^2)^{1/2} \tag{10.11}$$

$$s_{gap} = [(t_A/3)^2 + (t_B/3)^2 + (t_C/3)^2]^{1/2} \tag{10.12}$$

$$s_{gap} = [(0.015/3)^2 + (0.003/3)^2 + (0.020/3)^2]^{1/2} \tag{10.13}$$

$$s_{gap} = 0.008393 \text{ in.} \tag{10.14}$$

Transforming the random variable d_{gap} to the standard unit normal distribution z, we obtain a measure of the probability that the clearance is less than 0.015 in. We know that for normally distributed random variables we can find the probability that $x \leq x^*$ by numerically integrating the probability density function as follows:

$$p(x \leq x^*) = \int_{-\infty}^{x^*} f(x) \, dx = \int_{-\infty}^{x^*} \frac{1}{\sqrt{2\pi}\sigma} e^{-\frac{1}{2}\left(\frac{x^* - \mu}{\sigma}\right)^2} dx \tag{10.15}$$

Rather than integrating equation (10.15) for every problem, we can transform any normally distributed random variable x, to the unit normal distribution variable z, using equation (10.16). The table of probability values derived from equation (10.17) is shown in Table 10.9.

$$z^* = \frac{x^* - \bar{x}}{s} \tag{10.16}$$

$$p(z \le z^*) = \int_{-\infty}^{z^*} f(z)\, dz = \int_{-\infty}^{z^*} \frac{1}{\sqrt{2\pi}} e^{-\frac{1}{2}z^2}\, dz \tag{10.17}$$

The likelihood that the clearance is smaller than the required amount can be stated as $p(d \le d^*)$, for $d^* = 0.015$, as shown in equation (10.18).

$$p(x \le x^*) = p(d_{gap} \le 0.015) = p(z \le z^*) \tag{10.18}$$

$$z^* = \frac{x^* - \bar{x}}{s} = \frac{d^* - \bar{d}}{s_d} = \frac{0.015 - 0.05}{0.008393} = -4.170 \tag{10.19}$$

Therefore,

$$p(d_{gap} \le 0.015) = p(z \le -4.170) = \sim 0. \tag{10.20}$$

Using the unit normal distribution table, we find the probability that the clearance gap is smaller than 0.015 not even listed on the table because it is so small. Using a spreadsheet function to evaluate the unit normal distribution integral, we find the computed probability is 0.0000152. This is equivalent to 15.2 occurrences per million assemblies. We now see why the worst-case method is considered conservative. Based on this very small probability, the existing design is acceptable.

Using **sensitivity analysis**, we can examine the stack of tolerances to see which specific parts contribute the most variation (Ullman, 1997). We rearrange equation (10.8) as follows:

$$s^2 = s_1^2 + s_2^2 + \cdots s_n^2 \tag{10.21}$$

Then, by dividing through by s^2, we obtain equation (10.22), where P_i are the percentage contributions to the assembly variance.

$$1 = \frac{s_1^2}{s^2} + \frac{s_2^2}{s^2} + \cdots \frac{s_n^2}{s^2} = P_1 + P_2 + P_3 \cdots P_n \tag{10.22}$$

Using the values from the example, we can create a summary shown in Table 10.10. We see that part C contributes 63.1 percent of the variation and part A contributes 35.5 percent. Therefore, if we were out of tolerance, we would focus our attention on part C.

TABLE 10.9   Unit Normal Probability Distribution

Normal Probability Distribution

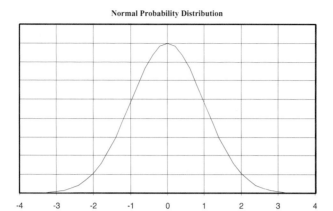

Z	0.00	0.01	0.02	0.03	0.04	0.05	0.06	0.07	0.08	0.09
−3.5	0.00023	0.00022	0.00022	0.00021	0.00020	0.00019	0.00019	0.00018	0.00017	0.00017
−3.4	0.00034	0.00032	0.00031	0.00030	0.00029	0.00028	0.00027	0.00026	0.00025	0.00024
−3.3	0.00048	0.00047	0.00045	0.00043	0.00042	0.00040	0.00039	0.00038	0.00036	0.00035
−3.2	0.00069	0.00066	0.00064	0.00062	0.00060	0.00058	0.00056	0.00054	0.00052	0.00050
−3.1	0.00097	0.00094	0.00090	0.00087	0.00084	0.00082	0.00079	0.00076	0.00074	0.00071
−3.0	0.00135	0.00131	0.00126	0.00122	0.00118	0.00114	0.00111	0.00107	0.00104	0.00100
−2.9	0.00187	0.00181	0.00175	0.00169	0.00164	0.00159	0.00154	0.00149	0.00144	0.00139
−2.8	0.00256	0.00248	0.00240	0.00233	0.00226	0.00219	0.00212	0.00205	0.00199	0.00193
−2.7	0.0035	0.0034	0.0033	0.0032	0.0031	0.0030	0.0029	0.0028	0.0027	0.0026
−2.6	0.0047	0.0045	0.0044	0.0043	0.0041	0.0040	0.0039	0.0038	0.0037	0.0036
−2.5	0.0062	0.0060	0.0059	0.0057	0.0055	0.0054	0.0052	0.0051	0.0049	0.0048
−2.4	0.0082	0.0080	0.0078	0.0075	0.0073	0.0071	0.0069	0.0068	0.0066	0.0064
−2.3	0.0107	0.0104	0.0102	0.0099	0.0096	0.0094	0.0091	0.0089	0.0087	0.0084
−2.2	0.0139	0.0136	0.0132	0.0129	0.0125	0.0122	0.0119	0.0116	0.0113	0.0110
−2.1	0.0179	0.0174	0.0170	0.0166	0.0162	0.0158	0.0154	0.0150	0.0146	0.0143
−2.0	0.0228	0.0222	0.0217	0.0212	0.0207	0.0202	0.0197	0.0192	0.0188	0.0183
−1.9	0.0287	0.0281	0.0274	0.0268	0.0262	0.0256	0.0250	0.0244	0.0239	0.0233
−1.8	0.0359	0.0351	0.0344	0.0336	0.0329	0.0322	0.0314	0.0307	0.0301	0.0294
−1.7	0.0446	0.0436	0.0427	0.0418	0.0409	0.0401	0.0392	0.0384	0.0375	0.0367
−1.6	0.0548	0.0537	0.0526	0.0516	0.0505	0.0495	0.0485	0.0475	0.0465	0.0455
−1.5	0.0668	0.0655	0.0643	0.0630	0.0618	0.0606	0.0594	0.0582	0.0571	0.0559
−1.4	0.0808	0.0793	0.0778	0.0764	0.0749	0.0735	0.0721	0.0708	0.0694	0.0681
−1.3	0.0968	0.0951	0.0934	0.0918	0.0901	0.0885	0.0869	0.0853	0.0838	0.0823
−1.2	0.1151	0.1131	0.1112	0.1093	0.1075	0.1056	0.1038	0.1020	0.1003	0.0985
−1.1	0.1357	0.1335	0.1314	0.1292	0.1271	0.1251	0.1230	0.1210	0.1190	0.1170
−1.0	0.1587	0.1562	0.1539	0.1515	0.1492	0.1469	0.1446	0.1423	0.1401	0.1379
−0.9	0.1841	0.1814	0.1788	0.1762	0.1736	0.1711	0.1685	0.1660	0.1635	0.1611
−0.8	0.2119	0.2090	0.2061	0.2033	0.2005	0.1977	0.1949	0.1922	0.1894	0.1867
−0.7	0.2420	0.2389	0.2358	0.2327	0.2296	0.2266	0.2236	0.2206	0.2177	0.2148
−0.6	0.2743	0.2709	0.2676	0.2643	0.2611	0.2578	0.2546	0.2514	0.2483	0.2451
−0.5	0.3085	0.3050	0.3015	0.2981	0.2946	0.2912	0.2877	0.2843	0.2810	0.2776
−0.4	0.3446	0.3409	0.3372	0.3336	0.3300	0.3264	0.3228	0.3192	0.3156	0.3121
−0.3	0.3821	0.3783	0.3745	0.3707	0.3669	0.3632	0.3594	0.3557	0.3520	0.3483
−0.2	0.4207	0.4168	0.4129	0.4090	0.4052	0.4013	0.3974	0.3936	0.3897	0.3859
−0.1	0.4602	0.4562	0.4522	0.4483	0.4443	0.4404	0.4364	0.4325	0.4286	0.4247
0.0	0.5000	0.4960	0.4920	0.4880	0.4840	0.4801	0.4761	0.4721	0.4681	0.4641

TABLE 10.10   Sensitivity Analysis of Three-Part Assembly

Part	t	s	s^2	P_i
A	0.015	0.00500	2.50E-05	35.5
B	0.003	0.00100	1.00E-06	1.4
C	0.020	0.00667	4.44E-05	63.1
				100.0

Example

A 5-inch-diameter pin will be assembled into a 5.005-inch journal bearing. The pin manufacturing tolerance is specified as 0.003 inch. A minimum clearance fit of 0.001 inch is needed. Determine the tolerance required of the hole such that 0.1 percent of the mates will not meet the minimum clearance. Assume that the manufacturing variations are normally distributed.

Using the unit normal probability table, we find that for a 0.1 percent, $z = -3.09$. The mean value of the clearance gap can be found from:

$$\overline{d}_{gap} = \overline{d}_{hole} - \overline{d}_{pin} = 5.005 - 5.000 = 0.005 \text{ in.}$$

We can then find the standard deviation of the assembly as:

$$z^* = \frac{x^* - \overline{x}}{s} = \frac{d^* - \overline{d}}{s_d} = \frac{0.001 - 0.005}{s_d} = -3.09$$

$$s_d = \frac{0.001 - 0.005}{-3.09} = 0.00129 \text{ in.}$$

$$s_d = (s_{hole}^2 + s_{pin}^2)^{1/2} = 0.00129 \text{ in.}$$

Then we can find the standard deviation of the hole from:

$$s_d = (s_{hole}^2 + (0.003/3)^2)^{1/2} = 0.00129 \text{ in.}$$

$$s_{hole}^2 = (0.00129)^2 - (0.003/3)^2 = 0.676(10)^{-6} \text{ in.}^2$$

$$s_{hole}^2 = (t_{hole}/3)^2 = 0.676(10)^{-6} \text{ in.}^2$$

Then, solving for the tolerance, we obtain: $t_{hole} = 0.0025$ in.

10.4.3   Tolerance Design Guidelines

The following guidelines can be used to make tolerance design decisions.

- Consider which dimensions, if any, are critical to the fit and function of the assembly. Not all dimensions will have an impact. Quantitatively examine tolerance stacks using worst-case or statistical tolerancing methods.
- Consider whether any critical parts could be redesigned to become non-critical.
- Consider the feasibility of selective assembly.
- Consider revising nominal dimensions before specifying tighter tolerances.

- If a tighter tolerance is needed, use sensitivity analysis to identify prospects.
- Specify commercial tolerances when possible.
- Reduce the number of fine-tolerance specifications.

10.5   DESIGN FOR THE ENVIRONMENT

Our personal health and well-being is intimately linked to the environment in which we live. We are dependent on the air we breathe, the water we drink, and the raw materials we consume. As the world population increases, more people share the same environment, encouraging us to find better ways to preserve and protect the environment. We cannot continue to pollute as we have in the past. As trustees for the company we work for and as citizens of the world we share, it is our moral obligation. And as a member of the engineering profession, it is our ethical responsibility.

Design for Minimal Use of Materials. Let's examine the effects of material waste, by starting with the finished product. Was it packaged carefully to protect it during shipping? Broken or damaged products are often disposed of, that is, wasted. As the product was assembled, were any parts damaged, or thrown away because they were not within specifications? Could these parts have been designed better? During the fabrication of the parts, how much waste material was manufactured as scrap? Was the scrap recyclable? Upon receipt of the raw materials, were any damaged in transit, or spoiled owing to aging? Imagine if we could reduce our raw material usage 10 percent. Imagine the impact that would have on fuel used to ship the raw materials. Imagine the reduction of air pollution and highway repairs because of the reduction in raw materials shipping, and so on. The bottom line is that we can make a difference. Materials usage has a ripple effect, just like the ripples produced by dropping a single stone into a still pond.

Design for Recyclability. Assuming that we have reduced the usage of materials as much as reasonably possible, the next effort should focus on returning **recyclable** materials back into the product stream. Nonrecyclable materials, of course, are sent to waste disposal sites and should be designed to reduce volume or toxicity. Polymers, such as polyethylenes and polystyrenes, and metals, such as steel, copper, silver, and brass, are very recyclable if they can be economically separated from the nonrecyclable materials. A separate field called design for disassembly considers aspects such as the appropriate use of fasteners that facilitate disassembly, as well as decreasing separation labor time (Otto and Wood, 2001). We should consider ways to improve access and removal of recyclable materials and consider the effective use of removable fasteners.

Design for Remanufacture. Not all products are composed of components that simultaneously fail. The typical laser printer, for example, has a replaceable toner cartridge that can be reused if appropriately remanufactured. **Remanufacturing** includes disassembly, inspecting, cleaning, and replacement of worn or broken parts. Product design efforts should focus on facilitating these activities, especially design for disassembly.

Design for Energy Efficiency. Products should be designed to be manufactured and used with less energy. Less electrical use during manufacture means less heat generated by inefficient motors and devices. Also, using less electricity means less fossil fuel consumed, less air pollution, and possibly less global warming. Similarly, energy efficiency during usage also improves the environment owing to reductions in air, noise, and thermal pollution.

Design for the Workplace. We also need to recognize and remedy the exposure of our co-workers and community to the environmental hazards associated with manufacturing, such as air pollution, noise pollution, and exposure to toxic and/or hazardous chemicals. Close adherence to OSHA regulations is a must. We should also consider alternative materials or processes, whenever possible, that are more environmentally friendly.

10.6 SUMMARY

- FMEA is used to identify potential failure modes, causes, and effects.
- FMEA is used to eliminate or reduce the occurrence of failures and/or mitigate the severity of the effects of failures.
- Hazards that can injure people or damage property exist in the manufacture, use, and retirement of products.
- Engineers have legal and ethical responsibilities regarding the safety, health, and welfare of employees, customers, and the general public.
- The safety hierarchy is an effective design-for-safety procedure.
- Natural variations occur in material properties, manufacturing processes, and product usage.
- Tolerance design ensures part interchangeability as well as satisfactory fit and function.
- Worst-case tolerance design is a simple and conservative method, but may be costly.
- Statistical tolerance design makes effective use of fundamental probability mathematics and manufacturing process data to specify tolerances that are more realistic but less conservative.
- Design for the environment methods consider minimal use of materials, recyclability, remanufacture, energy efficiency, and workplace conditions.

REFERENCES

Blinn, K. W. 1989. *Legal and Ethical Concepts in Engineering.* Englewood Cliffs, NJ: Prentice Hall.

Creveling, C. M. 1997. *Tolerance Design: A Handbook for Developing Optimal Specifications.* Reading, MA: Addison-Wesley Longman.

Himmelfarb, D. 1985. *A Guide to Product Failures and Accidents*. Lancaster, PA: Technomic
    Publishing Company.
Lindbeck, J. R. 1995. *Product Design and Manufacture*. Upper Saddle River, NJ: Prentice Hall.
McDermott, R. E., R. J. Mikulak, and M. R. Beauregard. 1996. *The Basics of FMEA*. New
    York: Quality Resources.
MIL-STD-1472F. 1999. *U.S. Department of Defense, Design Criteria Standard, Human
    Engineering.*
Otto, K. P., and K. L. Wood. 2001. *Product Design: Techniques in Reverse Engineering and
    New Product Development*. Upper Saddle River, NJ: Prentice Hall.
Pahl, G., and W. Beitz. 1996. *Engineering Design*, 2d ed. New York: Springer-Verlag.
Stoll, H. W. 1999. *Product Design Methods and Practices*. New York: Marcel Dekker.
Ullman, D. G. 1997. *The Mechanical Design Process*, 2d ed. New York: McGraw-Hill.

KEY TERMS

Civil actions	Failure modes and effects	Safe-life design
Clearance fit	analysis (FMEA)	Safety hierarchy
Criminal actions	Failure mode	Selective assembly
Design for the environment	Hazard	Sensitivity analysis
Design for safety	Impact hazards	Severity rating
Design for X	Interference fit	Statistical tolerance
Detection	Occupational safety	design method
Detection rating	Occurrence rating	Tolerance
Dimension	Products liability	Tolerance design
Ejection hazards	Recyclable	Worst-case tolerance
Entanglement hazards	Redundant design	design
Entrapment hazards	Remanufacturing	
Fail-safe design	Risk priority number	

EXERCISES

Self-Test. Write the letter of the choice that best answers the question

1. _____ In the failure modes and effect analysis method, the reason why a part fails owing to a defect in design or manufacture is called the:

    a. effect

    b. cause

    c. severity

    d. detection

2. _____ The adverse consequences that the customer might experience is called:

    a. severity

    b. cause

    c. effect

    d. detection

3. _____ The risk priority number for a part failure mode is the product of all the following factors except:

   a. severity rating

   b. deflection rating

   c. occurrence rating

   d. detection rating

4. _____ All of the following are examples of prevalent part failure modes except:

   a. binding

   b. rotating

   c. leaking

   d. fracturing

5. _____ What "likelihood" would rate a 10 on the occurrence rating chart?

   a. occasional

   b. remote

   c. expected

   d. improbable

6. _____ At what stage in the life of a product may injuries occur to employees or customers:

   a. manufacture

   b. use

   c. retirement

   d. all of these

7. _____ Injuries received by touching hot surfaces, sharp edges, or electrically charged parts are caused by:

   a. contact hazards

   b. entrapment hazards

   c. impact hazards

   d. entanglement hazards

8. _____ Injuries received from hammer blows to the hand are caused by:

   a. contact

   b. impact

   c. ejection

   d. entanglement

9. _____ The design method that establishes priorities for risk reduction is called:

   a. safety alternatives

   b. hazard reduction

   c. safety hierarchy

   d. hazard hierarchy

10. _____ The fundamental design principles used to design safe products and systems include all of the following except:

    a. dual use

    b. safe- life

    c. redundant design

    d. fail-safe

2: c, 4: b, 6: d, 8: b, 10: a

11. Explain the difference between a failure mode and the cause of a failure.

12. Consider the potential design or manufacturing defects in the typical steam iron sold for home use. List some failure modes, causes, and effects. Describe the severity of the effects.

13. Prepare an FMEA chart for the power switch of the steam iron in the previous question.

14. How can risk priority numbers be used?

15. When should FMEA be used?

16. Briefly list the six basic types of hazards and give an example of a product that you own or have used.

17. Connect to http://www.fmeainfocentre.com/download_area.htm or www.fmeca.com. What is the purpose of the Web site? How can it be used?

18. Consider the design of a new weed whacker that uses a small gasoline engine to power a rotating cutter blade 6 inches long by 2 inches wide by $\frac{1}{4}$ inch thick. Use the safety hierarchy to make recommendations to protect users from injury or property damage.

19. Explain the difference between safe-life design, fail-safe design and redundant design. Give an example of each.

20. Describe the difference between an interference fit and a clearance fit. Give an example of each. Then explain how inadequate tolerance specifications in your examples could lead to poor fit or function.

21. Explain how minimizing the use of materials is friendly to the environment.

22. How does design for recyclability differ from remanufacture?

CHAPTER 11

Human Factors/Ergonomics

When you have completed this chapter you will be able to

- Explain the role of human factors in engineering design
- Describe the human-machine system model
- Apply sound design principles for visual and auditory displays
- Select appropriate control devices
- Specify human limitations for applying forces and torques
- Specify size and range of motion limitations
- Describe and apply three strategies for design for fit (human)

11.1 INTRODUCTION

Human factors is a term used to describe the abilities, limitations, and other physiological or behavioral characteristics of humans that affect the design and operation of tools, machines, systems, tasks, jobs, and environments. **Ergonomics** is a synonym frequently used by practitioners in Europe.

By careful consideration of human factors we strive to achieve user-friendliness, convenience, effectiveness, efficiency, and increased productivity. We also hope to enhance desirable human values such as improved safety, re-duced fatigue or stress, increased comfort, greater user acceptance, increased job satisfaction, and improved quality of life (Sanders and McCormick, 1993).

The following three scenarios involve humans using a different product. Let's consider the physical interactions and the decisions made when perform-ing the various tasks.

Making a pot of coffee with an electric coffeemaker. The consumer:

- grasps the pot, removes the pot top, and fills it with cold water,
- replaces the top, and pours the water into the coffeemaker water reservoir,
- removes the basket, inserts a new filter along with fresh coffee grounds,
- reinserts the basket into the coffeemaker, and
- pushes the power switch.

Changing the oil on a motorcycle. The repair person:

- starts and runs the engine for a few minutes to warm the oil and entrain particles that have settled to the bottom of the oil sump,
- places a suitable basin underneath the oil drain plug,
- removes ratchet wrench and socket from toolbox,
- loosens and removes the drain plug,
- inspects the plug and O-ring seal for dirt or wear,
- waits for oil to drain,
- replaces the plug finger-tight,
- tightens plug to a specified torque with torque wrench,
- removes the oil refill cap and pours in correct amount of fresh oil, and
- pours the waste oil into a suitable container for transport to recycle site.

Riding a bicycle. The rider:

- removes the bicycle from storage,
- climbs onto the bike and pedals down the street,
- pushes the handlebar to steer around a dog in the street,
- slows down by grasping hand brake lever and squeezing,
- steers to the side of the road to let car pass by,
- pushes the shift lever to a lower gear decreasing pedal force to climb hill,
- squeezes brake lever to stop at the top of the hill, dismounts, and
- walks the bike to roadside and enjoys the view.

First, we may notice that the user physically senses each product by seeing, hearing, and touching various components. Second, the user makes decisions, sometimes based on the sensory input, such as determining the amount of coffee grounds, how long to run the engine for warmup, and steering left or right around the dog. Finally, the user outputs or exerts forces and/or torques on parts of the machine. For example, the user grips the basket, pours water, pushes power switch, twists drain plug, pushes on handlebar to steer, pushes gearshift lever, and squeezes brake lever. The product, in turn, uses the human output to control its response.

We can model the interactions between the user and product as a **human-machine system**, as shown in Figure 11.1. The system includes the human operator,

Human-machine system

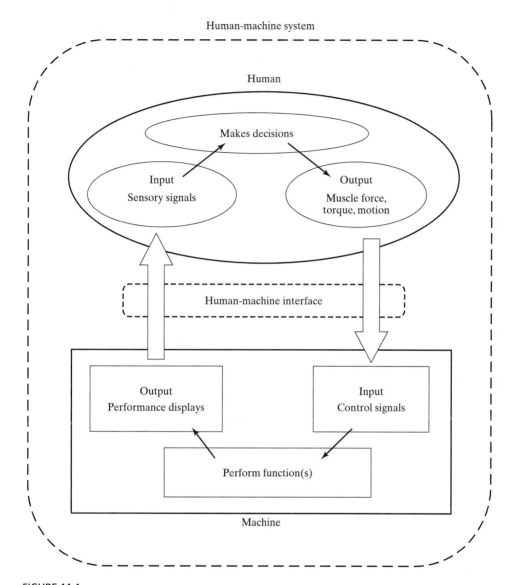

FIGURE 11.1

The human-machine system involves input and output interactions by means of interfaces.

the machine, and the surrounding environment. The "machine" could be a consumer product such as a coffeemaker, motorcycle, or bicycle. The machine could also represent larger pieces of equipment, systems, and facilities such as a milling machine, a potato-chip-packaging system, or a nuclear-power plant.

Let's now use the model to understand why the study of human factors is so important in the design of a machine.

Humans receive information about the current state of the machine and the environment via **sensory inputs** such as by seeing, hearing, touching, smelling, and tasting. They use this sensory information to make decisions. Then, they implement their decision by applying an output force or torque to the machine. Similarly, the machine uses these outputs as control inputs, to perform the desired changes in function, and displays new outputs.

Most important, the system depends upon the interfaces between the human and the machine. When the user pushes a switch to turn on power to the coffeemaker, the switch lights up to indicate its "on" state. If the user loses grip of the bicycle handlebar interface, the bicycle will not "know" how to respond. To reinstall the oil drain plug, the user must be strong enough to supply the correct amount of torque to the plug (interface). An interface is any part of the machine that the human touches with a part of his or her body or is exposed to. An interface is also any component that the human sees, hears, or is exposed to. For example, we hear the muffler, or we see a grinding wheel rotating. We may also be exposed to radiation or dangerous chemical gases or vapors. Interfaces are the means by which we interact with the product (Stanton, 1998).

As we design and manufacture products for people to use we must consider how humans interface with machines. We need to design machines that fit the size of human hands, feet, arms, legs, and chests. We need to identify and accommodate human limitations in detecting and interpreting sensory inputs. We need to accommodate the capacity of a human to correctly process information and make timely decisions such as reducing reaction time. We need to acknowledge human limits with regard to applying forces or torques. We also need to design our product so that our customer may use it in ideal weather or foul, or during various phases of use including setup, operate, maintain, repair, take-down, disassemble, recycle, and dispose.

11.2   SENSORY INPUT LIMITATIONS

Human senses include sight, hearing, touch, smell, and taste. The human body can also sense acceleration by the vestibular sacs in the ear. This sense also contributes to balance. The body also has a kinesthetic sense, which uses receptors to feel joint and muscle movement, permitting us to feel the motion of our arms and legs. Each of these senses, however, is limited in its abilities. Let's examine our senses of sight, hearing, and touch in particular.

11.2.1   Sight

Humans differ in their ability to see. Some may not be able to focus on near (or far) objects. Some of us may not be able to distinguish certain colors. All of us, on the other hand, need a minimum of illumination to read or perform various tasks. In fact, as we age, even more illumination is needed. Some objects may move so fast that we cannot "see" them. Even when we can "see" an object, we may not perceive it correctly. For example, objects may actually be nearer than they appear when looking into a rearview mirror. Or a poorly designed visual display could be confusing or difficult to interpret. Some displays may even cause "optical illusions."

In some cases we need to see and perceive a product's "visual displays" in order to operate it. For example, we frequently look at our car's instrument panel as we drive on the superhighway. If we see that the oil warning light is "on," we can stop the engine before permanent damage occurs. If we see that the water temperature gauge is within limits, we need not take action. A properly designed instrument panel displays abnormal operating conditions in a manor that can be quickly scanned and correctly interpreted. Poorly designed machinery, however, may require the replacement of a part that is hidden or visually obscured during routine maintenance.

Visual display types include:

- *indicator lights*, such as an automobile oil pressure warning light,
- *continuous readout gauges*, such as the fuel gauge,
- *digital counters*, such as the odometer or digital radio tuner, and
- *graphical panels*, such as a railroad switchyard control center, or a power plant.

Woodson (1981) lists a number of principles that enhance the effectiveness of warnings, signals, labels, and other displays:

- *Conspicuity.* The display should be conspicuous in that it should be prominently located, novel, and relevant.
- *Emphasis.* Important words should be visually emphasized.
- *Legibility.* Character fonts, size, and contrast should be exploited.
- *Intelligibility.* Succinctly tell the operator what the hazard is and how to fix it.
- *Visibility.* The display should be visible in all lighting conditions, including day or night.
- *Maintainability.* The display should resist aging, wear, and vandalism.
- *Standardization.* Display standard words and symbols whenever possible.

11.2.2 Hearing

Humans hear frequencies between 20 and 20,000 Hz. We are more sensitive to higher frequencies, however. We make use of our hearing to sense the operating condition of machinery. When operating a car brake, we can hear worn brake pads scraping on the rotor, indicating replacement. We may also hear the loud noise of a leaking muffler, or the repetitive thumping of a flat tire, or the hissing of an overheated radiator. A significant problem related to machinery in the workplace, however, is the **masking** of one sound by others. An audible alarm, for example, may go undetected if surrounding noise masks it.

A variety of auditory displays can be used in machine designs, including: bells, buzzers, horns, sirens, tones, and electronic devices that speak single words or standard messages. Auditory displays should be avoided in noisy applications or when the operator is receiving too many other audio signals.

11.2.3 Touch/Kinesthetic/Vestibular

Humans are very sensitive to the tactile stimulation of skin tissues. We can feel sharp points and rough and smooth surfaces, as well as detect hot or cold temperatures or excessive electric voltages. As we operate a motorcycle we may feel excessive vibration through the handlebar, indicating a potential problem. We may detect a bearing failure in a machine by feeling how hot the surface is. We can feel the heat being radiated by a hot furnace. And we can inspect the surface of a part by feeling its roughness. Using our **kinesthetic sense**, we can "feel" how a golf club or baseball bat responds as we swing it, or the improved performance of a new pair of downhill skis or a new fishing rod. Similarly, using our **vestibular sense** we can feel the centrifugal acceleration of a roller coaster in our skin, our muscles and joints, and our middle ear.

11.2 HUMAN DECISION-MAKING LIMITATIONS

Humans are limited in their ability to make decisions, especially when saturated with conflicting signals that are difficult to interpret. For example, let's consider the case when we are driving at night on a dark country road in a rainstorm with fog when a deer jumps in front of us. What do we do? Brake, steer, or pray? A well-trained driver will know where the brake pedal and steering wheel are and won't have to think about them first. And most likely, the driver will coordinate braking and steering to avoid the animal. Not all products are as user-friendly, however.

When we make a decision, we receive sensory input information, interpret it, develop a set of choices, predict the outcomes associated with those choices, evaluate the pros and cons, and then select the "best" choice. In many cases we can take our time deciding. In other cases, such as in driving a car, a quick decision or short reaction time may be required.

Sanders and McCormick (1993) describe reaction time as a practical measure of performance. They define **simple reaction time** as the time to initiate a response when only one particular stimulus occurs and the same response is always required. Touching a hot surface, and jerking away, would be an example. Most decisions in real life, however, deal with multiple choices. **Choice reaction time** is the time when several stimuli are presented, each of which requires a different response. Choice reaction times are longer because of the decision making or thinking that we do to.

Our role as product designers should be to make our machines display appropriate visual and auditory signals clearly and quickly. The user should be trained to react to these stimuli in a safe and predetermined manner, by actuating "the" control that is clearly identified for the specific situation.

11.3   HUMAN MUSCLE OUTPUT

Humans implement their decisions by outputting a force (i.e., push or pull) or torque (i.e., twist) on a part of the machine. In some cases we apply a force or torque to a "control." For example, we step on the brake to slow down, push on a handlebar to steer our bicycle, twist the control on our radio to adjust its volume, twist the handle of a water faucet to control its flow, and squeeze the brake lever on our motorcycle to slow down. In other cases we hold and move the whole product to perform a task, as in trimming a hedge with an electric trimmer, or in shaving with an electric razor, or when cutting a ham with an electric carving knife.

Our ability to apply a force or torque depends on a number of **muscle strength** factors, including whether we are sitting, standing, or lying down, what part of the body is applying the force (i.e., the hand, foot, leg, or arm, etc.), and the direction and duration of the force. Hand, thumb-finger, and arm strength for 5th-percentile males adults is shown in Figure 11.2. Lifting strength for men and women is shown in Figure 11.3. Intermittently applied leg strength for men is shown in Table 11.1.

Note that the data have been compiled using percentiles. This method is more useful than having just an arithmetic mean. Percentiles tell us about the distribution of sizes among the population. For example, the value for the 5th percentile tells us that 5 percent or less of the population has a dimension of that value. Similarly, the 95th percentile tells us that 95 percent of the population has dimensions smaller than that listed. The average is a statistical measure, an arithmetic mean, which may or may not exist in reality.

As we apply a force through a displacement we perform translational work. As we apply a torque through an angular displacement we perform rotational work. Human motion depends on the **range of motion** for each of the joints, as shown in Figures 11.4–11.6.

We know from physics that work requires energy and energy expended per unit of time gives us power. When a human operator mentions "ease of use," we might question whether he or she is talking about the amount of physical effort, or energy expenditure needed to use the product, or the eventual muscle fatigue that can occur. **Energy expenditure** depends upon a number of factors, including the magnitude of the force or torque and the amount of displacement, its duration, number and length of rest periods, and whether the worker is standing, sitting, or lying, as well as environmental conditions, such as temperature and humidity. The mean value daily energy expenditures for a few occupations are given in Table 11.2.

Human muscle output is applied to the machine at a machine-input interface. Quite often, the interface is a control. Controls include levers, cranks, hand wheels, handles, pedals, push buttons, rotary knobs, rotary selector switches, toggle switches, rocker switches, slides, joysticks, and roll balls.

Researchers have studied the adequacy of various controls with respect to the required input force or torque, operational characteristics, and assigned control tasks, as shown in Figures 11.7–11.9 (Salvendy, 1987). The controls are

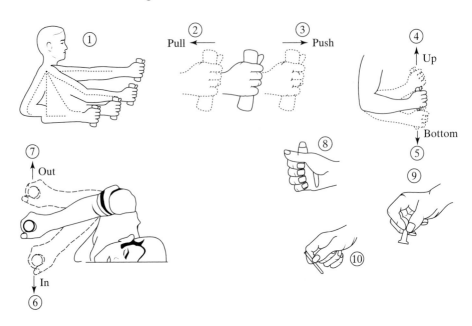

Arm strength (N)													
(1)	(2)		(3)		(4)		(5)		(6)		(7)	
Degree of elbow flexion (deg)	Pull		Push		Up		Down		In		Out	
	L*	R*	L	R	L	R	L	R	L	R	L	R
180	222	231	187	222	40	62	58	76	58	89	36	62
150	187	249	133	187	67	80	80	89	67	89	36	67
120	151	187	116	160	76	107	93	116	89	98	45	67
90	142	165	98	160	76	89	93	116	71	80	45	71
60	116	107	98	151	67	89	80	89	76	89	53	76
Hand and thumb-finger strength (N)												
	(8)			(9)			(10)					
	Hand grip			Thumb-finger grip (Palmer)			Thumb-finger grip (Tips)					
	L	R										
Momentary hold	250	260		60			60					
Sustained hold	145	155		35			35					

*L = Left; R = Right

FIGURE 11.2

Muscle strength of the arm, hand, and thumb for 5th-percentile males.
(Adapted from MIL-STD-1472F. Courtesy of the U.S. Department of Defense.)

A. **Standing two-handed pull: 38 cm (15.0) level.** Standing with feet 45 cm (17.7 in) apart and knees bent; bending at the waist, grasping both sides of a 45 cm (17.7 in) handle located directly in front, 38 cm (15.0 in) above standing surface, and pulling, using primarily arms, shoulders, and legs

Strength measurements	5th percentile		95th percentile	
	Male	**Female**	**Male**	**Female**
Mean force (N)	737.5	330.9	1354.5	817.6
Mean force (lbf)	(165.80)	(74.39)	(304.50)	(183.80)
Peak force (N)	844.7	396.9	1437.2	888.3
Peak force (lbf)	(189.90)	(89.23)	(323.10)	(199.70)

B. **Standing two-handed pull: 50 cm (19.7 in) level.** Standing with feet 45 cm (17.7 in) apart and knees straight; bending at the waist, grasping both sides of a 45 cm (17.7 in) handle located directly in front, 50 cm (19.7 in) above standing surface, and pulling, using primarily arms and shoulders

Strength measurements	5th percentile		95th percentile	
	Male	**Female**	**Male**	**Female**
Mean force (N)	758.0	326.1	1341.6	840.7
Mean force (lbf)	(170.41)	(73.31)	(301.60)	(189.00)
Peak force (N)	830.9	374.1	1441.7	905.2
Peak force (lbf)	(186.79)	(84.10)	(324.11)	(203.50)

C. **Standing two-handed pull: 100 cm (39.4 in) level.** Standing erect with feet 45 cm (17.7 in) apart, grasping both sides of a 45 cm (17.7 in) handle located directly in front, 100 cm (39.4 in) above the standing surface, and pulling, using the arms

Strength measurements	5th percentile		95th percentile	
	Male	**Female**	**Male**	**Female**
Mean force (N)	444.4	185.0	931.0	443.0
Mean force (lbf)	(99.91)	(41.59)	(209.30)	(99.59)
Peak force (N)	504.0	218.0	988.4	493.3
Peak force (lbf)	(113.30)	(49.01)	(222.20)	(110.90)

FIGURE 11.3

Static muscle strength for vertical pull.
(Reprinted from *Human Factors Design Guide* by Wagner, Snyder, and Duncanson. Courtesy of the Federal Aviation Administration, U.S. Department of Commerce.)

TABLE 11.1 Horizontal Push and Pull Forces That Can Be Exerted Intermittently by Males (adapted from MIL-STD-1472F)

Exertable horizontal force	Applied with	Coefficient of friction
110 N (24.7 lbf) push or pull	both hands or one shoulder or the back	with low traction $0.2 < \mu < 0.3$
200 N (45.0 lbf) push or pull	both hands or one shoulder or the back	with medium traction $\mu = 0.6$
240 N (54.0 lbf) push or pull	one hand	If braced against a vertical wall, 510–1,520 mm (20.08–59.84 in.) from and parallel to the push panel
310 N (70.0 lbf) push	both hands or one shoulder or the back	With high traction $\mu > 0.9$
490 N (110.2 lbf) push or pull	both hands or one shoulder or the back	If braced against a vertical wall, 510–1,520 mm (20.08–59.84 in.) from and parallel to the push panel or if anchoring the feet on a perfectly nonslip ground (like a footrest)
730 N (164.1 lbf) push	the back	If braced against a vertical wall, 580–1,090 mm (22.83–42.91 in.) from and parallel to the push panel or if anchoring the feet on a perfectly nonslip ground (like a footrest)

separated into three categories: rotating, translating, and swiveling. The basic shape of the control, its approximate dimensions, and permissible actuating forces and torques are given along with ratings as to how well the control performs various functions. Note that the • symbol means recommended, the ∘ means not suitable, and the half-filled circle means acceptable.

Example

Assume that we have been hired by Micron Industries, Inc. to design a custom machine tool to inspect raw materials as the factory receives them. A 300-hp motor drives the tool at one of four speeds. The control will electronically select one of the four speeds. It will be located on a control panel near the machine. Use Figures 11.7–11.9 to determine feasible controls for the application. Discuss some of the functional characteristics of each control.

The principal criterion for selection is that the control needs to energize one of three choices. Scanning the figures and selecting only those recommended we find five possible controls:

Rotary selector switch is recommended for having tactile feedback, a visible setting, and marginally recommended for a precise adjustment. It has an acceptable rating for accidental actuation and large force actuation. It is not recommended for continuous or quick adjustment.

Handle (slide) is also recommended for large-force actuation and marginally recommended for precise adjustment. It is acceptable for quick adjustment, tactile feedback, and setting visible. It is rated not suitable due to accidental actuation.

D-handle is also recommended for large-force actuation. It is acceptable for precise adjustment and quick adjustment. It is rated not quite suitable for tactile feedback and setting visible. It is rated not suitable due to accidental actuation.

Lever is also recommended for large-force actuation. It is marginally recommended for precise adjustment. It is rated acceptable for quick adjustment, tactile feedback and setting visible. It is rated not suitable due to accidental actuation.

Joystick is also recommended for precise adjustment. It is rated acceptable for quick adjustment, tactile feedback, and setting visible. It is marginally suitable for large-force application. It is rated not suitable due to accidental actuation.

The rotary selector switch appears to be quite acceptable for the application. It is marginally better than the others because of its acceptable rating on accidental actuation.

11.4 PHYSICAL SIZE LIMITATIONS

Humans vary in size such as the length and width of body parts, including fingers, hands, arms, legs, and torsos. Some differences are due to sex and age. For example, men are generally taller than women, and children are smaller yet. Most humans grow taller as they grow to adulthood and then slightly shorten in their later years.

To fit the user, a product should not be too big or too small. A leaf rake that is too short might make the user stoop uncomfortably. A bicycle handlebar that is too large in diameter might not be graspable by a child. A sofa chair that is too narrow might squeeze the user. A coffeepot handle similarly has to fit the users. And the access area to a motorcycle drain plug must be large enough to fit the user's hand.

What are the sizes that we should consider? **Anthropometrics** is a field of human factors that deals with the measurements of the human form such as height and/or reach (Kroemer et al., 2001). Researchers around the world routinely compile and categorize such information by age, sex, and/or country. The U.S. armed forces has compiled a significant amount of data and has prepared a number of documents to aid in the design of machines, equipment, systems, computers and facilities (MIL-STD-1472F, 1999; MIL-HDBK-759C, 1997; Wagner et al., 1996). Static human physical characteristics are given for adult males and females in Figures 11.10–11.17.

11.4.1 Designing for Fit

How should we design our product, machine, or system to fit our users? Should we design for the average person? Probably not, since the average person really doesn't exist. Anthropometric data appear to be normally distributed. Therefore, designing for an average person may eliminate many people slightly larger or smaller.

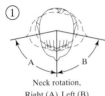

Neck rotation,
Right (A), Left (B)

	Range of motion (degrees)			
	Males		Females	
Joint movement	5th percentile	95th percentile	5th percentile	95th percentile
Neck, rotation right	73.3	99.6	74.9	108.8
Neck, rotation left	74.3	99.1	72.2	109.0

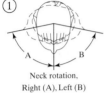

Neck extension (A)
Flexion (B)

	Range of motion (degrees)			
	Males		Females	
Joint movement	5th percentile	95th percentile	5th percentile	95th percentile
Neck, flexion	34.5	71.0	46.0	84.4
Neck, extension	65.4	103.0	64.9	103.0

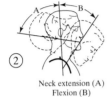

Neck lateral bend
Right (A), Left (B)

	Range of motion (degrees)			
	Males		Females	
Joint movement	5th percentile	95th percentile	5th percentile	95th percentile
Neck, lateral right	34.9	63.5	37.0	63.2
Neck, lateral left	35.5	63.5	29.1	77.2

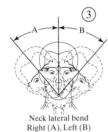

Horizontal adduction (A)
Horizontal adduction (B)

	Range of motion (degrees)			
	Males		Females	
Joint movement	5th percentile	95th percentile	5th percentile	95th percentile
Shoulder, abduction	173.2	188.7	172.6	192.9

Shoulder rotation,
Lateral (A), Medial (B)

	Range of motion (degrees)			
	Males		Females	
Joint movement	5th percentile	95th percentile	5th percentile	95th percentile
Shoulder, rotation lat	46.3	96.7	53.8	85.8
Shoulder, rotation med	90.5	126.6	95.8	130.9

FIGURE 11.4

Range-of-motion.
(Reprinted from *Human Factors Design Guide* by Wagner, Snyder, and Duncanson. Courtesy of the Federal Aviation Administration, U.S. Department of Commerce.)

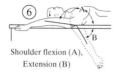

Shoulder flexion (A),
Extension (B)

	Range of motion (degrees)			
	Males		Females	
Joint movement	5th percentile	95th percentile	5th percentile	95th percentile
Shoulder, flexion	164.4	210.9	152.0	217.0
Shoulder, extension	39.6	83.3	33.7	87.9

Elbow flexion (B),
Extension (A)

	Range of motion (degrees)			
	Males		Females	
Joint movement	5th percentile	95th percentile	5th percentile	95th percentile
Elbow, flexion	140.5	159.0	144.9	165.9

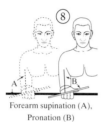

Forearm supination (A),
Pronation (B)

	Range of motion (degrees)			
	Males		Females	
Joint movement	5th percentile	95th percentile	5th percentile	95th percentile
Forearm, pronation	78.2	116.1	82.3	118.9
Forearm, supination	83.4	125.8	90.4	139.5

Wrist ulnar bend (A),
Radial bend (B)

	Range of motion (degrees)			
	Males		Females	
Joint movement	5th percentile	95th percentile	5th percentile	95th percentile
Wrist, radial	16.9	36.7	16.1	36.1
Wrist, ulnar	18.6	47.9	21.5	43.0

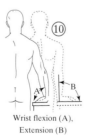

Wrist flexion (A),
Extension (B)

	Range of motion (degrees)			
	Males		Females	
Joint movement	5th percentile	95th percentile	5th percentile	95th percentile
Wrist, flexion	61.5	94.8	68.3	98.1
Wrist, extension	40.1	78.0	42.3	74.7

FIGURE 11.5

Range-of-motion continued.
(Reprinted from *Human Factors Design Guide* by Wagner, Snyder, and Duncanson. Courtesy of the Federal Aviation Administration, U.S. Department of Commerce.)

Hip flexion

| Joint movement | Range of motion (degrees) | | | |
| | Males | | Females | |
	5th percentile	95th percentile	5th percentile	95th percentile
Hip, flexion	116.5	148.0	118.5	145.0

Hip adduction (A), abduction (B)

| Joint movement | Range of motion (degrees) | | | |
| | Males | | Females | |
	5th percentile	95th percentile	5th percentile	95th percentile
Hip, abduction	26.8	53.5	27.2	55.9

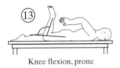

Knee flexion, prone

| Joint movement | Range of motion (degrees) | | | |
| | Males | | Females | |
	5th percentile	95th percentile	5th percentile	95th percentile
Knee, flexion	118.4	145.6	125.2	145.2

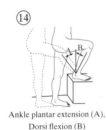

Ankle plantar extension (A), Dorsi flexion (B)

| Joint movement | Range of motion (degrees) | | | |
| | Males | | Females | |
	5th percentile	95th percentile	5th percentile	95th percentile
Ankle, planar	36.1	79.6	44.2	91.1
Ankle, dorsi	8.1	19.9	6.9	17.4

FIGURE 11.6

Range-of-motion continued.
(Reprinted from *Human Factors Design Guide* by Wagner, Snyder, and Duncanson. Courtesy of the Federal Aviation Administration, U.S. Department of Commerce.)

TABLE 11.2 Mean Value Energy Expenditure per Day for Various Occupations (adapted from Sanders and McCormick, 1993)

Occupation	Energy Expended (kcal/day)
Construction worker (male)	3,000
Steelworker (male)	3,280
Coal miner (male)	3,660
Housewives	2,090
University males/females	2,930/2,290

Hand- and foot-operated control devices and their operational characteristics and control functions; ● recommended; ○ not suitable. (*Handbook of Human Factors*; G. Salvendy (ed.); Copyright © 1987. This material is used by permission of John Wiley & Sons, Inc.)

Path of C. motion	Control	Dimension [mm]	Force F [N] Moment M [Nm]		2 positions	>2 positions	Continuous adjustment	Precise adjustment	Quick adjustment	Large force application	Tactile feedback	Setting visible	Accidental actuation
			D	**M**									
Turning movement	Handwheel	D : 160–800 d : 30–40	160–200 mm 200–250 mm	2–40 mm 4–60 mm	◐	◐	●	●	◐	●	○	○	◐
	Crank	Hand (Finger) D : <250 (<100) l : 100 (30) d : 32 (16)	**R** <100 mm 100–250 mm	0,6–3 Nm 5–14 Nm	◐	◐	●	●	●	◐	◐	◐	○
	Rotary knob	Hand (Finger) D : 25–100(15–25) h : > 20 (>15)	15–25 mm 25–100 mm	0,02–0,05 Nm 0,3–0,7 Nm	◐	◐	●	●	◐	○	○	○	◐
	Rotary selector switch	l : 30–70 h : >20 b : 10–25	30 mm 30–70 mm	0,1–0,3 Nm 0,3–0,6 Nm	◐	◔	◔	◔	◔	◐	●	●	◐
	Thumbwheel	b : > 8	0,4–5 N		◐	◐	●	●	●	○	○	○	◐
	Rollball	D : 60–120	0,4–5 N		○	○	●	●	◔	○	○	○	◐

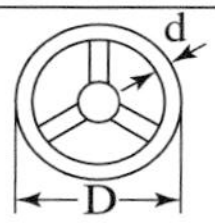

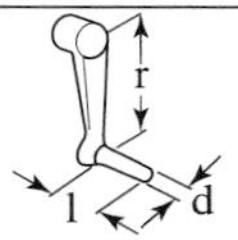

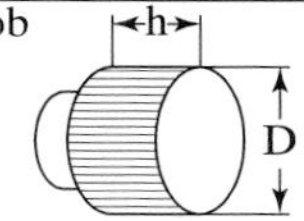

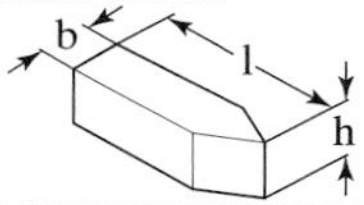

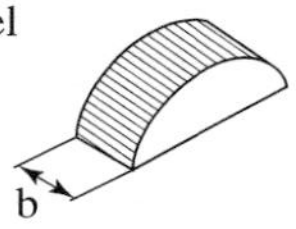

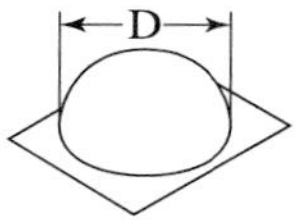

FIGURE 11.7

Path of C. motion	Control	Dimension [mm]	Force F [N] Moment M [Nm]	2 positions	>2 positions	Continuous adjustment	Precise adjustment	Quick adjustment	Large force application	Tactile feedback	Setting visible	Accidental actuation
Swiveling movement	Lever 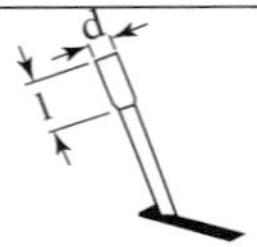	d : 30–40 l : 100–120	10–200 N	●	●	●	◔	◐	●	◐	◐	○
	Joystick 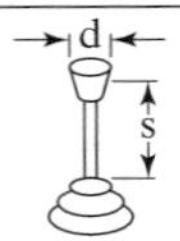	s : 20–150 d : 10–20	5–50 N	●	●	●	●	◐	◔	◐	◐	○
	Toggle switch 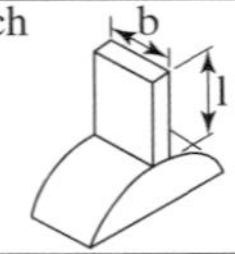	b : > 10 l : > 15	2–10 N	●	◐	○	○	●	●	●	●	○
	Rocker switch 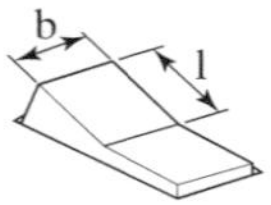	b : > 10 l : > 15	2–8 N	●	○	○	○	●	○	●	●	◐
	Rotary disk 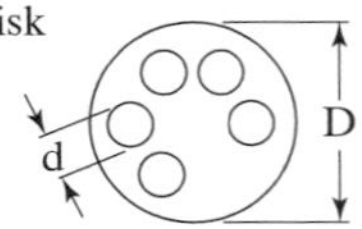	d : 12–15 D : 50–80	1–7 N	●	◔	◔	○	◔	○	○	○	◐
	Pedal 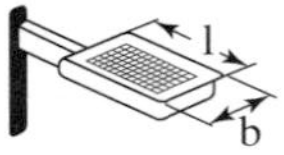	b : 50–100 l : 200–300 l : 50–100 (Forefoot)	Sitting : 16–100 N Standing : 80–250 N	◔	◔	◐	◐	●	●	◐	○	○

FIGURE 11.8

Hand- and foot-operated control devices and their operational characteristics and control functions: ● recommended; ○ not suitable. (*Handbook of Human Factors*; G. Salvendy (ed.); Copyright © 1987. This material is used by permission of John Wiley & Sons, Inc.)

Path of C. motion: Linear movement

Control	Dimension (mm)	Force F [N] Moment M [Nm]	2 positions	>2 positions	Continuous adjustment	Precise adjustment	Quick adjustment	Large force application	Tactile feedback	Setting visible	Accidental actuation
Handle (Slide) 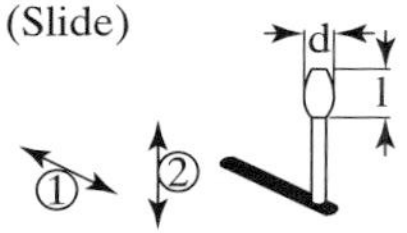	d : 30–40 \ l : 100–120	F_1 : 10–200 N \ F_2 : 7–140 N	●	◑	●	◑	◑	●	◑	◑	○
D-Handle 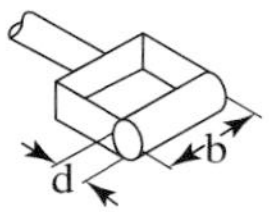	d : 30–40 \ b : 110–130	10–200 N	●	●	●	●	●	●	◔	◔	○
Push button 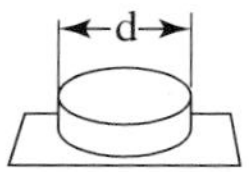	Finger : d > 15 \ Hand : d > 50 \ Foot : d > 50	Finger : F = 1–8 N \ Hand : F = 4–16 N \ Foot : F = 15–90 N	●	○	○	○	●	◑ ●	○	○	●[b] ○
Slide 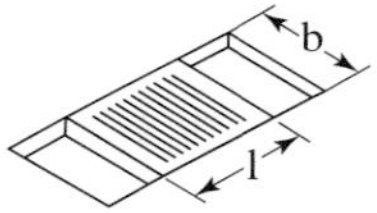	l : > 15 \ b : > 15	1–5 N (Touch grip)	●	●	●	●	●	○	○	●	●
Slide 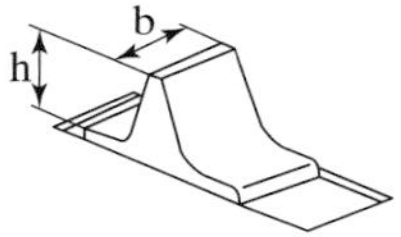	b : > 10 \ h : > 15	1–10 N (Thumb-finger grip)	●	●	●	●	●	●	○	●	●
Sensor key 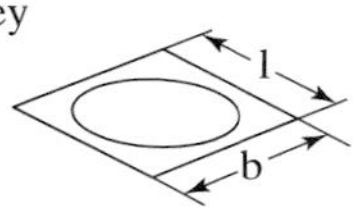	l : > 14 \ b : > 14		●	○	○	○	●	○	○	○	◑

[b] Recessed installation of the control

FIGURE 11.9

Hand- and foot-operated control devices and their operational characteristics and control functions; ● recommended; ○ not suitable. (*Handbook of Human Factors*; G. Salvendy (ed.); Copyright © 1987. This material is used by permission of John Wiley & Sons, Inc.)

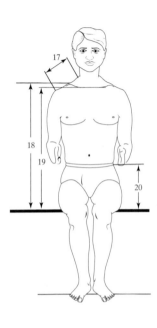

17 Shoulder length. The surface distance along the top of the shoulder from the junction of the neck and shoulder to the point of the shoulder (acromion).

	Sample		1st	5th	Percentiles 50th	95th	99th
A	Men	cm	12.4	13.3	15.0	16.9	17.7
		(in)	(4.9)	(5.3)	(5.9)	(6.7)	(7.0)
B	Women	cm	12.0	12.7	14.5	16.2	17.1
		(in)	(4.7)	(5.0)	(5.7)	(6.4)	(6.7)

18 Mid-shoulder height, sitting. The vertical distance from the sitting surface of the shoulder halfway between the neck and the point of the shoulder, measured with the subject sitting.

	Sample		1st	5th	Percentiles 50th	95th	99th
A	Men	cm	56.3	58.3	63.0	67.7	69.4
		(in)	(22.2)	(23.0)	(24.9)	(26.7)	(27.3)
B	Women	cm	52.3	53.9	58.4	63.1	64.7
		(in)	(20.6)	(21.2)	(23.0)	(24.8)	(25.5)

19 Trunk (suprasternale) height, sitting. The vertical distance from the sitting surface to the lowest point of the notch in the upper edge of the breast bone (suprasternale), measured with the subject sitting.

	Sample		1st	5th	Percentiles 50th	95th	99th
A	Men	cm	53.1	55.2	59.6	64.2	65.9
		(in)	(20.9)	(21.7)	(23.5)	(25.3)	(25.9)
B	Women	cm	49.8	51.1	55.3	59.6	61.2
		(in)	(19.6)	(20.1)	(21.8)	(23.5)	(24.1)

20 Waist height, sitting. The vertical distance from the sitting surface to the level of the waist (natural indentation), measured with the subject sitting.

	Sample		1st	5th	Percentiles 50th	95th	99th
A	Men	cm	24.8	26.0	28.7	31.5	32.9
		(in)	(9.8)	(10.2)	(11.3)	(12.4)	(13.0)
B	Women	cm	22.8	24.4	28.0	31.5	32.7
		(in)	(9.0)	(9.6)	(11.0)	(12.4)	(12.9)

FIGURE 11.10

Static human physical characteristics.
(Reprinted from *Human Factors Design Guide* by Wagner, Snyder, and Duncanson. Courtesy of the Federal Aviation Administration, U.S. Department of Commerce.)

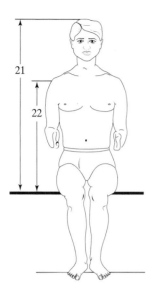

21   **Sitting height.** The vertical distance from the sitting
surface to the top of the head, measured with the
subject sitting.

	Sample		1st	5th	Percentiles 50th	95th	99th
A	Men	cm	82.8	85.5	91.4	97.2	99.1
		(in)	(32.6)	(33.7)	(36.0)	(38.3)	(39.0)
B	Women	cm	77.5	79.5	85.1	91.0	93.3
		(in)	(30.5)	(31.3)	(33.5)	(35.8)	(36.7)

22   **Shoulder (acromiale) height, sitting.** The vertical
distance from the sitting surface to the point of the
shoulder (acromion), measured with the subject
sitting.

	Sample		1st	5th	Percentiles 50th	95th	99th
A	Men	cm	129.9	134.2	144.2	154.6	158.4
		(in)	(51.1)	(52.8)	(56.8)	(60.1)	(62.4)
B	Women	cm	120.4	123.9	133.3	143.7	147.5
		(in)	(47.4)	(48.8)	(52.5)	(56.6)	(58.1)

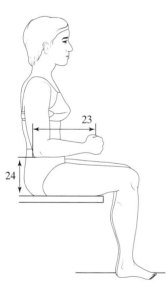

23   **Elbow-grip length.** The horizontal distance from the
back of the elbow to the center of the clenched fist.

	Sample		1st	5th	Percentiles 50th	95th	99th
A	Men	cm	32.3	33.2	35.9	39.1	40.3
		(in)	(12.7)	(13.1)	(14.1)	(15.4)	(15.9)
B	Women	cm	28.9	30.0	32.8	35.8	37.2
		(in)	(11.4)	(11.8)	(12.9)	(14.1)	(14.7)

24   **Elbow rest height.** The vertical distance from the
sitting surface to the bottom of the tip of the elbow,
measured with the subject sitting and the forearm
held horizontally.

	Sample		1st	5th	Percentiles 50th	95th	99th
A	Men	cm	16.8	18.4	23.2	27.4	29.2
		(in)	(6.6)	(7.2)	(9.1)	(10.8)	(11.5)
B	Women	cm	15.8	17.6	22.1	26.4	28.2
		(in)	(6.2)	(6.9)	(8.7)	(10.4)	(11.1)

FIGURE 11.11

Static human physical characteristics.
(Reprinted from *Human Factors Design Guide* by Wagner, Snyder, and Duncanson. Courtesy of the
Federal Aviation Administration, U.S. Department of Commerce.)

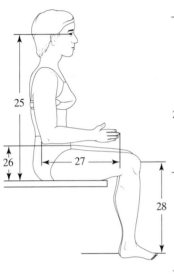

25 **Eye height, sitting.** The vertical distance from the sitting surface to the outer corner of the eye (ectocanthus), measured with the subject sitting.

Sample		1st	5th	Percentiles 50th	95th	99th
A Men	cm	71.2	73.5	79.2	84.8	86.9
	(in)	(28.0)	(28.9)	(31.2)	(33.4)	(34.2)
B Women	cm	66.4	68.5	73.8	79.4	81.6
	(in)	(26.1)	(30.0)	(29.1)	(31.2)	(32.1)

26 **Thigh clearance.** The vertical distance from the sitting surface to the highest point of the thigh, measured with the subject sitting.

Sample		1st	5th	Percentiles 50th	95th	99th
A Men	cm	14.1	14.9	16.8	19.0	20.1
	(in)	(5.6)	(5.9)	(6.6)	(7.5)	(7.9)
B Women	cm	13.4	14.0	1.8	18.0	19.0
	(in)	(5.3)	(5.5)	(6.2)	(7.1)	(7.5)

27 **Elbow-fingertip length.** The horizontal distance from the back of the elbow to the tip of the middle finger, with the hand extended.

Sample		1st	5th	Percentiles 50th	95th	99th
A Men	cm	43.4	44.8	48.3	52.4	54.2
	(in)	(17.1)	(17.6)	(19.2)	(20.6)	(21.3)
B Women	cm	39.1	40.6	44.2	48.3	49.8
	(in)	(15.4)	(16.0)	(17.4)	(19.0)	(19.6)

28 **Knee height, sitting.** The vertical distance from the footrest surface to the top of the knee, measured with the subject sitting.

Sample		1st	5th	Percentiles 50th	95th	99th
A Men	cm	49.7	51.4	55.8	60.6	62.3
	(in)	(19.6)	(20.2)	(22.0)	(23.9)	(24.5)
B Women	cm	45.4	47.4	49.8	56.0	57.8
	(in)	(17.9)	(18.7)	(20.2)	(22.0)	(22.8)

FIGURE 11.12

Static human physical characteristics.
(Reprinted from *Human Factors Design Guide* by Wagner, Snyder, and Duncanson. Courtesy of the Federal Aviation Administration, U.S. Department of Commerce.)

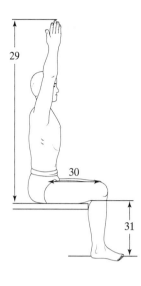

29   **Vertical reach, sitting.** The vertical distance from the sitting surface to the tip of the middle finger, measured with the subject sitting and the arm, hand, and fingers extended vertically.

	Sample		1st	5th	Percentiles 50th	95th	99th
A	Men	cm	129.3	133.8	143.3	153.2	156.7
		(in)	(50.1)	(52.7)	(56.4)	(60.3)	(61.7)
B	Women	cm	119.7	123.3	132.7	141.8	145.4
		(in)	(47.1)	(48.5)	(52.2)	(55.8)	(57.2)

30   **Abdominal depth, sitting.** The depth of the abdomen, with the subject sitting.

	Sample		1st	5th	Percentiles 50th	95th	99th
A	Men	cm	18.6	19.9	23.6	29.1	31.4
		(in)	(7.3)	(7.8)	(9.3)	(11.5)	(12.4)
B	Women	cm	17.3	18.5	21.9	27.1	29.5
		(in)	(6.1)	(7.3)	(8.6)	(10.7)	(11.6)

31   **Popliteal height, sitting.** The vertical distance from the footrest surface to the underside of the lower leg, measured with the subject sitting.

	Sample		1st	5th	Percentiles 50th	95th	99th
A	Men	cm	37.8	39.5	43.3	47.6	49.5
		(in)	(14.9)	(15.6)	(17.1)	(18.7)	(19.5)
B	Women	cm	33.7	35.1	38.9	42.9	44.6
		(in)	(13.3)	(13.8)	(15.3)	(16.9)	(17.6)

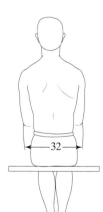

32   **Forearm-forearm breadth, sitting.** The horizontal distance across the body between the outer surfaces of the forearms, measured with the forearms flexed and held against the body.

	Sample		1st	5th	Percentiles 50th	95th	99th
A	Men	cm	45.1	47.79	54.5	62.1	65.3
		(in)	(17.8)	(18.8)	(21.5)	(24.5)	(25.7)
B	Women	cm	39.4	41.5	46.7	52.8	56.0
		(in)	(15.5)	(16.3)	(18.4)	(20.8)	(22.1)

FIGURE 11.13

Static human physical characteristics.
(Reprinted from *Human Factors Design Guide* by Wagner, Snyder, and Duncanson. Courtesy of the Federal Aviation Administration, U.S. Department of Commerce.)

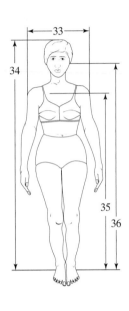

33 **Shoulder (bideltoid) breadth.** The horizontal distance across the upper arms between the maximum bulges of the deltoid muscles; the arms are hanging and relaxed.

Sample			1st	5th	Percentiles 50th	95th	99th
A	Men	cm	43.4	45.0	49.1	53.5	55.2
		(in)	(17.1)	(17.7)	(19.3)	(21.1)	(21.7)
B	Women	cm	38.0	39.7	43.1	47.2	49.2
		(in)	(15.0)	(15.6)	(17.0)	(18.6)	(19.4)

34 **Stature.** The vertical distance from the floor to the top of the head.

Sample			1st	5th	Percentiles 50th	95th	99th
A	Men	cm	160.3	164.7	175.5	186.7	190.9
		(in)	(63.1)	(64.8)	(69.1)	(73.5)	(75.2)
B	Women	cm	148.3	152.8	162.7	173.7	178.0
		(in)	(58.4)	(60.2)	(64.1)	(68.4)	(70.1)

35 **Suprasternale height.** The vertical distance from the floor to the lowest point of the notch in the upper edge of the breast bone (suprasternale).

Sample			1st	5th	Percentiles 50th	95th	99th
A	Men	cm	130.2	134.3	143.7	153.7	157.5
		(in)	(51.3)	(52.9)	(56.6)	(60.5)	(62.0)
B	Women	cm	120.7	124.1	132.9	142.5	146.4
		(in)	(47.5)	(48.9)	(52.3)	(56.1)	(57.6)

36 **Tragion height, standing.** The vertical distance from the floor to the tragion, the cartilaginous notch at the front of the ear.

Sample			1st	5th	Percentiles 50th	95th	99th
A	Men	cm	147.4	151.9	162.4	173.4	177.5
		(in)	(58.0)	(59.8)	(63.9)	(68.3)	(69.9)
B	Women	cm	136.3	140.7	150.4	161.2	165.4
		(in)	(53.7)	(55.4)	(59.2)	(63.5)	(65.1)

FIGURE 11.14

Static human physical characteristics.
(Reprinted from *Human Factors Design Guide* by Wagner, Snyder, and Duncanson. Courtesy of the Federal Aviation Administration, U.S. Department of Commerce.)

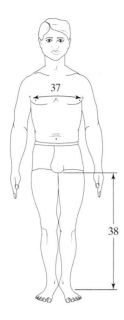

37   Chest (bust) circumference. The circumference of the torso measured at the level of the nipples.

Sample			1st	5th	**Percentiles** 50th	95th	99th
A	Men	cm	84.5	88.6	98.7	111.3	116.8
		(in)	(33.3)	(34.9)	(38.9)	(43.8)	(50.0)
B	Women	cm	78.1	81.4	90.1	102.2	107.7
		(in)	(30.8)	(32.1)	(35.5)	(40.2)	(42.4)

38   Crotch height. The vertical distance from the floor to the midpoint of the crotch.

Sample			1st	5th	**Percentiles** 50th	95th	99th
A	Men	cm	73.2	76.4	83.5	91.6	94.6
		(in)	(28.8)	(30.1)	(32.9)	(36.1)	(37.2)
B	Women	cm	67.0	70.0	77.0	84.6	88.1
		(in)	(26.4)	(27.6)	(30.3)	(33.3)	(34.7)

39   Waist circumference (natural indentation). The horizontal circumference of the torso at the level of the natural indentation of the waist.

Sample			1st	5th	**Percentiles** 50th	95th	99th
A	Men	cm	69.9	73.0	83.4	97.1	102.9
		(in)	(27.5)	(28.7)	(32.8)	(38.2)	(40.5)
B	Women	cm	60.7	63.7	71.7	84.3	91.0
		(in)	(23.9)	(25.1)	(28.2)	(33.2)	(35.8)

40   Waist circumference (omphalion). The horizontal circumference of the torso at the level of the navel (omphalion).

Sample			1st	5th	**Percentiles** 50th	95th	99th
A	Men	cm	70.0	73.3	85.6	101.6	107.7
		(in)	(27.6)	(28.9)	(33.7)	(40.0)	(42.4)
B	Women	cm	64.4	67.6	78.1	94.6	102.6
		(in)	(25.4)	(26.6)	(30.8)	(37.2)	(40.4)

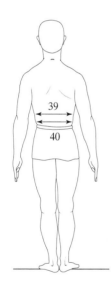

FIGURE 11.15

Static human physical characteristics.
(Reprinted from *Human Factors Design Guide* by Wagner, Snyder, and Duncanson. Courtesy of the Federal Aviation Administration, U.S. Department of Commerce.)

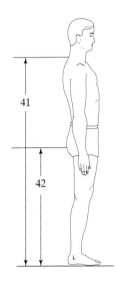

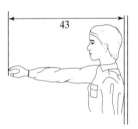

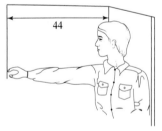

41 Cervicale height. The vertical distance from the floor to the cervicale, the tip of the spine of the seventh cervical vertebra at the base of the neck.

	Sample		Percentiles				
			1st	5th	50th	95th	99th
A	Men	cm	137.4	141.8	151.8	162.4	166.1
		(in)	(54.1)	(55.8)	(59.8)	(63.9)	(65.4)
B	Women	cm	127.3	131.4	140.6	150.8	154.8
		(in)	(50.1)	(51.7)	(55.4)	(59.4)	(60.9)

42 Buttock height. The vertical distance from the floor to the maximum posterior protrusion of the buttock.

	Sample		Percentiles				
			1st	5th	50th	95th	99th
A	Men	cm	78.4	81.5	88.5	96.9	100.5
		(in)	(30.9)	(32.1)	(34.8)	(38.1)	(39.6)
B	Women	cm	73.9	76.7	83.7	91.5	94.9
		(in)	(29.1)	(30.2)	(33.0)	(36.0)	(37.4)

43 Functional (thumb-tip) reach. The horizontal distance from the wall to the tip of the thumb, measured with the subject's shoulders against the wall, the arm extended forward, and the index finger touching the tip of the thumb.

	Sample		Percentiles				
			1st	5th	50th	95th	99th
A	Men	cm	72.0	73.9	80.0	86.7	89.7
		(in)	(28.4)	(29.1)	(31.5)	(34.1)	(35.3)
B	Women	cm	65.8	67.7	73.4	79.7	82.4
		(in)	(25.9)	(26.7)	(28.9)	(31.4)	(32.4)

44 Functional (thumb-tip) reach, extended. Measured similarly to functional (thumb-tip) reach, except that the right shoulder is extended forward as far as possible, while the left shoulder is kept pressed firmly against the wall.

	Sample		Percentiles				
			1st	5th	50th	95th	99th
A	Men	cm	77.9	80.5	87.3	94.2	97.7
		(in)	(30.0)	(31.7)	(34.4)	(37.1)	(38.5)
B	Women	cm	71.2	73.5	79.6	86.2	89.0
		(in)	(28.0)	(28.9)	(31.3)	(33.9)	(35.0)

FIGURE 11.16

Static human physical characteristics.
(Reprinted from *Human Factors Design Guide* by Wagner, Snyder, and Duncanson. Courtesy of the Federal Aviation Administration, U.S. Department of Commerce.)

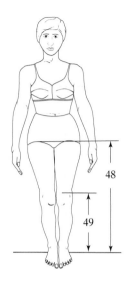

48 Hip (trochanteric) height. The vertical distance from the floor to the level of the maximum posterior protrusion of the greater trochanter of the femur (trochanterion).

Sample			1st	5th	50th	95th	99th
					Percentiles		
A	Men	cm	82.1	85.3	92.7	100.9	104.0
		(in)	(32.3)	(33.6)	(36.5)	(39.7)	(40.9)
B	Women	cm	76.1	78.9	86.0	93.8	97.5
		(in)	(30.0)	(31.1)	(33.9)	(36.9)	(38.4)

49 Knee height, midpatella. The vertical distance from the footrest surface to the top of the knee, measured with the subject sitting.

Sample			1st	5th	50th	95th	99th
					Percentiles		
A	Men	cm	44.3	46.1	50.4	55.2	56.8
		(in)	(17.4)	(18.2)	(19.8)	(21.7)	(22.4)
B	Women	cm	39.9	41.7	45.8	50.3	52.3
		(in)	(15.7)	(16.4)	(18.0)	(19.8)	(20.6)

FIGURE 11.17

Static human physical characteristics.
(Reprinted from *Human Factors Design Guide* by Wagner, Snyder, and Duncanson. Courtesy of the Federal Aviation Administration, U.S. Department of Commerce.)

One strategy is to design for the **extreme case**, that is, design to fit the smallest or largest person. For example, if we want an access doorway to be wide enough to handle the "largest" person, we would consider the shoulder width of the 95th-percentile male. That size would accommodate 95 percent of all adult males, a higher percentage of most females, and almost 100 percent of all children. If, in a different case, we need to determine the arm reach distance for a control, we would want the "smallest" arm length and therefore consider the 5th-percentile female.

A **close fit** may be required in some cases, however. A close fit is when we try to match all our differently sized customers. Clothing manufacturers accommodate a close fit by producing a collection of sizes to obtain more or less "close fits" for all of their customers. It is somewhat expensive, however, considering the costs of producing and maintaining different inventories.

Finally, we may need to design for **adjustability**. How would we design an automobile seat, for example? If we used the 95th-percentile male it would not fit most women or adolescent drivers. In this case the driver needs to be in contact with the steering wheel and foot controls. Therefore, for cases such as these we design for adjustability. Other examples of design for adjustability include office chairs, microphone stands, stereo headphones, and bicycle seat posts.

Sunday, May 11, 2003

Designers have moms on the mind

By Anita Lienert / Special to The Detroit News

DEARBORN—Early in the development of the 2004 Mercury Monterey minivan, Jared Glaspell's supervisor at Ford Motor Co. suggested that he strap on an "Empathy Belly" and don a frumpy denim smock designed to make him look and feel like a pregnant woman.

The request came as the 26-year-old Glaspell and 15 other, mostly male, engineers were standing around a clay model of the minivan, searching for ways to improve the vehicle for female consumers.

The $900 patented pregnancy simulator—one of two owned by Ford—includes breasts, a large water-filled "belly," a tight corset that restricts breathing and two 7-pound metal balls that emulate fetal movement.

"There were 'aha' moments," Glaspell said of wearing the suit. "I became more aware of where the instrument panel controls were placed. I wanted more grab handles to get in and out of the vehicle. Other guys wore it and said they began to appreciate how much your body can change when you are pregnant. But after about an hour in the pregnancy suit, we'd all be begging for mercy."

The contraption highlights a growing trend in the auto industry of creating cars and trucks not only aimed at women, but targeted specifically at making pregnant women—a once-overlooked segment of the market—more comfortable behind the wheel.

Birthways Inc. of Vashon Island, Wash., created the Empathy Belly. The company claims that after about 10 minutes, wearers feel a weight gain of about 30 pounds, their breathing grows more shallow and they're tired.

The suit originally was designed as a tool for expectant fathers and adolescents in sex-education programs. At Ford, it's a critical tool used in many new product programs. The 2004 Monterey, which goes on sale in September, is the first minivan Ford has developed using lessons learned from the Empathy Belly.

Still, some industry observers question the value of taking such extreme measures to get into the minds—and bodies—of the consumer.

"Anything that connects the people engineering the cars to the people who buy the cars is good," said auto consultant James N. Hall of AutoPacific Inc. in Troy.

"But to think this device is simulating pregnancy is a male fantasy."

Even if they aren't strapping on pregnancy suits, engineers at other automakers also are focusing on the needs of pregnant drivers and passengers.

At the New York auto show last month, Gene Stefanyshyn, who's in charge of midsize cars for General Motors Corp., showed off the 2004 Malibu Maxx hatchback and described how it was designed with pregnant women in mind.

"Historically, we've designed cars around the 50th percentile male figure," Stefanyshyn said. "With the Malibu Maxx, we were thinking about pregnant women. The whole flexibility of the vehicle's interior speaks to that effort."

Opening the driver's door, Stefanyshyn pointed out features that engineers considered necessary for making a pregnant driver's life easier.

The vehicle, which goes on sale in January, boasts standard adjustable pedals, standard vertical height adjustment on the driver's seat and a standard adjustable steering wheel—all fairly unusual on a vehicle expected to start at under $20,000. But they help give a pregnant woman an almost custom-fit cockpit.

Back at Ford, Julie Dils, a 39-year-old mother of two and a prototype planning supervisor, stands in front of an early version of the Monterey pointing out pregnancy-friendly features.

The Monterey's second-row seats easily fold and tumble forward—a first for a Mercury minivan. The third-row seat folds flat—another first. An articulating arm adds length to the strap used to pull the third-row seat into the floor. Dils said that will make it easier for a woman in the final three months of pregnancy to pull the heavy seat into a deep well in the floor.

Like the Malibu Maxx, the Monterey has standard adjustable pedals, which make it easier for a pregnant woman to find a comfortable spot behind the wheel.

The Monterey will be available with a power liftgate, too. Dils said management was planning to drop the option until the mothers on the Monterey team lobbied hard for its inclusion. They argued that a power liftgate is crucial for a mother loaded down with groceries and kids who is trying to access the rear of the minivan.

Dils is a believer in the pregnancy suit. But she said the 40 working mothers on the Monterey team—mainly engineers—probably did more to ensure that the minivan would be friendly to pregnant women.

"Wearing the suit is an enlightenment," the eight-months-pregnant Dils said, holding her 6-year-old daughter, Madison. "The men noticed so many things a normal male engineer wouldn't notice. But to make it absolutely realistic, they need to sleep in it." *You can reach Anita Lienert at conseye@aol.com.*

Historically design engineers have used anthropometric data of the 50 percentile male to design the interior of an automobile. To satisfy the needs of the adult female market, however, male design engineers are now using an "empathy belly" to simulate the human factors associated with pregnancy. (By permission of The Detroit News.)

11.5 WORKSPACE CONSIDERATIONS

When we design larger machines and systems such as processing or manufacturing facilities, we should consider the effects of climate, illumination, noise, and motion on worker performance, safety, and job satisfaction.

Climate. The design of a workspace should take into consideration the physical exertion levels of its occupants in relation to acceptable temperature and humidity ranges. In addition, thought should be given to airborne particulate filtration, odor control, and ventilation of harmful vapors.

Illumination. Being able to see equipment displays and actuate machine controls is directly linked to how well the workspace is illuminated. Important factors include the intensity of the light, the color content, glare, and reflection.

Noise. Excessive noise levels can lead to hearing loss as well as impact job performance. Important factors include sound level, frequency, duration, and the fluctuations in level or frequency.

Motion. The ability to operate machinery may depend on whether the operator is subjected to vibration or acceleration. Depending on the frequency and intensity, workers subjected to whole-body vibration may feel discomfort. In addition, humans are affected by excessive accelerations with respect to head motion in the "footward," forward, backward, and lateral directions (Sanders and McCormick, 1993). Symptoms include discomfort, disorientation, and motion sickness.

11.6 SUMMARY

- Human factors affect the design and operation of tools, machines, systems, tasks, jobs, and environments.

- The human-machine system model provides a handy tool to consider interface interactions when designing a product.

- Visual and auditory displays require considerations of conspicuity, emphasis, legibility, intelligibility, visibility, maintainability, and standardization.

- Humans have output limitations with respect to force, torque, work, and power.

- Design for human-fit strategies include extreme fit, close fit, and adjustable fit.

REFERENCES

Kroemer, K. H. E., H. B. Kroemer, and K. E. Kroemer-Elbert. 2001. *Ergonomics: How to Design for Ease and Efficiency.* Upper Saddle River, NJ: Prentice Hall.
MIL-HDBK-759C. 1997. *Human Engineering Design Guidelines.* U.S. Department of Defense.
MIL-STD-1472F. 1999. *Design Criteria: Human Engineering.* U.S. Department of Defense.
Salvendy, G. (ed.). 1987. *Handbook of Human Factors.* New York: John Wiley & Sons.

Sanders, M. S., and E. J. McCormick. 1993. *Human Factors in Engineering and Design*, 7th ed. New York: McGraw-Hill.

Stanton, N. 1998. *Human Factors in Consumer Products*. London: Taylor & Francis.

Wagner, D., J. A. Birt, D. Snyder, and P. Duncanson. 1996. *Human Factors Design Guide*. U.S. Department of Transportation, Federal Aviation Administration, New Jersey.

Woodson, W. E. 1981. *Human Factors Design Handbook*. New York: McGraw-Hill.

KEY TERMS

Adjustability	Extreme case	Muscle strength
Anthropometrics	Human factors	Range of motion
Choice reaction time	Human-machine system	Sensory inputs
Close fit	Human senses	Simple reaction time
Energy expenditure	Kinesthetic sense	Vestibular sense
Ergonomics	Masking	

EXERCISES

Self-Test. Write the letter of the choice that best answers the question.

1. _____ Touching a hot stove and jerking away is an example of:
    a. choice reaction time
    b. simple reaction time
    c. relevance reaction time
    d. stimulus reaction time

2. _____ A subfield of human factors that deals with the measurement of the human form such as height and reach is called:
    a. biometrics
    b. anthropometrics
    c. biomechanics
    d. animatronics

3. _____ We sense joint and muscle movement by our:
    a. tactile sense
    b. sense of touch
    c. kinesthetic sense
    d. biorhythms

4. _____ Display and warning signals can be enhanced by:
    a. legibility
    b. visibility
    c. conspicuity
    d. intelligibility
    e. all of these

5. _____ Examples of design for adjustability include all of the following except:

   a. microphone stands

   b. driver car seats

   c. stereo head phones

   d. computer printers

6. _____ A term used to describe the abilities, limitations, and other physiological or behavioral characteristics of humans that affect design and operation of tools, machines, systems, tasks, jobs, and environments is:

   a. ergonomics

   b. human factors

   c. biofeedback

   d. both a and b

7. _____ All of the following parts are an example of a human-machine interface except:

   a. car steering wheel

   b. motorcycle handlebar

   c. TV remote

   d. flower garden

8. _____ If your main screening criterion was to prevent accidental actuation, which of the following control devices would you not consider:

   a. rollball

   b. thumbwheel

   c. slide

   d. crank

9. _____ If you were designing an elastic tool belt for both men and women, what waste circumference range would you consider:

   a. 60.7–102.9 cm

   b. 69.9–102.9 cm

   c. 69.9–91 cm

   d. 60.7–91 cm

10. _____ All of the following are examples of a close fit strategy except:

   a. shoes

   b. gloves

   c. roller skates

   d. wallet

2: b, 4: e, 6: d, 8: d, 10: d

11. Considering the five activities discussed for making a pot of coffee, give two or more specific examples of

    a. human input,

    b. decision making, and

    c. human output

12. Using terms and concepts form the human-machine model, create five design-for-maintenance recommendations.

13. What three recommendations would you make for the design of control switches in an army tank?

14. Explain the difference between a conspicuous display and a visible display.

15. Describe the difference between a legible display and an intelligible display.

16. What would you do to prevent masking of an auditory display?

17. Do we sense the balance or "feel" of a computer keyboard by the sense of touch or kinesthetic sense?

18. Use the figures and/or tables to recommend weight limitations for the total weight of a toolbox. Explain your reasons.

19. Use the figures and/or tables to recommend the maximum pull force on a hand-actuated lever. Explain your reasons.

20. Use the figures and/or tables to recommend the maximum leg force to actuate a brake pedal. Explain your reasons.

21. Use the figures and/or tables to support the claim that females are more flexible than males. Discuss your reasons.

22. Calculate the amount of work required to shovel a 1,000-pound gravel pile from the ground to a truck bed four feet higher than the ground. Compare your answer with Table 11.2.

23. Many power lawn mowers use a "lever" to control the engine speed. How do the operational characteristics from Figure 11.8 compare to your experience?

24. Discuss the operational characteristics of a pedal from Figure 11.8 with respect to controlling rocket motor thrust in a spacecraft.

25. Use the figures and/or tables to recommend the height of assembly workbench (standing position). Explain your reasons.

26. Use the figures and/or tables to recommend the height for the seat of a chair. Explain your reasons.

27. Use the figures and/or tables to recommend the width of the seat of a comfortable folding chair. Explain your reasons.

28. Use the figures and/or tables to recommend the maximum reach and height (from the floor) for a wall-mounted push button. Explain your reasons.

29. What is the purpose of an empathy belly?

CHAPTER 12

Introduction to Engineering Economics

L E A R N I N G
O B J E C T I V E S

When you have completed this chapter you will be able to

- Understand and apply simple- and compound-interest formulas
- Apply present-worth, future-worth, uniform-series, and gradient formulas
- Evaluate engineering alternatives using a variety of methods
- Describe and apply fundamental breakeven equations
- Determine the breakeven point

12.1 INTRODUCTION

Evaluating the merits of different alternatives may be as simple as comparing annual costs. For example, candidate A will cost $45 versus candidate B at $55. Usually, however, alternatives include other aspects, such as complex stream cash flows over time. For example, let's assume that we wish to purchase one of two machine tools. Machine tool A will last three years and cost $45,000 to replace. Machine B initially costs $30,000, performs as well as the other machine, but needs to be replaced every two years. Given that the interest rate is 6 percent per year, how would we evaluate the economic merits of each machine?

This problem includes time as an important factor. Intuitively we know that a dollar today can be worth more than a dollar three years from now. We call that the **time value of money** because we can put the money to work and earn more money over the next three years. So, we might be inclined to choose Machine B, which has the lower initial cost, saving us the most initial cost. We also find that the machines have different lives. Is there a way that we can evaluate these alternatives on a common basis? The short answer is yes. We can and we will solve this problem later in the chapter.

First, however, we need to develop a few evaluation tools. In the next sections, we define the basic concepts of interest, principal, compound interest, present value, future value, and uniform-series payments. Then, we will describe and apply five methods for evaluating engineering alternatives, including present worth, future worth, equivalent uniform annual worth, rate of return, and discounted payback period. Finally, we will model the revenues and costs associated with producing a product to determine the breakeven point of production.

12.2   FUNDAMENTAL CONCEPTS

Interest is the amount we pay for the use of money. For example, if we borrow an amount of money called the **principal** amount, interest is the difference between the total amount owed and the principal.

$$\text{Interest} = \text{total amount owed} - \text{principal amount} \qquad (12.1)$$

A relative measure of how much the interest is with respect to the principal is called the **interest rate**.

$$\text{Interest rate } (\%) = \frac{\text{interest}}{\text{principal}} \times 100 \qquad (12.2)$$

The length of time between interest payments or receipts is called the interest period. We often find the interest period to be on a yearly basis, such as 8 percent per year. For car loans and home mortgage loans, we often find that lenders require the interest period to be monthly, such as 0.5 percent per month.

Example

Assume that our company borrowed $50,000 for one year. At the end of the year, the company paid back $51,500. Determine the interest charged during the one-year interest period and the interest rate.

Using equation 12.1, we can substitute $50,000 as the principal and $51,500 as the total amount to obtain:

$$\text{Interest} = \text{total amount} - \text{principal}$$

$$\text{Interest} = \$51,000 - 50,000 = \$1,500$$

$$\text{Interest rate } (\%) = \frac{\text{interest}}{\text{principal}} \times 100 = \frac{1,500}{50,000}(100) = 3.0\%$$

Example

ARW, Inc. borrowed $1,500,000 for one year. The interest rate is 5 percent per year. Determine the interest to be paid and the total amount to be paid.

Rearranging equation 12.2 we obtain:

$$\text{Interest amount} = \text{interest rate (\%)} \times \text{principal}/100$$

$$= 5\% \ (\$1,500,000)/100 = \$75,000$$

Total amount owed = original loan + interest = $1,500,000 + 75,000 = $1,575,000.

Simple Interest Interest amounts can be earned as either **simple interest** or **compound interest**. Simple interest is the amount earned per interest period ignoring any amounts earned during prior interest periods. We designate the principal as P, the interest rate as i, and the interest period as n; therefore:

$$\text{Interest} = (\text{principal})(\text{number of interest periods})(\text{interest rate})$$

$$= Pni \tag{12.3}$$

Example

Assume that we borrow $10,000 for five years from our Aunt Martha at 7 percent per year simple interest, how much would we owe her at the end of five years?

$$\text{Total amount} = \text{principal} + \text{interest} = \$10,000 + Pni$$

$$= \$10,000 + \$10,000(5)(7/100)$$

$$= \$10,000 + \$3,500$$

$$= \$13,500$$

The following symbols are used to represent quantities in time value of money equations:

n = number of interest periods, for example, years, months

I = interest rate per period, for example, percent per year, percent per month

P = monetary value at the present, dollars

F = monetary value at a future time, dollars

A = series of monetary amounts, equal in value, for example, dollars/year, dollars/month

I = interest earned during an interest period

We can illustrate the complex nature of some problems by using a cash flow diagram. We use upward-pointing arrows to represent cash receipts, or inflows, and downward-pointing arrows for cash disbursements, or outflows. The length of each arrow is proportional to the amount. Along the abscissa, we show the number of interest periods as well as the interest rate per period.

Example

Illustrate the cash flows shown in the table at the left using a cash flow diagram. The interest rate is 5 percent per year.

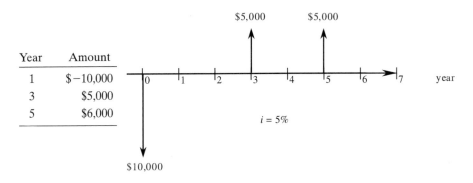

Year	Amount
1	$-10,000
3	$5,000
5	$6,000

12.3   TIME VALUE OF MONEY

Interest is compounded when it is allowed to accumulate from one period to the next. The future value F in period one and two includes the principal and accumulated interest as follows:

$$I_1 = iP$$

$$F_1 = P + I_1 = P + iP = P(1 + i)$$

$$I_2 = i(F_1) = iP(1 + i)$$

$$F_2 = P + I_1 + I_2 = P(1 + i) + iP(1 + i) = P(1 + i)(1 + i)$$
$$= P(1 + i)^2$$

For n interest periods, the compounded amount will equal

$$F = P(1 + i)^n \tag{12.4}$$

This is the law of compound interest.

12.3.1   Single-Payment Compound-Amount Factor

Dividing both sides of the equation by P, we obtain the single-payment compound-amount factor F/P, fully noted as $(F/P, i\%, n)$ shown in equation 12.5. The factor has been computed for different values of interest and number of periods in Appendix B.

$$F/P = (1 + i)^n \tag{12.5}$$

Example

What is the compound amount of a $3,000 single payment at the end of the fifth year given that it earns 8 percent per year compounded annually?

$$F = P(1 + i)^n = \$3,000(1 + .08)^5 = \$3,000(1.469328) = \$4,407.98$$

Example

Use the $(F/P, i\%, n)$ factor from the appendix to solve the previous example. We find that for 8 percent, for five years, the factor $(F/P, 8\%, 5) = 1.4693$. Therefore, $F = P(F/P) = P(F/P, 8\%, 5) = \$3,000(1.4693) = \$4,407.90$. The eight-cent difference is due to rounding.

12.3.2   Single-Payment Present-Worth Factor

The **present worth** of an amount received or paid in the future is the reciprocal of the single-payment compound-amount factor and is designated as P/F, or as $(P/F, i\%, n)$. It is called the single-payment present-worth factor, and is shown in equation 12.6. Values for the factor have been tabulated in Appendix B.

$$P/F = (F/P)^{-1} = (1 + i)^{-n} \tag{12.6}$$

Example

Determine the present worth of a machine that will be sold 15 years from now for $10,000. Assume the given interest rate is 6 percent per year.

$$P = F(P/F) = F(F/P, 6\%, 15) = \$10,000(1 + 0.06)^{-15}$$
$$= \$10,000(0.417265) = \$4,172.65$$

12.3.3   Uniform-Series Present-Worth Factor

The present worth P of a uniform series of amounts A can be illustrated on the cash flow diagram shown in Figure 12.1. The present value is calculated using equation 12.7, the uniform-series present-worth factor P/A (Blank and Tarquin, 1998). Values for the uniform-series present-worth factor $(P/A, i\%, n)$ have been tabulated in Appendix B.

$$P/A = \frac{(1 + i)^n - 1}{i(1 + i)^n} \tag{12.7}$$

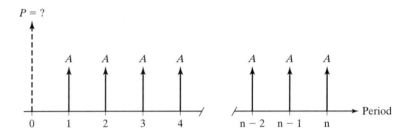

FIGURE 12.1

Uniform series present worth.

Example

Determine the present value of a machine tool that saves $500 a year for 25 years. Assume that the interest rate is 6 percent per year over the life of the tool.

$$P = A(P/A) = A\left[\frac{(1 + i)^n - 1}{i(1 + i)^n}\right] = \$500\left[\frac{(1 + 0.06)^{25} - 1}{0.06(1 + 0.06)^{25}}\right]$$

$$= \$500(12.7834) = \$6,391.70$$

or, we can use the tables to conveniently obtain:

$$P = A(P/A, 6\%, 25) = \$500[12.7834] = \$6,391.70$$

Example

What interest rate would be needed for us to invest $25,000 in year 0, to obtain a uniform series of $5,000 per year for 10 years?

Using the simpler notation we find:

$$P = A(P/A, i\%, n) = \$5,000(P/A, i\%, 10) = \$25,000$$

Rearranging the equation, we find:

$$(P/A, i\%, 10) = \$25,000/\$5,000 = 5.000$$

Looking in Appendix B along the 10-period row, we find the $P/A = 5.2161$ for 14 percent and 4.8332 for 16 percent. Interpolating, we can estimate the interest rate as:

$$\frac{5.2161 - 5.000}{5.2161 - 4.8332} = \frac{14 - i}{14 - 16}$$

$$i = 15.13\%$$

Using GOALSEEK in an Excel spreadsheet we find a more accurate value of 15.10 percent.

12.3.4   Capital-Recovery Factor

To determine a uniform series A, given a present worth P, we use the capital-recovery factor A/P, which is the reciprocal of the uniform-series present-worth factor, shown as equation 12.8. Table values for $(A/P, i\%, n)$ are given in Appendix B.

$$A/P = \frac{i(1 + i)^n}{(1 + i)^n - 1} \qquad (12.8)$$

Example

What uniform series amount per year needs to be saved by investing in a new machine that costs $6,391.70 and will last for 25 years? Assume that the interest rate is 6 percent. Use the table values for the capital-recovery factor.

$$A = P(A/P) = P(A/P, 6\%, 25) = \$6391.70(0.0782) = \$499.83$$

Note that this amount is almost the same as found in the previous example, except for rounding.

Nominal versus Effective Rate of Interest   The **nominal interest rate** r, is the interest rate per year, typically quoted in interest-related loan agreements. The **effective interest rate** i, is the interest rate per interest period. The effective interest rate i is equal to the ratio of the nominal rate r, divided by the number of interest periods per year m.

$$i = r/m \qquad (12.9)$$

Similarly, the total number of interest periods n, for the total number of years y, is

$$n = my \qquad (12.10)$$

Example

DCF Banks of America, Inc. has offered to finance the construction of our new $550,000 warehouse. The loan will have a term of 15 years with a nominal interest rate of 6 percent compounded monthly. Determine the effective interest rate, number of interest periods, and monthly loan payment.

Since the interest is compounded monthly $m = 12$ periods per year, the effective rate is

$$i = 8\%/12 = 0.5\% \text{ per month}$$

$$n = 12(15) = 180 \text{ periods}$$

$$A = P(A/P) = P(A/P, 0.5\%, 180) = P\frac{i(1 + i)^n}{(1 + i)^n - 1} = \$550,000\frac{0.005(1 + 0.005)^{180}}{(1 + 0.005)^{180} - 1}$$

$$A = \$550,000\frac{0.005(1 + 0.005)^{180}}{(1 + 0.005)^{180} - 1} = \$550,000(0.008438531) = \$4,641.19/\text{month}$$

12.3.5 Uniform-Series Compound-Amount Factor

To determine the compound amount F, given a uniform-series amount A, for n periods, at an interest rate per period i, we find the future worth of the present value P that produces the uniform series as follows. F/A is called the uniform-series compound-amount factor. The cash flow stream is shown in Figure 12.2.

$$F/A = (F/P)(P/A) = (1 + i)^n\frac{(1 + i)^n - 1}{i(1 + i)^n} \tag{12.11}$$

$$F/A = \frac{(1 + i)^n - 1}{i} \tag{12.12}$$

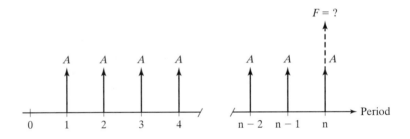

FIGURE 12.2

Uniform-series compound-amount cash flow diagram.

12.3.6 Uniform-Series Sinking-Fund Factor

Note that the reciprocal of the equation produces the uniform-series sinking-fund factor A/F shown in equation 12.13. Table values for the uniform-series compound-amount and the uniform-series sinking-fund factors are given in Appendix B.

$$A/F = \frac{i}{(1 + i)^n - 1} \tag{12.13}$$

Example

ARW, Inc. would like to set up a sinking-fund account to replace a machine tool 15 years from now. The future cost of the machine is estimated to be $45,000. Assume that the interest rate per period is 8 percent. Use Appendix B to estimate the annual deposits required.

$$A = F(A/F) = F(A/F, 8\%, 15) = \$45,000(0.0368) = \$1,656/\text{year}$$

12.3.7 Gradient-Series Factors

In some cases we find cash flows that increase or decrease in a regular fashion, as shown in Figure 12.3. Note that the uniform series has the gradient series superposed on top. To find the present value for the combined cash flows we will need to add the *P/A* component to the *G/A* component.

We can find the present worth *P*, future worth *F*, and uniform series *A* for gradient-series cash flows from the following gradient factors (Thuesen and Fabrycky, 1993):

$$P/G = \frac{(1 + i)^n - 1}{i^2(1 + i)^n} - \frac{n}{i(1 + i)^n} \tag{12.14}$$

$$F/G = \frac{(1 + i)^n - 1}{i^2} - \frac{n}{i} \tag{12.15}$$

$$A/G = \frac{1}{i} - \frac{n}{(1 + i)^n - 1} \tag{12.16}$$

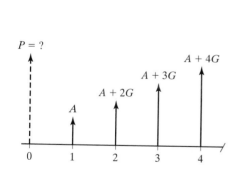
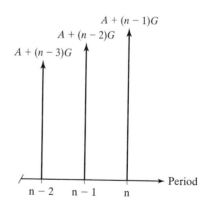

FIGURE 12.3

Uniform series *A*, with superposed gradient *G*.

Example

A new piece of equipment will require $150 worth of maintenance at the end of the first year. The costs will increase $50 to $200 for the second year, $250 for the third, and so on through year 10, at which time the equipment will be retired. Sketch a cash flow diagram and use the table values to determine the amount of money the company should put away now to cover these expenses. Assume the interest rate per period is 8 percent.

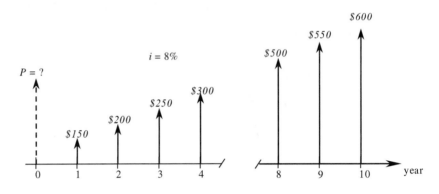

$$P = A(P/A, 8\%, 10) + G(P/G, 8\%, 10) = \$150(6.7107) + \$50(25.977) = \$2,305.46$$

Interest factor formulas are summarized in Table 12.1. Values for various interest rates and periods are given in Appendix B.

TABLE 12.1 Summary Table of Interest Factors

Name of Factor	Formula	Symbol
Single-Payment Compound-Amount	$F/P = (1 + i)^n$	$(F/P, i\%, n)$
Single-Payment Present-Worth	$P/F = (1 + i)^{-n}$	$(P/F, i\%, n)$
Uniform-Series Present-Worth	$P/A = \dfrac{(1 + i)^n - 1}{i(1 + i)^n}$	$(P/A, i\%, n)$
Uniform-Series Compound-Amount	$F/A = \dfrac{(1 + i)^n - 1}{i}$	$(F/A, i\%, n)$
Single-Payment Uniform-Series (Capital-recovery)	$A/P = \dfrac{i(1 + i)^n}{(1 + i)^n - 1}$	$(A/P, i\%, n)$
Compound-Amount Uniform-Series (Sinking-fund)	$A/F = \dfrac{i}{(1 + i)^n - 1}$	$(A/F, i\%, n)$
Uniform-Gradient Present-Worth	$P/G = \dfrac{(1 + i)^n - 1}{i^2(1 + i)^n} - \dfrac{n}{i(1 + i)^n}$	$(P/G, i\%, n)$
Uniform-Gradient Future-Worth	$F/G = \dfrac{(1 + i)^n - 1}{i^2} - \dfrac{n}{i}$	$(F/G, i\%, n)$
Uniform-Gradient Uniform-Series	$A/G = \dfrac{1}{i} - \dfrac{n}{(1 + i)^n - 1}$	$(A/G, i\%, n)$

12.4   EVALUATING ECONOMIC ALTERNATIVES

The previous sections provide us with the tools to evaluate economic alternatives having different cash flows over time. We will use one of five methods to determine whether alternatives have equivalent economic value: present-worth, future-worth, equivalent-uniform annual-worth, rate-of-return, and payback period.

12.4.1   Present-Worth Method

The present worth PW of each alternative is calculated. The alternative with the most positive PW is the best alternative. Assuming that we have a stream of cash flows CF_j at the end of year j for n years, assuming interest rate i, we find the PW as

$$PW = CF_0 + CF_1(P/F, i\%, 1)$$
$$+ CF_2(P/F, i\%, 2) + \cdots + CF_n(P/F, i\%, n) \qquad (12.17)$$
$$PW = \Sigma_j CF_j(P/F, i\%, j) \qquad (12.18)$$

Many companies will make an initial investment outlay in year 0, expecting to receive inflows over time. In such cases the present worth is usually called the net present worth or **net present value**.

Example

Our company has received a vendor quote stipulating that it will supply 5,000 parts per year for the next three years at $25,000 per year. Our manufacturing engineering department, however, estimates that if the company invests in a new machine tool for $10,000, we can produce the same 5,000 parts per year for an annual cost of $18,000 per year. Assume that the interest rate is 8 percent per year; which alternative is more economical, make or buy?

The present worth of the "buy" alternative is

$$PW_{buy} = \$-25{,}000(P/F, 8\%, 1) - \$25{,}000(P/F, 8\%, 2) - 25{,}000(P/F, 8\%, 3)$$

$$PW_{buy} = \$-25{,}000(0.9259) - 25{,}000(0.8573) - 25{,}000(0.7938) = \$-64{,}425$$

The present worth of the "make" alternative is

$$PW_{make} = -\$10{,}000 - \$18{,}000(P/A, 8\%, 3)$$

$$PW_{make} = -\$10{,}000 - \$18{,}000(2.5771) = \$-56{,}388$$

The "make" alternative would be the best choice based upon estimated present worth.

Example

Machine tool A will last three years and cost $45,000 to replace. Machine B costs $30,000, performs as well as the other machine, but need to be replaced every two years. Which machine is better assuming the interest rate is 6 percent per year?

This problem is a case of unequal lives. The *PW* method assumes that each alternative will last the same number of years. For cases such as these, we find the least common multiple of years that would produce equivalent life spans. Often, this will be the product of the two lives. For this problem the least common multiple is 2(3), or 6 years. The equivalence of the common multiple is shown in Figure 12.4.

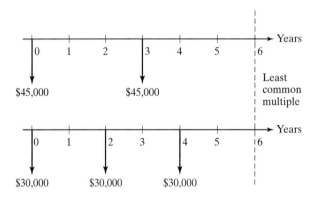

FIGURE 12.4

Least common multiple is used for alternatives with unequal lives.

Machine A

$$PW_A = \$-45{,}000 + \$-45{,}000(P/F, 6\%, 3)$$

$$PW_A = \$-45{,}000 + \$-45{,}000(0.8396) = \$-82{,}782$$

Machine B

$$PW_B = \$-30{,}000 + \$-30{,}000(P/F, 6\%, 2) + \$-30{,}000(P/F, 6\%, 4)$$

$$PW_B = \$-30{,}000 + \$-30{,}000(0.8900) + \$-30{,}000(0.7921) = \$-80{,}463$$

Therefore, machine B is the more economical choice.

12.4.2 Future-Worth Method

The **future worth** (*FW*) of each alternative is calculated. The alternative with the most positive *FW* is the best alternative. Assuming that we have a stream of cash flows CF_j at the end of year j for n years, assuming interest rate i, we find the *FW* as

$$FW = CF_0(F/P, i\%, n) + CF_1(P/F, i\%, n-1) + \cdots + CF_n \quad (12.19)$$

Therefore:

$$FW = \Sigma_j CF_j(F/P, i\%, n - j) = PW(F/P, i\%, n) \qquad (12.20)$$

We can also compute the FW by compounding the PW forward in time as

$$FW = PW(F/P, i\%, n) \qquad (12.21)$$

Example

An equipment supplier would like to offer a service contract to perform overhauls on our machine at the end of year 2 and 4 costing \$1,500 and \$3,500, respectively. However, our maintenance department says that they can keep the machine in perfect working order for \$1,500 a year for the same four years. Assume that the interest rate is 8 percent per year and use the future-worth method to determine which alternative is more economical: we repair, or they repair.

The future worth of the "they repair" alternative is

$$FW_{they} = \$-1,500(F/P, 8\%, 2) - \$3,500$$

$$FW_{they} = \$-1,500(1.1664) - \$3,500 = \$-5,250$$

The future worth of the "make" alternative is

$$FW_{we} = \$-1,500(F/A, 8\%, 4)$$

$$FW_{we} = \$-1,500(4.5061) = \$-6,759$$

Based on the future worth of each alternative, we find that the "make" decision is not economical.

12.4.3   Equivalent-Uniform Annual-Worth Method

The **equivalent-uniform annual worth** $EUAW$, of each alternative is calculated. The alternative with the most positive $EUAW$ is the best alternative. Assuming that we have cash flows CF_j at the end of year j for n years and interest rate i, we find the $EUAW$ as

$$EUAW = PW(A/P, i\%, n) = \Sigma_j CF_j(P/F, i\%, j)(A/P, i\%, n) \quad (12.22)$$

If there is a salvage value SV at the end of the nth year, it is converted to a uniform series and subtracted as follows:

$$EUAW = PW(A/P, i\%, n) - SV(A/F, i\%, n) \qquad (12.23)$$

A major advantage of the EUAW method is that alternatives that have different lives do not have to be compared using the least common multiple of the lives.

Example

A new machine initially costs $40,000 and $10,000 per year to operate. The machine has a 10-year life and $1,000 salvage value. Assume that the interest rate is 10 percent per year. As an alternative, the company can lease the machine for $7,500 per year for 10 years. Use the EUAW method to determine which alternative is most economical.

The *EUAW* of the purchase option is obtained from

$$EUAW = PW(A/P, i\%, n) - SV(A/F, i\%, n)$$
$$= \$-40{,}000(0.1627) - (\$-1{,}000)(0.0627)$$
$$EUAW = \$-6{,}508.00 + \$62.70 = \$-6445.3$$

The *EUAW* of the lease option is $\$-7{,}500$.
Therefore, the purchase option is more economical.

12.4.4 Rate-of-Return Method

The **rate of return** (ROR) of each alternative is calculated. The alternative with the greatest rate of return is the best alternative. Assuming that we have a stream of cash flows CF_j at the end of year j for n years, we find the ROR as that specific interest rate i^* such that the present value of the cash flows is zero.

$$PW = CF_0 + CF_1(P/F, i^*\%, 1) + \cdots + CF_n(P/F, i^*\%, n) = 0.0 \quad (12.24)$$

Example

The redesign of an existing product will require a $50,000 investment but will return $16,719 each year for years 1 through 5. The company has a minimum attractive rate of return of 25 percent. Use the ROR method to determine whether to complete the redesign.

To determine the *ROR* we need to find the i^* such that the $PW = 0$.

$$PW = \$-50{,}000 + \$16{,}719(P/A, i^*\%, 5) = 0.0$$

$$(P/A, i^*\%, 5) = \$50{,}000/\$16{,}719 = 2.991$$

Looking through the tables we find that $i^* = 20\%$, which is inadequate to go ahead with the redesign.

If the *P/A* factor had been a value between two tables, we would have interpolated.

12.4.5 Payback Period

The payback period is the amount of time that it takes for the cash flows to recover the initial investment. The **simple payback period** is equal to the investment divided by the annual savings as

$$SPP = \frac{\text{investment}}{\text{annual savings}} \quad (12.25)$$

While the SPP is a convenient method, it ignores the time value of money. The more preferred measure is the **discounted payback period** (DPP), which is the number of years for the discounted cash flow to equal zero. It is the same as the number of years for the net present value to break even.

$$0 = -P + \sum_{1}^{j} CF_j(P/F, i\%, j) \qquad (12.26)$$

Example

Calculate the simple payback period for a new machine tool that costs $245,000 and will save the company $95,000 per year.

$$SPP = \frac{\text{investment}}{\text{annual savings}} = \frac{\$245,000}{\$95,000/\text{year}} = 2.58 \text{ years}$$

Example

Use a spreadsheet to determine the DPP for a new product that will result in the following disbursements and receipts shown in the table at the left.

				Discounted Cash Flow		
Year	Cash Flow CF_j	Year	CFi ($)	(P/F, 8%, n)	CFi(P/F, 8%, n) ($)	Sum ($)
0	$-750,000	0	-750,000	1.00000	-750,000	-750,000
1	125,000	1	125,000	0.92593	115,741	-634,259
2	190,000	2	190,000	0.85734	162,894	-471,365
3	350,000	3	350,000	0.79383	277,841	-193,524
4	565,000	4	565,000	0.73503	415,292	221,768
5	345,000	5	345,000	0.68058	234,801	456,569
6	255,000	6	255,000	0.63017	160,693	617,263
7	125,000	7	125,000	0.58349	72,936	690,199
8	100,000	8	100,000	0.54027	54,027	744,226

We find that the cumulative discounted cash flow zeros out between the end of the third year and the end of the fourth year. Interpolating we find the DPP = 3.46 years.

12.5 BREAKEVEN ECONOMICS

We can develop a simplified mathematical model to show the relationships between the costs and revenues of manufacturing and selling a product. If we designate the revenue per unit as r, and the quantity of product sold as q, then the total revenues R is obtained from

$$R = rq \qquad (12.27)$$

The costs of manufacturing can be broken down into **variable costs** VC and **fixed costs** FC. Variable costs refer to items that depend upon the amount of product manufactured, such as raw materials and production labor. Fixed costs refer to cost elements that do not vary with production, such as administrative labor, advertising, legal, and building depreciation.

If we designate FC as the fixed costs and v as the variable costs per unit of production, we can obtain the total cost of production TC as

$$TC = FC + VC \qquad (12.28)$$

$$TC = FC + vq \qquad (12.29)$$

Profit is the difference between revenues and costs. It can be determined from

$$\text{Profit} = R - TC = rq - FC - vq \qquad (12.30)$$

The **breakeven point** q^* is that specific volume of production such that the profit is zero; that is, the revenues equal the total costs. Note that when total costs exceed revenues, a **loss** occurs.

$$\text{Breakeven profit} = rq - FC - vq = 0 \qquad (12.31)$$

$$q^* = \frac{FC}{r - v} \qquad (12.32)$$

These linear relationships are shown in Figure 12.5. Note that for production quantities less than q^*, the company suffers a loss, and that for $q > q^*$, the company earns a profit. The figure also illustrates the sensitivity of profits to fixed costs.

Example

The company receives $5 per unit sold and the variable costs to make each unit are $3.50. Determine the fixed costs such that the company will break even when making 20,000 units.

$$q^* = \frac{FC}{r - v}$$

$$20{,}000 = \frac{FC}{5 - 3.5}$$

$$FC = \$1.5/\text{unit} \ (20{,}000 \ \text{units}) = \$30{,}000$$

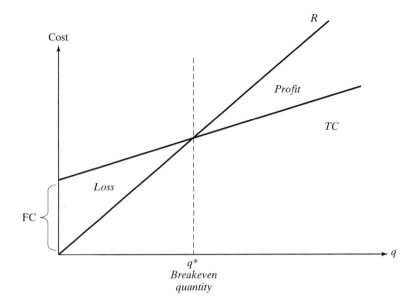

FIGURE 12.5

Breakeven point analysis for linearly related variable costs and revenues.

12.6 SUMMARY

- Engineering economic analyses need to consider the time value of money.
- Interest factor formulas and tables are useful in evaluating alternatives.
- The PW and FW methods can be used to evaluate alternatives having different lives by using the least common multiple of years.
- The EUAW method is preferred because it has the advantage of not requiring the use of the least common multiple.
- The breakeven point is that level of production (and sales) that results in a zero profit.

REFERENCES

Blank, L. T., and A. J. Tarquin. 1998. *Engineering Economy*, 4th ed. New York: McGraw-Hill.
Thuesen, G. J., and W. J. Fabrycky. 1993. *Engineering Economy*, 8th ed. Englewood Cliffs, NJ: Prentice Hall.

KEY TERMS

Breakeven point	Future worth	Profit
Compound interest	Interest	Rate of return
Discounted payback	Interest rate	Simple interest
period	Loss	Simple payback period
Effective interest rate	Net present value	Time value of money
Equivalent-uniform	Nominal interest rate	Variable costs
annual worth	Present worth	
Fixed costs	Principal	

EXERCISES

Self-Test. Write the letter of the choice that best answers the question.

1. _____ The time between interest payments or receipts is the:

    a. loan period

    b. contract term

    c. interest period

    d. collection period

2. _____ The original amount borrowed or invested is the:

    a. loan value

    b. principal

    c. payment schedule

    d. interest rate

3. _____ When interest is earned at the end of each interest period and allowed to accumulate from one period to the next, it is called:

    a. nominal interest

    b. compound interest

    c. simple interest

    d. effective interest

4. _____ The amount of time it takes for the cash flow to recover the initial investment is called:

    a. payback period

    b. recovery period

    c. investment period

    d. compound period

5. _____ The amount paid for the use of money is called:

    a. investment

    b. cost factor

    c. interest

    d. revenue

6. _____ The interest rate per interest period is called:

    a. nominal interest rate

    b. the rate of return

    c. compound interest rate

    d. effective interest rate

7. _____ All of the following methods can be used to determine whether alternatives have equivalent economic value except:

    a. profit

    b. present worth

    c. rate of return

    d. payback period

8. _____ When alternatives having different lives do not have to be compared using the least common multiple is a major advantage of:

    a. present-worth method

    b. equivalent-uniform annual-worth method

    c. rate-of-return method

    d. future-worth method

9. _____ Cost items that depend upon the amount of product manufactured, such as raw materials and production labor, are called:

    a. production costs

    b. fixed costs

    c. future costs

    d. variable costs

10. _____ The specific quantity of production such that the revenues are just enough to cover the total costs is called:

    a. breakeven point

    b. crossover point

    c. payback point

    d. profit point

2: b, 4: a, 6: a, 8: b, 10: a

11. Determine the interest due on a $5,000 loan after three years. Assume that the loan accrues simple interest at 6 percent per year. Determine the total amount to be paid.

12. Uncle George received a $1,500 interest payment after five years for a loan to his niece. Given that the interest rate was 10 percent per year, what was the principal?

13. After 10 years, what is the compound amount for a single payment amount of $500 given that the amount earned 5 percent interest per year, compounded annually.

14. An IOU is purchased for $895. It will pay $1,000 three years from the date of purchase. Determine the interest rate per period.

15. Redesign of an existing product will produce an additional $1,500,000 and $2,500,000 at the end of the next two years. What is the present value of the cash flows given that the interest rate per period is 10 percent.

16. The company is considering purchasing a new CAD system for the engineering department that will cost $150,000 to install. Given that the cost of money is 7 percent, what uniform-series amount will be required each year, for the next five years, to pay for the new system?

17. A new machine tool is expected to cost $450,000 to acquire and install. The manufacturing engineering department estimates that the machine will save $65,000 each year for the next 10 years. Determine the interest rate per period. Is the machine a good investment if the company's borrowing rate is 15 percent?

18. ARW, Inc. would like to buy a new materials-handling system for its factory. Overhauls, costing $50,000 and $75,000, will be required at the end of the fifth and tenth year of operation, respectively. The new system will save $5,000 per year for its 15-year life. What is the present value of the cash flows, given that the interest rate per period is 6 percent?

19. In 15 years a machine tool will need to be replaced at a future cost of $95,000. What annual amount should the company deposit to a sinking fund that can earn 10 percent?

20. The company would like to finance the $250,000 construction of a new assembly line. The bank is willing to provide a five-year loan at 6 percent per year, compounded monthly. What is the monthly payment amount?

21. An engineering student would like to buy a used car. He has $2,000 available to use as a down payment and he will finance the balance over four years at 12 percent per year compounded monthly. Determine the most expensive car he can buy.

22. Use the present-worth method to determine whether receiving $4,000 annually for 10 years is equivalent to receiving $5,000 annually for eight years, if the interest rate is 8 percent per year.

23. In nine years a machine tool will have a $12,488 salvage value. If $1,000 can be saved each year until then, what interest rate would it have to earn?

24. Use the present-worth method to determine which machine the company should choose, assuming the interest rate is 10 percent per year.

	Machine G	Machine H
Initial cost	$35,000	$30,000
Annual operating expenses	500	1,500
Salvage value	1000	0
Life	10	10

25. Use the future-worth method to determine which machine the company should choose, assuming the interest rate is 15 percent per year.

	Machine K	Machine L
Initial cost	$55,000	$45,000
Annual operating expenses	500	1,500
Salvage value	4,500	3,000
Life	15	15

26. Use the equivalent-uniform annual-worth method to determine which machine the company should choose, assuming the interest rate is 12 percent per year.

	Machine K	Machine L
Initial cost	$105,000	$125,000
Annual operating expenses	10,000	9,500
Salvage value	14,500	5,000
Life	25	15

27. Use the equivalent-uniform annual-worth method to determine which alternative is more economical, assuming the interest rate is 14 percent per year.

	I	J
Initial cost	$8,500,000	$50,000,000
Annual operating expenses	8,000	7,000
Salvage value	5,000	2,000
Life	5	500

28. Determine the rate of return on a machine that will cost $575,000 to purchase and install and will produce $45,000 of annual savings for the next 15 years.

29. Determine the rate of return for problem 28, but assume that the machine has 1 $100,000 salvage value at the end of the fifteenth year.

30. Determine the rate of return for the purchase of a $5,000 security that will return $6,500 at the end of 8 years.

31. Develop a spreadsheet to determine the rate of return for the following stream of cash flows.

Year	Cash Flow
0	$−750,000
1	120,000
2	185,000
3	350,000
4	575,000
5	345,000
6	245,000
7	125,000
8	95,000

32. Develop a spreadsheet to determine the discounted payback period for the following stream of cash flows.

Year	Cash Flow
0	$-550,000
1	20,000
2	75,000
3	150,000
4	275,000
5	145,000
6	65,000
7	25,000
8	15,000

33. Determine the breakeven point for a company that has monthly fixed costs of $50,000, revenue per unit of $125, and variable cost per unit of $55.

34. Assume that the fixed costs in problem 33 decrease by $10,000. What is the new breakeven point? Calculate the percent change in fixed costs and the percent change in breakeven quantity.

CHAPTER   13

Detail Design

L   E   A   R   N   I   N   G
O   B   J   E   C   T   I   V   E   S

When you have completed this chapter you will be able to

- Describe the essential inputs and outputs of the detail-design process.
- Describe concurrent engineering-team responsibilities
- Discriminate drawing types and uses
- Describe illustration graphics
- Describe different forms of written documentation
- List the topics in a design-project report
- List the contents of a project progress report
- List the steps in preparing an oral presentation
- Describe guidelines for giving an oral presentation
- Explain product documentation and management

13.1   INTRODUCTION

During the parametric design phase we will have determined most of the form of our product. We will have found optimal values for the design variables such as dimensions and tolerances, and selected materials and primary manufacturing processes. Also, we will have prepared a collection of analyses predicting the expected performance of the product and have an estimate of the customer's satisfaction. In some cases we will also have prototype test results supporting our "final" design. These inputs to the detail-design phase are shown in Figure 13.1.

During detail design we use these data to make the remaining decisions on specific details that can no longer be deferred. For example, we might find that during parametric design of a transmission a shaft diameter between 0.5 and 0.65 inch would be "optimal." Knowing that stock shaft materials are available in

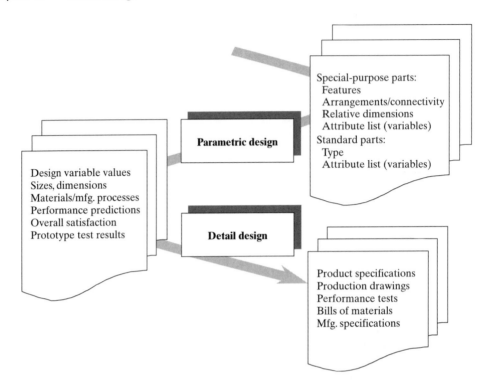

Design variable values
Sizes, dimensions
Materials/mfg. processes
Performance predictions
Overall satisfaction
Prototype test results

Parametric design

Detail design

Special-purpose parts:
    Features
    Arrangements/connectivity
    Relative dimensions
    Attribute list (variables)
Standard parts:
    Type
    Attribute list (variables)

Product specifications
Production drawings
Performance tests
Bills of materials
Mfg. specifications

FIGURE 13.1

Detail-design phase includes the generation of product specifications, production drawings, performance-test summaries, bills of materials, and manufacturing-process specifications.

$\frac{1}{2}$-inch and $\frac{5}{8}$-inch sizes, we need to determine which to specify for manufacture and assembly. Assuming that our development team is concurrently engineering, it has a purchasing representative that would investigate cost and availability of both sizes. The team would make its final decision using that information. Using the final shaft size, the design-engineering-team member would reanalyze the expected performance. A subsequent beta prototype might also be built and tested, if needed.

13.1.1 Product Realization

When we consider the overall product realization process we recognize that a number of related activities are needed before a product is finally manufactured, including: production design, manufacturing, distribution, sales, service, and disposal. Departments responsible for these activities will also be making decisions as they prepare for full production. Assuming that we have been communicating with these "downstream" departments, we should not expect any surprises.

At the completion of the detail-design phase most, if not all, of the design decisions have been made, thereby "freezing" the design of the product. The resulting detail-design documents provide the required information necessary to manufacture the product (Hales, 1993). These include production drawings (e.g., layout, detail, and assembly), bill(s) of materials, owner's manual(s), and product performance specification sheet(s).

13.1.2 Design Is Iterative and Nonlinear

We should note that the design phases have been presented as a linear sequence of decision-making activities. This was done to keep the explanations simple and concise. Engineering practice, on the other hand, is rarely linear. Owing to differences in product, industry, or management style, some design activities will be completed "out of sequence."

A large equipment manufacturer, for example, may freeze the design of a large casting because of long-lead-time requirements, even though other subassembly designs are not finished. Another manufacturer may select another manufacturing process because beta testing shows that the intended process is not capable of maintaining satisfactory tolerances.

Companies often specialize their product development processes to take advantage of unique circumstances and consequently involve nonlinear bypasses or iterations of redesign, and possibly respecification. In general, however, product development activities involve aspects of formulation, concept design, configuration design, parametric design, and detail design.

In the next sections, we will examine the types of decisions that remain to be made during the detail-design phase or later, and how they are appropriately documented so that the whole manufacturing organization will know who will be doing what and when.

13.2 MAKING DETAIL-DESIGN DECISIONS

We might ask, "When is the detail design completed?" Is the design frozen when the fabrication and assembly lines are up and running, and finished goods are being shipped to our warehouses? Before that? After that? Another question that we might also ask, "Who is responsible for making these final decisions?" Is it design engineering, or manufacturing engineering, or production?

Let's examine some of the "details" needed by some of these departments, and gain a better understanding of what the words "detail design" mean. Table 13.1 is an example of a concurrent-engineering team with seven members representing the critical departments in its company.

For companies that concurrently engineer, the answer to these questions is that the "team" is responsible for the timing and quality of these decisions. Perhaps this answer is somewhat oversimplified. But, when the team manager is a high-level officer in the company, the team will have the authority and responsibility to make things happen.

Recall that the team is composed of representatives from all the critical functional departments of the organization. Therefore, we find that in practice,

TABLE 13.1 Example Concurrent Engineering Team Membership and Responsibilities Related to the Detail Design, Manufacture, and Distribution of the Finished Product

Department Representative	Team Responsibilities
Sales/marketing	Product warranty
	Shipping
	Warehousing
	Advertising campaign
	Product literature
	Owner's manual
	Product launch
Industrial design	Product trim details
	Finish details
	Ergonomic refinements
	Product packaging
Design engineering	Detail-design performance analyses
	Preproduction prototype performance tests
	Manufacturing process specifications
	Owner manual(s)
	Layout drawing
	Detail drawings
	Assembly drawings
	Bills of materials
	Engineering change notices
	Patents, trademarks, copyrights
Industrial engineering	Facility layout/remodeling
	Material-handling equipment
	Inventory warehousing
	Product flow
	Assembly planning
Manufacturing engineering	Fixture design/fabrication
	Tool design/fabrication
	Process equipment refurbishment/adaptation
	Process equipment acquisition/installation
	Process planning
	Make or buy,
Purchasing	Vendor qualification, selection, negotiation
	Outsourcing parts/subassemblies
	Materials planning
	Raw materials
Production	Quality control
	Tooling changeover
	Worker training
	Acceptance testing

the whole organization is involved in getting the product designed, manufactured, and delivered to the customer. The idea that "a detail-design package" is prepared and forwarded to some hypothetical department is somewhat meaningless.

For simple products, such as a single-part product, the detail-design package is sometimes frozen before the product is launched. The product works as it was tested and manufactured and delivered to the customer. For more complex products, such as a home vacuum cleaner, we might discover design or manufacturing defects during initial production runs. We would change the design or manufacturing specifications to fix the problem using an engineering-change notice (ECN) form to advise company personnel. Finally, for highly complex products, such as a gas turbine electric generation facility, the detail design is completed in the field, as the facility is constructed, shaken-down, acceptance-tested, and commissioned.

Therefore, a collection of documents emerges from multiple departments throughout the company in the final weeks or months before launch. It is this final collection of documents that we might loosely call the "detail design." The design of a product does not freeze at once, but rather progressively, heterogeneously, in different departments throughout the organization. Since the design is not "thrown over the wall" a distinct transition of responsibility does not really occur. Ideally, the detail-design phase ends and the manufacturing phase begins in a seamless fashion.

In the next section we will examine how employees communicate with each other as the product design freezes.

13.3   COMMUNICATING DESIGN AND MANUFACTURING INFORMATION

As we have seen in the preceding chapters, a team will make many decisions during the design of a new product about the product's form and how it will satisfy the customer's desired function(s). Also, we recognize that the whole organization is involved in those decisions. How, then, do we keep everyone "in the loop"?

The answer is quite simple, but rather difficult to accomplish in practice—but not if we communicate to all the stakeholders often, thoroughly, and clearly. Frequent communications by e-mail, letter, phone calls, and written reports are instrumental in informing stakeholders. Keeping everyone informed helps to maintains coordination among the various departments. By communicating thoroughly and clearly we save ourselves work in the long run, by not having to spend time answering a multitude of follow-up questions or correct a lot of misunderstandings.

We next examine three principal modes of communication: (1) graphic documents, (2) written documents, and (3) oral presentations.

13.3.1   Graphic Documents

Production drawings are used to communicate a product design in sufficient detail to enable its manufacture. Production drawings are also called working drawings and include: detail drawings and assembly drawings with bill of materials or parts list.

Detail Drawings   **Detail drawings** are orthographic projection views (front, side, or top) of a part showing its geometric features drawn to scale along with full dimensions, tolerances, manufacturing-process notes, and title block (Bertoline et al., 1997). Detail drawings sometimes include section views and oblique-plane views. Sometimes a pictorial view (isometric or oblique) is shown. Detail drawings do not have a bill of materials. An example of a detail drawing is shown in Figure 13.2.

Assembly Drawings   **Assembly drawings** show the components that make up a product or subassembly. They illustrate the location of each part in relation to the other parts and, in some cases, illustrate the operation of the product. The parts are shown to scale in orthographic and/or pictorial views. Balloon notes are used to identify each part and cross-reference it to the bill of materials also on the drawing. **Sectioned assembly drawings** cut away a portion of the assembly to expose or illustrate the details of an interior portion of the assembly. A completed title block is added along with manufacturing and assembly notes. No dimensions are shown on assembly drawings. The **bill of materials (BOM)** is a table of part information organized with column headings for: part number, part name, material, quantity used in assembly, and other special notes as required. In some companies, the bill of materials is placed on a separate sheet with its own drawing number. An example assembly drawing is shown in Figure 13.3.

Illustrations   Graphics are also used in business reports, owner's manuals, and oral reports to illustrate parts, assemblies, information flow, decision-making, and numerical data. **Charts** are used to portray relationships among numerical data, for example, sales versus time. **Schematics** are diagrams of electrical or mechanical systems using abstract symbols, for example, a piping schematic or electronics schematic. **Figures** can be defined as diagrams that illustrate textual material. **Sketches** are hand-drawn preliminary, or rough, "drawings" drawn without the use of drawing instruments. **Diagrams** are drawings intended to explain how something works or the relationship between the parts. For example, we use free-body diagrams to analyze static-equilibrium forces and moments. A complete free-body diagram shows a set of axes, the object free of its surroundings, relevant geometry, its orientation, forces, and moments, and motion, if any. A complete free body is shown as Figure 8.6.

13.3.2 Written Documents

Written documents can be as brief as a one-page letter or as long as a 100-page technical report. They can be written using the personal first-person "We" or "I," or the impersonal third person, such as "it has been determined …" Let's examine the content and format of some typical written documents including: letters, memoranda, test reports, project progress reports, research reports, and owner's manuals (Beer and McMurry, 1997).

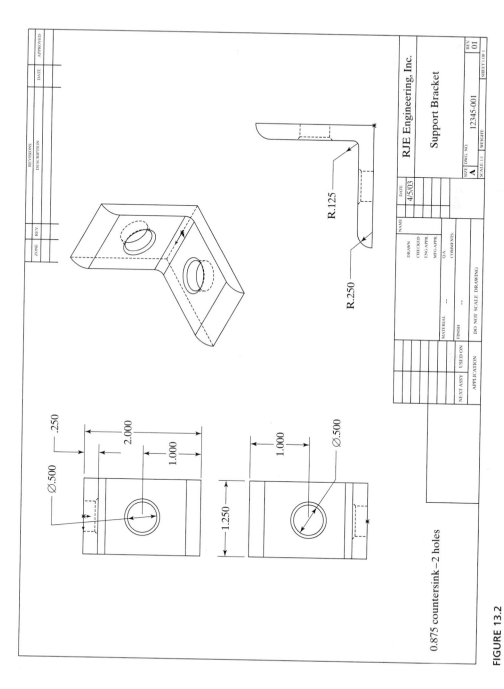

RJE Engineering, Inc.

Support Bracket

SIZE	DWG. NO.	REV
A	12345-001	01
SCALE 1:1	WEIGHT:	SHEET 1 OF 1

FIGURE 13.2

Detail drawing of a support bracket showing dimensioned features, drawn to scale.

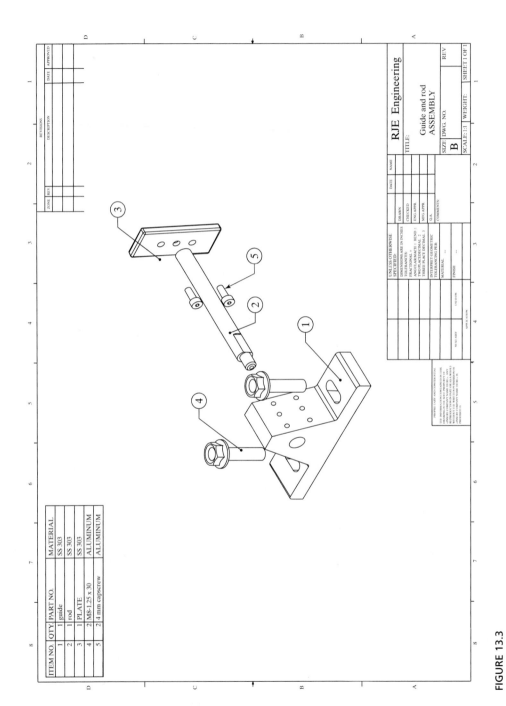

ITEM NO.	QTY	PART NO.	MATERIAL
1	1	guide	SS 303
2	1	rod	SS 303
3	1	PLATE	SS 303
4	2	M8-1.25 x 30	ALUMINUM
5	2	4 mm capscrew	ALUMINUM

FIGURE 13.3

Assembly drawing showing parts of an assembly, their relationship, and a bill of materials.

Letters   Letters are brief communications, usually less than a page in length, sent to a few selected individuals on a specific topic with which the readers are somewhat familiar. E-mail "letters" are informal in format and usually very brief.

Memoranda   **Memoranda** are typically longer, from three to nine pages, sent to a broader audience, and cover more topics in greater depth than a letter. Memoranda are often transmitted via the Internet as attachments to a transmittal letter. A transmittal letter is a cover letter that introduces an accompanying report or memorandum.

Test Reports   **Test reports** are technical reports detailing engineering or scientific tests on materials, prototypes, and/or products. Test reports can vary in length from a few pages to a couple hundred pages. The contents usually include sections on: test objectives, test procedures, data/results, summary, and recommendations.

Project Progress Reports   A project is a unique sequence of activities or work tasks, undertaken once, to achieve a specific set of objectives. Project **progress reports**, sent to clients and other stakeholders, communicate project status regarding the workscope, schedule, and budget. Project status reports are usually brief, from three to five pages in length. For a large-budget project, such as the development of a new fighter plane, the status reports can be a couple hundred pages. Status reports are prepared weekly, monthly, quarterly, and annually depending upon the budget, schedule, and stakeholders' desires. An example project progress report template is shown in Figure 13.4.

Design-Project Report   A formal report is most often prepared at the conclusion of a design project. The report summarizes the work tasks undertaken and discusses the recommended design in detail. A **design-project report** usually includes content on: the nature of the design problem, design formulation, concept design, configuration design, parametric design, prototype testing, and detail-design description and performance. A sample design-project-report outline is shown in Figure 13.5.

Research Reports   **Research reports** include sections similar to test reports, but are longer in length and broader in coverage, including additional sections such as: an abstract, background, literature review, laboratory/test program description, and bibliography.

Owner's Manuals   **Owner's manuals** include sections on: setting up or installing the product, how to operate the product, how to maintain it (i.e., cleaned, lubricated, and adjusted), and how to repair it, if necessary. Owner's manuals can vary in length from one page for a simple consumer product, such as a manual can opener, to hundreds of pages for more complex products, such

> To: Mr. Adam Smith, Client contact title
> cc: All the other stakeholders
>
> From: Mr. John Doe, Project Manager
>
> Re: Design Project Progress Report #XXX-Y
>
> Current Project Status for period __/__/__ through __/__/__
>   brief discussion of work completed since last report (scope)
>   technical issues resolved, and those unresolved
> Future Work
>   work to be performed until next report
>   technical issues to be resolved
>   reminders of deadlines, requested information
>   changes in scope (i.e., Engineering Change Notices)
>   impacts on schedule, milestone dates, deliverables
> Project Budget and Schedule
>   brief discussion of progress with respect to schedule
>   cost variances, schedule variances, earned-value analysis
> Action Items
>   recommendations
>   response(s) needed from client, customers, others,
>   e.g., "Please review and approve within two weeks"

FIGURE 13.4

Project-progress-report template indicating updates on the project's scope of work, schedule, and budget.

as a motor vehicle, or thousands of pages for complex facilities, such as a nuclear power plant.

Engineering-Change Notices   **Engineering-change notices (ECN)** are brief descriptions of changes made to a product. They are detailed on a company-approved form that is subsequently authorized and distributed to all the critical departments in a manufacturing organization.

Patents, Trademarks, and Copyrights   Drawings, illustrations, and textual descriptions are forms of "intellectual property." They represent investment of the company's funds, and as assets, they can be protected by patent, trademark, and/or copyright laws.

13.3.3   Oral Presentations

When we have the opportunity to make an oral presentation before our management, customers, or colleagues, we should honor and respect their willing participation and, consequently, put forth our best efforts. In so doing, we will

1. Introduction
    1.1 Background
    1.2 Problem statement
    1.3 Mission statement
2. Design Problem Formulation
    2.1 Initial site visit
    2.2 House of quality
    2.3 Design objectives
    2.4 Engineering design specifications
    2.5 Solution evaluation parameters
    2.6 Satisfaction curves
3. Project Engineering
    3.1 Work breakdown structure
    3.2 Workscope
    3.3 Schedule
    3.4 Budget
    3.5 Earned-value analysis
    3.6 Work task responsibility table
    3.7 Project team organization chart
4. Concept Design
    4.1 Generation of alternatives
    4.2 Analyses
    4.3 Evaluation and recommendations
5. Configuration Design
    5.1 Generation of alternatives
    5.2 Analyses
    5.3 Design for assembly
    5.4 Design for manufacture
    5.5 Evaluation and recommendations
6. Parametric Design
    6.1 Analytical/experimental methods used
    6.2 Calculations summary
    6.3 Product safety/failure modes and effects analysis
    6.4 Design refinement/optimization
    6.5 Results and recommendations
7. Prototype Tests
    7.1 Prototype fabrication
    7.2 Test results
8. Final Design
    8.1 Product specifications
    8.2 Product cost estimates
    8.3 Drawings (layout, detail, assembly)
    8.4 Bill of materials
9. Recommendations and conclusions
    Appendices

FIGURE 13.5

Example design-project report summarizing the major work tasks undertaken to develop a new product.

increase our chances of achieving our objectives and, in most cases, also make good first impressions. In this section we will first examine three characteristics of presentations (Martin, 1993). Then we will consider how to prepare for a presentation. Finally, we conclude by discussing some guidelines on giving a presentation.

Presentation Characteristics   We will likely make a number of presentations during our engineering careers, both formal and informal, including: design reports, progress reports, technical proposals, sales reports, prototype testing reports, and training sessions. Presentations differ in three characteristics:

Time. On one occasion, we may be given a few minutes to make a minipresentation as part of a meeting. On another occasion, we may have 30 minutes. And on another, we may have hours available, to train employees, for example.

Topic. Given the wide scope of engineering responsibilities and activities in the product-realization process, we can expect presentations on various topics, including: sales, marketing, product development, industrial design, concept design, configuration design, parametric design, detail design, prototype testing, production planning, manufacturing engineering, purchasing, and service.

Temperament. An informal presentation can often be accomplished with a few well-constructed sentences. In other cases, an accompanying visual aid such as a drawing, diagram, chart, or schematic is used to refer to specific details. In contrast, we give comprehensive formal presentations to thoroughly address topics that are important to customers, upper management, and fellow employees. Therefore, formal presentations usually involve detailed information and require the utmost in preparation and execution.

Preparing a Presentation   When we give a presentation, it is our responsibility to be understood. We cannot assume that our audience has understood us. They may have heard what we said, but did they understand us? To achieve this understanding we need to prepare and deliver a clear and engaging presentation.

An excellent presentation requires excellent preparation.

We might consider the following steps in preparing for an "excellent" presentation:

Step 1. Plan	Consider who our audience will be, what we wish to communicate, why we are giving the presentation, and how long it should be.
Step 2. Outline	Prepare a draft outline of the topics. If we are giving a group presentation, we need to agree upon responsibilities. Estimate the time to be devoted to each topic. Break up longer topics into smaller chunks. Combine or eliminate incidental topics. Discuss the draft outline with your co-workers. Confirm the draft outline with your immediate supervisor.
Step 3. Compose	Use the outline to compose 3×5 (or 5×7) note cards. Write clear and concise statements for major ideas and facts. Number each card in succession. Compose clear overhead slides or PowerPoint slides, use font >20 pt. Prepare videos using CAD animations or camcorder movies. Prepare posters, 35-mm slides, working models, or demos.
Step 4. Rehearse	Practice saying the note card phrases. Give the draft presentation to some friendly co-workers. Rehearse using the intended room and audiovisual aids. Videotape and critically evaluate our delivery and use of visual aids.
Step 5. Refine	Revise or rewrite the note cards. Eliminate confusing visual aids. Refine the visual aids. Revise presentation room layout or equipment.

Preparing a Formal Design-Project Presentation We will often have the opportunity to give a formal presentation at the conclusion of each design phase. These occur at various milestones in the product-development schedule. In some cases, we might give a presentation on the work that was completed during the whole project. An example outline is presented in Table 13.2.

Questions, Answers, and Follow-up At the conclusion of a presentation we have the opportunity to learn whether our listeners understood the presentation. Typically an engaged audience asks questions and makes comments. We

TABLE 13.2 Sample Outline for a Formal Presentation of a Design Project

1. Project title, participants
2. Design-problem background
3. Design objectives
4. Alternative designs
5. Analyses and evaluations
6. Design for assembly
7. Design for manufacture
8. Prototype-test results
9. Recommended design
10. Expected performance
11. Summary
12. Acknowledgments

can be ready by anticipating some likely questions and preparing a few concise answers. A slide that was eliminated from a lengthy draft presentation could be held in reserve, then shown. We can even ask the audience questions. For example, "Do you agree that our design is economical?" "Did we demonstrate that the fabrication and assembly concerns have been resolved?" "Can we have your approval to go ahead and schedule production as soon as possible?" Finally, if we don't have an answer for a listener, we should get one and personally contact the listener at a later date. Follow-up is a natural, usual, and necessary part of business presentations.

Guidelines for Giving the Presentation   The following guidelines are recommended for oral presentations:

1. Make the listeners physically comfortable. Provide adequate seating and lighting. Ensure that the room temperature, noise level, and ventilation are agreeable.
2. Expect that we will be somewhat nervous. Accept it. Convert nervousness to enthusiasm.
3. Take a deep breath and relax for a moment before beginning.
4. Start on time, stick to the presentation schedule, and finish on time; do not go over.
5. Pronounce words clearly, with sufficient volume and at a relaxed pace.
6. Vary the pitch or tone of voice occasionally.
7. Add enthusiasm to the delivery.
8. Use visual aids judiciously.
9. Use appropriate gestures and avoid annoying mannerisms.
10. Make frequent eye contact with the audience.
11. Use a pointer when appropriate.
12. Relax and "enjoy the ride."

13.4   PRODUCT DATA MANAGEMENT

Graphic and written documents are instrumental to a company's success. Product-related documents include items such as: CAD drawings (*.dwg), CAD solid models (*.prt), design specifications, design data, manufacturing-process plans, numerical-control programs, product-performance analyses, raw materials, and finished-goods-inventory data.

Considering the product-realization process, computer applications generally relate to engineering, manufacturing, and administration as shown in Figure 13.6.

Computer applications related to product design and manufacture can be categorized into the following groups:

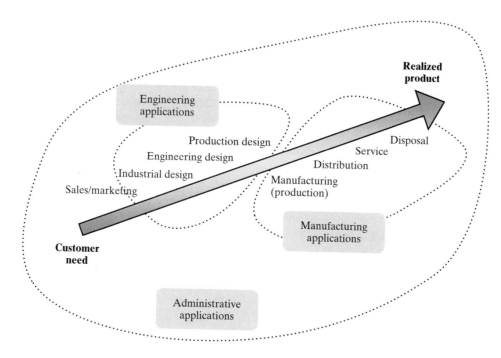

FIGURE 13.6

Computer applications used in the product-realization process for engineering, manufacturing, and administrative needs.

Engineering

CAD. **Computer-aided design (CAD)** software is used to prepare two-dimensional and three-dimensional drawings of product parts and assemblies. Full-featured solid-modeling packages can also render parts as if artistically painted or photographed. In addition, assembly components can be animated, thus illustrating various motions.

CAE. **Computer-aided engineering (CAE)** refers to any computer package that aids the engineering of a product. This includes analysis software typically relating to loads, stresses, vibrations, motions, heat transfer, fluid mechanics, and thermodynamics.

BOM. Bill(s) of materials are documents that list the components, quantity, and part numbers used in a product. Often a BOM is an official drawing that is automatically generated by the company's computer-aided-design software.

Manufacturing

AS/RS. **Automatic storage and retrieval system (AS/RS)**'s are integrated materials handling hardware and software used to automatically store and retrieve raw materials or finished goods from inventory locations in the factory.

CAT. **Computer-aided tolerancing (CAT)** software is used to analyze and specify part tolerances with respect to economic production and reliable product function.

MRP. **Material-requirements planning (MRP)** systems use computers to "explode" a bill of materials to determine the total amount of materials required for a production run and when they are required. Materials can then be ordered just-in-time (JIT) for production, rather than stored in inventory. JIT reduces handling and carrying costs.

CAPP. **Computer-aided process planning (CAPP)** systems establish the type and sequence of manufacturing processes for each part produced. CAPP is also used to plan or balance production among the various machine tools in a factory.

CNC. **Computer numerical control (CNC)** refers to software and hardware used to automatically generate parts, such as a CNC milling machine or NC lathe.

CAI. **Computer-aided inspection (CAI)** systems assist in obtaining and recording raw material and finished-goods-inspection data.

CMM. **Coordinate measuring machines (CMM)** refers to hardware and software that obtains and records measurements of parts.

SPC. **Statistical process control (SPC)** software obtains and analyzes process characteristic statistics. The data are used to alert machine operators and management when a manufacturing machine or process is producing out-of-tolerance parts.

Administrative. Administrative computer applications include accounting, payroll, sales, marketing, finance, and human resources.

The use of computers has facilitated electronic-document preparation, storage, retrieval, and revision. It has also automated the distribution process. Copies of a drawing or textual document can be sent instantly to many recipients, compared to the mailing of hard-copy reproduction of "blueprints" or "whiteprints."

To help manage product-development information some companies are using **product-data management** (PDM) systems, which are special computer software packages written just for that purpose (Rehg, 1994). A generalized version of a product-data-management system is shown in Figure 13.7.

Management of product information includes the initial approval and filing, or archiving and subsequent revisions, of original "master" documents. Revision control refers to authorizing engineering-change notices, which then require revisions to the master documents.

Proper document control is very important. Each document communicates detailed information that evolves during the design and manufacture of a product. The advantages of using a corporate intranet or the world wide Internet, are

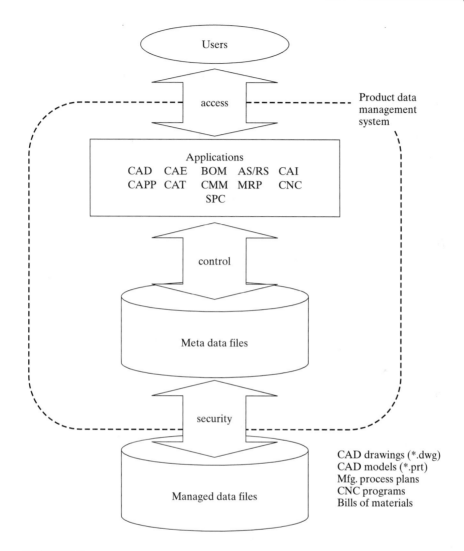

FIGURE 13.7

Product-data-management systems provide access to the users in the company while securely isolating the master data files.

many. However, we must be aware of security issues as well, by installing fire-walls and virus checkers, thereby protecting our company's intellectual property. The improper management or control of a company's documents can lead to:

- out-of-date, incomplete, or incorrect drawings being reproduced and distributed,
- documents being sent to the wrong individuals,

- loss or destruction of very important information, and even
- distribution of proprietary information to competitors.

By using a product-data-management system, an authorized engineer or technician "checks out" a virtual master drawing from the database. At this point anyone attempting to retrieve the drawing will be advised that the master is currently being revised, and that only unofficial copies will be permitted. After the approved revisions have been made, the master is "returned" to the database with a new revision number. At this point, individuals in the approval chain are automatically notified that a revised document is ready for their approval. After each individual has electronically approved the revised drawing, it is made accessible to authorized employees or customers.

13.5 SUMMARY

- Detail-design information includes detail and assembly drawings, manufacturing-process specifications, and performance-test data.
- The freezing of a product design occurs progressively, heterogeneously, throughout the whole manufacturing company when concurrently engineered.
- Graphic documentation includes detail drawings, assembly drawings, and illustrations.
- Written documentation includes e-mails, letters, memoranda, test reports, project reports, and research reports.
- Oral presentations vary in time, topics, and temperament.
- Preparation steps for an oral presentation include: plan, outline, compose, rehearse, and refine.

REFERENCES

Beer, D., and D. McMurry. 1997. *A Guide to Writing as an Engineer.* New York: Wiley.

Bertoline G. R., E. N. Wiebe, C. L. Miller, and J. L. Mohler. 1997. *Technical Graphics Communication.* New York: McGraw-Hill.

Hales, C. 1993. *Managing Engineering Design.* Essex, England: Longman Scientific.

Martin, J. C. 1993. *The Successful Engineer.* New York: McGraw-Hill.

Rehg, J. A. 1994. *Computer Integrated Manufacturing.* Englewood Cliffs, NJ: Prentice Hall.

KEY TERMS

Automatic storage and
  retrieval system (AS/RS)
Assembly drawings
Bill of materials (BOM)
Computer-aided design
  (CAD)
Computer-aided
  engineering (CAE)
Computer-aided inspection
  (CAI)
Computer-aided process
  planning (CAPP)
Computer-aided
  tolerancing (CAT)

Computer numerical
  control (CNC)
Coordinate measuring
  machines (CMM)
Chart
Design-project report
Detail drawings
Diagram
Engineering-change notice
  (ECN)
Figure
Memoranda
Material-requirements
  planning (MRP)

Owner's manual
Product-data management
Progress report
Research report
Sectioned assembly
  drawing
Schematic
Sketch
Statistical process control
  (SPC)
Test report

EXERCISES

Self-Test. Write the letter of the choice that best answers the question.

1. _____ Production drawings are also called:
    a. detail drawings
    b. graphic drawings
    c. working drawings
    d. design drawings

2. _____ Orthographic projection views of a part, showing its geometric features drawn
    to scale along with full dimensions, tolerances, manufacturing-process notes,
    and title block, are called:
    a. assembly drawings
    b. working drawings
    c. detail drawings
    d. graphic drawings

3. _____ A table of part information, organized with column headings for part number,
    part name, material, quantity used in assembly, and other special notes, is called:
    a. parts bill
    b. piece invoice
    c. resource invoice
    d. bill of materials

4. _____ Diagrams of electrical or mechanical systems using abstract symbols are called:
    a. charts
    b. figures
    c. schematics
    d. sketches

5. _____ Illustrations that portray relationships among numerical data, for example, sales versus time, are called:

   a. charts

   b. figures

   c. schematics

   d. sketches

6. _____ Hand-drawn preliminary, or rough, drawings, drawn without the use of drawing instruments, are called:

   a. schematics

   b. sketches

   c. figures

   d. charts

7. _____ Communications, three to nine pages long, sent to a broad audience, and covering many topics are called:

   a. test reports

   b. progress reports

   c. memoranda

   d. letters

8. _____ Technical reports detailing engineering or scientific tests on materials, prototypes, and/or products are called:

   a. memoranda

   b. test reports

   c. letters

   d. progress reports

9. _____ Brief communications sent to a few selected individuals on a specific topic are called:

   a. letters

   b. memoranda

   c. progress reports

   d. test reports

10. _____ A formal report prepared at the conclusion of a design project, summarizing the work tasks undertaken and discussing their recommended design in detail, is called:

    a. progress report

    b. design-project report

    c. test report

    d. design report

11. _____ A technical report, similar to a test report, but broader in coverage, including additional sections such as an abstract, literature review, laboratory/test program description, and bibliography, is called:

    a. design-project report

    b. progress report

    c. research report

    d. project-progress report

2: c, 4: c, 6: b, 8: b, 10: b

12. Explain the differences between a detail drawing, an assembly drawing, and a sketch.

13. Give an example of a nonlinear design sequence.

14. Who is responsible for producing the owner's manual of a product?

15. What are the essential components of a "complete" free-body diagram?

16. Prepare a "complete" free-body diagram of a block resting on an inclined plane.

CHAPTER 14

Projects, Teamwork, and Ethics

LEARNING OBJECTIVES

When you have completed this chapter you will be able to

- Develop a project problem and mission statement
- Prepare a work breakdown structure diagram and scope of work
- Prepare a project plan, including a schedule and budget
- Complete an earned-value analysis including cost and schedule variances
- Describe the difference between a team and teamwork
- Explain and apply effective listening skills
- Apply consensus and voting decision-making methods
- Describe forming, storming, norming, and performing stages
- Develop a complete agenda for an effective meeting
- Describe and apply guidelines for an effective meeting
- Explain and apply the fundamental canons of a code of ethics
- Resolve an ethical dilemma

14.1 INTRODUCTION

The design and manufacture of a product is often accomplished as a project performed by a concurrent engineering team. As the group members participate in the development of a project plan, including a problem statement, mission statement, scope of work, schedule, budget, and responsibilities table, they begin to work as a team. Often successful projects can be traced back to effective teamwork. Similarly, effective teamwork can be traced back to thorough project planning.

We can draw an analogy between project planning and a football team during practice sessions, learning new plays on the board during the "chalk-talk," and then taking the field to drill individual assignments and hone timing. When team members know what they are supposed to do, and how to do it, they have a good chance at being successful. Similarly, project teams can establish effective teamwork through communication, decision making, collaboration, and self-management.

Finally, we will examine the fundamentals of professional ethics and how they relate to our project work tasks and everyday decision making.

14.2 PROJECTS

A **project** is a unique sequence of work tasks, undertaken once, to achieve a specific set of objectives. Design projects, in particular, focus on solving a specific design problem. Other projects include construction or software development. We perform the work tasks only once in a project. This by itself helps to distinguish a project from daily **work tasks**. A sales engineer, for example, performs the same work task by contacting a dozen customer prospects as part of his or her normal daily responsibilities.

The four major elements of a project are the scope of work (i.e., set of work tasks to be performed), the cost for labor, materials, and other resources to accomplish the scope, the time allotted, and the quality of the work performed (Lewis, 2002). The four elements are often interdependent. We can illustrate the coupling of the four elements as three legs of a triangle and the area of the triangle, as shown in Figure 14.1. If, for example, the scope of work is expanded to include more work tasks, the costs, quality, or time allotted will be revised. If we expand the scope without changing the cost or time, quality will usually suffer.

A project has three main phases: planning, executing, and closing. During planning we determine the what, who, how much, and when. We determine what the work tasks are, who the team members are, which tasks they will complete,

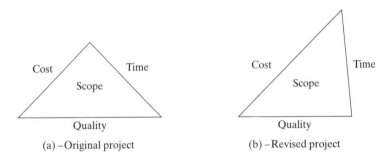

(a) – Original project   (b) – Revised project

FIGURE 14.1

Scope, cost, time, and quality are interrelated. Change any leg of the triangle or the area of the triangle and the others change also.

how much time, money, and other resources are needed, and when the tasks need to be completed. As we execute our work tasks, we also monitor and control our progress to determine if and when any corrective measures need to be taken. During the closing of a project we reflect and document on our successes and failures, such that we and others in the company can benefit.

14.2.1 Planning a Project

During the project-planning phase, team members prepare a package of key items collectively called the **project plan**. The plan includes a:

1. Problem statement
2. Mission statement
3. Project objectives
4. Work breakdown structure
5. Scope of work
6. Responsibilities table
7. Organization chart
8. Budget
9. Schedule
10. Risk assessment

The amount of effort that the team will spend preparing the project plan usually depends on the extent and type of project. Some design problems require hundreds or thousands of man-hours and take weeks, months, or years to solve. However, a simple selection design problem that is expected to take three man-hours would hardly require such a project plan.

A project plan helps team members control the scope, budget, and schedule.

A fundamental project management principle is that those who will work on a project should participate in planning it. That is because the person who will do the work usually knows best as to which specific tasks and deliverables need to be completed. He or she also knows how much time and/or other resources may be needed. Also, since a team usually includes members that have different skills, we can take advantage of their diverse knowledge and abilities.

1. Problem Statement. A **problem statement** is a simple sentence or two describing what the problem is, that is, what the desired end state is, what currently prevents it, and how we know when we have achieved the desired outcome.

2. Mission Statement. A **mission statement** is a paragraph of two or three sentences that states what we are going to do, for whom, and how we will go about doing it.

Example

Bob Bear, Inc. successfully produces a garden tractor, selling about 50,000 tractors per year. However, the company does not make a snow blower attachment for it. Recognizing that the company has underutilized manufacturing capacity, it would like to develop, manufacture, and sell an attachment that would make better use of the factory and enhance company profits. Write a problem statement and mission statement for a design project to remedy this situation

Problem Statement

Bob Bear, Inc. does not make a snow blower attachment for its garden tractor product. A project is needed to design, prototype, and test a new snow blower attachment that is easy to use, safe, and cost-effective.

Mission Statement

The project team will design, prototype, and test a new snow blower attachment for its existing tractor line. The team will formulate customer and company requirements, develop alternative snow blower concepts, configure snow blower attachments alternatives, establish feasible design variable values, prototype components, and test the final preproduction assembly.

3. Project Objectives. The mission statement is simple and broad. To better understand the amount of work in a project, however, we need to develop **project objectives**, which are specific outcomes that we expect to produce. In a design project, for example, we expect to complete a package of "deliverables," including production drawings, prototype test reports, bills of materials, manufacturing process specifications, and performance simulation reports. Project outcomes should be specific as to quantity and quality. Also, outcomes should be realistic in that they consider available time, personnel, and equipment. An example of some project objectives is shown in Table 14.1.

By detailing project objectives, we are determining specific "what"s that will be completed, not the "how." The "how" is determined when the work tasks are detailed, as discussed later.

4. Work Breakdown Structure. A **work breakdown structure** (WBS) is a block diagram that illustrates the major work tasks to be completed during a project. It is a graphical "outline" of sorts. The primary work task categories are shown at the top level, with subordinate tasks underneath each major task. In one glance, the reader gains an overview of the work to be completed. An example is shown in Figure 14.2.

5. Scope of Work. A **scope of work** is a written list of tasks to be completed during the project. Each major task is subdivided into level 2 tasks. For large projects, level 3 and level 4 task descriptions can be used. Alternatively, the list can be expanded into narrative sentences and paragraphs. Deliverable items, such as progress reports, test reports, physical prototypes, and computer programs, are also mentioned, in addition to the task description. Successful teams use the opportunity to jointly develop the workscope, thereby clarifying and

---

TABLE 14.1 Partial List of Project Objectives for a Design Project.

1. Obtain detailed understanding of customer and company requirements
   1.1 Completed survey of 1,000 customers
   1.2 Benchmark study of 5 competitive or substitute products
   1.3 House-of-quality diagram
   1.4 Engineering design specifications

2. Establish alternative concept designs
   2.1 Minimum of five feasible concept designs
   2.2 Sketches illustrating the physical principles, working geometry, and material
   2.3 Completed product function decomposition diagrams
   2.4 Completed morphological structure table
   2.5 Weighted-rating evaluation table

3. Develop alternative configuration designs
   3.1 Three feasible product architecture diagrams
   3.2 Special-purpose configuration sketches
   3.3 Completed design for assembly checklist
   3.4 Completed design for manufacture checklist

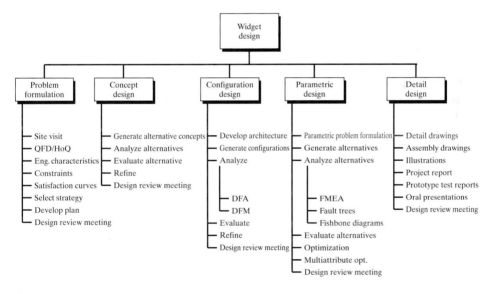

FIGURE 14.2

Work breakdown structure for a typical new product design project.

resolving potential misunderstandings before the project begins. An example scope of work is shown in Table 14.2.

6. Responsibilities Table. The individuals responsible for each task, as well as others who will be assisting, are listed in a project **responsibilities table**. Note that those in charge of a task have "R" next to their work time estimate.

TABLE 14.2   Example Scope of Work for Section 1

1.0 Design problem formulation
1.1 Visit site
        Meet with management/clients to discuss design problem
1.2 Contact consumers
        Make phone calls to determine pros and cons of current unit
        Set an appointment time to witness existing product in operation
1.3 Conduct benchmarking
        Research existing products that are currently available
        Contact manufacturers and request brochures
        Analyze the competition for functionality and performance
1.4 Complete QFD/HOQ
        Determine requirements, engineering characteristics
1.5 Determine parameters
        Define problem definition parameters, design variables
        Define solution evaluation parameters/performance parameters
1.6 Estimate satisfaction levels
        Estimate satisfaction levels for performance parameter
1.7 Prepare engineering design specifications
        List in-use purposes for the product
        List customer and company requirements
1.8 Finalize scope of work
        Refine work breakdown structure
1.9 Determine schedule
        Assign a time value to each task
        Prepare Gantt chart
1.10 Prepare budget
        Determine total number of engineering hours
        Determine total number of expert faculty hours
        Sum all hours and material cost
1.11 Prepare for and conduct design review meeting with management

The table helps to identify key team members for all project participants. An example responsibilities table is given in Table 14.3.

7. Organization Chart. An **organization chart** includes the name and function of each team member. The chart helps to facilitate communications and foster accountability. An example chart for a concurrent engineering design team is shown in Figure 14.3.

8. Budget. The planning for each work task is assigned to one or more individuals. They, in turn, estimate the man-hours, and other resources, necessary to complete the assigned task. The resources necessary to complete the work tasks described in the scope of work are summarized by task in the project **budget**. An example budget is shown in Table 14.4.

9. Schedule. A project **schedule** is a chart that illustrates when each task begins and ends. In addition, key events are denoted, including interim review meetings and report deliverables. The **Gantt chart** format shows work tasks as horizontal bars, as shown on the example in Figure 14.4. Special events, called **milestones** and deliverables are often shown as small diamonds, triangles, or circles.

TABLE 14.3   Project Responsibilities Table

Project Name: Widget Design   Date: 7/11/04

Task	Smith		Johnson		Tully		Hughs		n-th Person		Hours
1.1	6	R	1		1		2		2		12
1.2	3		3	R	2		3		3		14
1.3	1		2		3		6		6	R	18
1.4	2		1		2	R	2	→	4		11
1.5	4		1		1		3	R	5		14
1.6	3		2		2	R	2		2		11
1.7	2		1		2		5	R	3		13
↓											
m-th task											
Total hours	21		11		13		23		25		93

R - Responsible engineer in-charge

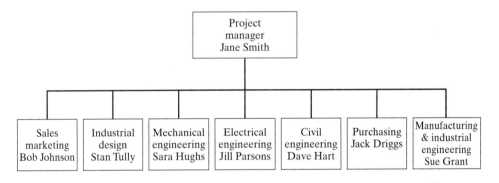

FIGURE 14.3

Example organization chart for design project illustrating the key project personnel. Some charts show other stakeholders, such as management or client personnel.

TABLE 14.4   Example Project Budget Listing Major Work Tasks, Time, and Expenses Required to Complete the Project Tasks

Project Name: Snow Blower Attachment Design   Date: 2/7/04

Project Budget

Task	Description	Sr. Engineers	Admin.	Hours	$
1.0	Design Problem Formulation	91	2	93	1900
2.0	Conceptual Design	96	6	102	2160
3.0	Configuration Design	97	8	105	2260
4.0	Parametric Design	160	9	169	3560
5.0	Detail Design	202	10	212	4440
	Total Hours	646	35	681	
	Rate: $/Hour	20	40		
	Total Labor Cost	$ 12,920	$ 1,400		$ 14,320
	Materials/Supplies				$200
			Total Costs:		$ 14,520.00

Project: New snowblower attachment									
	Week 1	Week 2	Week 3	Week 4	Week 5	Week 6	Week 7	Week 8	Week 9
Task	1/22–1/26	1/27–2/2	2/3–2/9	2/10–2/16	2/17–2/23	2/24–3/2	3/3–3/9	3/10–3/16	3/17–3/23
Design problem formulation									
1.1 Visit site	▨								
1.2 Contact customers		▨							
1.3 Benchmarking		▨	▨						
1.4 Complete QFD/HOQ		▨	▨						
1.5 Determine parameters			▨						
1.6 Estimate satisfaction levels			▨						
1.7 Prepare design specifications			▨	▨					
1.8 Finalize scope of work				▨					
1.9 Determine schedule				▨					
1.10 Prepare budget				▨					
1.11 Design review				◆					
Concept design									
2.1 Generate concepts						▨	▨		
2.2 Determine physical principles							▨	▨	
2.3 Conceptual drawings								▨	
2.4 Evaluate concepts									▨
2.5 Design review									◆

FIGURE 14.4

Example project schedule showing a portion of a project.

The major drawback of the Gantt chart is that succeeding tasks are not "connected" to preceding tasks. It is difficult to determine what effect a delay in a preceding task will have on succeeding tasks and the overall project completion date.

An advanced method, called activity network diagramming, overcomes Gantt chart drawbacks. Project activities are connected by arrows illustrating the network of dependencies. Combinations of parallel and sequential activities are thereby effectively managed. A critical path through the network can be also identified using some simple rules and basic arithmetic. The critical path determines the shortest time to complete the whole network of activities. Activity network diagramming is beyond the scope of this book. More information on this method and other aspects of project management can be found in Badiru and Pulat (1995) and Stub et al. (1994).

10. Risk Assessment. The last section of a project plan involves an assessment of the likely risks that face the project and possible contingency plans to overcome such situations. We call this activity **risk assessment**. First, we determine what risks might exist. Two questions that we might ask are "What is likely to go wrong with the project?" and "What could prevent us from achieving our objectives?"

Construction projects, for example face considerable risks, such as building-permit-approval delays, inclement weather, construction-material-delivery delays, and worker strikes. Design projects face risks such as the accidental death or sickness of key personnel, glitches with computer software, delays with fabricating prototypes, testing accidents, and vendor quote delays. The second aspect of the risk assessment is to consider and recommend a variety of contingency plans. Recognizing key risks in a project and having a couple of ready fall-back plans is just smart business.

With the plan, the team has a road map of what it should be working on, when it should have tasks completed, and how much time and resources to devote to each. Even though the plan is based on estimates that, in actuality, will be under or over budget, or ahead or behind schedule, the plan helps the team to keep under control. Without a plan the team has no idea if it is "on-track" or not. Without a road map the team may not even know where it is going.

During the planning phase of a project a significant amount of information begins to be generated and collected. Project notebooks can organize and protect data, analyses, and observations. It behooves each team member to maintain his/her own project notebook. In addition, a "team notebook" is sometimes used to store originals of key documents. Notebooks should be organized into convenient sections as suggested in Table 14.5.

As part of the planning phase, a project team may prepare a project proposal for review and approval by upper-level management. A project proposal is a written document that describes the problem, mission, objectives, scope of work, budget, schedule, and risks. A project proposal template is shown in Table 14.6.

14.2.2   Executing a Project

As we execute a project we complete our own individual work tasks. We also work in groups, meeting with our teammates to discuss important developments, coordinate work efforts, and make team decisions when required.

We also monitor actual project expenditures and compare these actual expenses to our budgeted expenses to determine whether corrective measures need be taken.

Earned-Value Analysis   **Earned-value analysis** is a widely used project control method that compares actual versus budgeted expenses on a period-by-period basis.

The first step in preparing an earned-value analysis is to calculate the **budgeted cost of work scheduled (BCWS)** for each time period. If, for example, we choose weekly time periods, we would take the total amount budgeted for the work task and distribute it over the weeks that we are scheduled. For example, Table 14.7 shows a project that has three tasks, A, B, and C. Task A is budgeted for $3,000. Let's assume that the project team members will work evenly over the three weeks during which it is scheduled. We therefore enter $1,000 per week for weeks 1, 2, and 3, under the column labeled work task A. Similarly, let's assume our schedule for work task B begins in week 2 and finishes at the end of week 4. Again let's assume that our work efforts are evenly divided among the three weeks. We therefore distribute the $10,500 in $3,500 increments over weeks 2, 3, and 4. We similarly distribute for work task C, which is scheduled to begin on week 3 and end at the end of week 8.

TABLE 14.5 Example Project Notebook Sections

Identification sheet
    Project name
    Team member name
    Telephone/e-mail addresses

Design problem formulation
    Engineering design specifications
    Customer notes
    House of quality
    Prior art (library research, Web)/benchmarks

Alternative generation, analyses, and evaluation
    Analysis plan
    Computations, experiments
    Citations for equations, data
    Spreadsheets
    Sketches, figures, schematics, drawings
    Evaluation plan
    Evaluations/computations
    References/bibliography

Project engineering
    Project meeting notes
    Work breakdown structure
    Scope of work
    Project schedule and updates
    Budget, earned-value analyses
    Risk assessments
    Time sheet—log of work/team mtg hours
    Punchlists of things to be done

Vendor information
    Telephone numbers, addresses
    Phone conversations notes
    Web site printouts
    Product/vendor literature

We determine the weekly budget by adding across the columns, arriving at a total budget of $1,000 for the first week, $4,500 for the second week, and so on. The BCWS is shown in Table 14.7 and is an accumulation of the weekly amounts. At any week, we can look at the BCWS column to know what we should have spent in total. The BCWS is a measure of the planned amount of work.

The next step is to determine the **actual cost of work performed (ACWP)**. Fortunately, the accounting department prepares these amounts as weekly expenses. The accounting department tallies payroll, material, and other costs that are actually expended each week as the project continues. These amounts are entered weekly into the weekly column labeled ACWP. Let's assume that the total expenses for the project are $600. This amount is shown in Table 14.8.

The next step is to calculate the **budgeted cost of work performed (BCWP)** by estimating the **percent complete** for each task and multiplying it by

TABLE 14.6 Example Template for a Design Project Proposal

Design Project Proposal
    Cover letter
    Title page
    Table of contents

Introduction
    Problem statement
    Mission statement
    Engineering design specifications (QFD)
    Project objectives

Scope of work
    Work breakdown structure (WBS), 2-level diagram
    Work scope describing work tasks
    Project deliverables associated with tasks

Schedule
    Gantt chart
    Critical path network diagram
    Milestones

Budget
    Responsibilities table
    Budget
    Other resource requirements

Project management
    Organization chart of project stakeholders
    Project budget and schedule control system
    Risk assessment
    Design change notice (DCN's) procedure

Appendix
    Site visit data

TABLE 14.7 Earned-Value-Analysis Table before the Project Starts (Week #0)

	Earned-Value Analysis						End of Week:		#0
	Work Task						Schedule		Cost
Week	A	B	C	Weekly	BCWS	BCWP	Variance	ACWP	Variance
1	1000			1000	1000				
2	1000	3500		4500	5500				
3	1000	3500	2800	7300	12800				
4		3500	2800	6300	19100				
5			2800	2800	21900				
6			2800	2800	24700				
7			2800	2800	27500				
8			2800	2800	30300				
Total	3000	10500	16800	30300					
% Complete	0	0	0						
BCWP	0	0	0	0					

TABLE 14.8   Earned-Value-Analysis Table as of the End of Week #1

| | Earned-Value Analysis | | | | | | End of Week: | | #1 |
| | Work Task | | | | | | Schedule | | Cost |
Week	A	B	C	Weekly	BCWS	BCWP	Variance	ACWP	Variance
1	1000			1000	1000	750	−250	600	150
2	1000	3500		4500	5500				
3	1000	3500	2800	7300	12800				
4		3500	2800	6300	19100				
5			2800	2800	21900				
6			2800	2800	24700				
7			2800	2800	27500				
8			2800	2800	30300				
Total	3000	10500	16800	30300					
% Complete	25	0	0						
BCWP	750	0	0	750					

the total task budget amount. The result is a measure of the amount of the budget that was planned for the work actually completed. Assume that at the end of week 1, we have completed about 25 percent of task A; 25 percent of $3,000 is $750. We enter this value in the first row of the BCWP column. Note that we actually completed $750 worth of task A.

Next we compute the **schedule variance** and the **cost variance**, for week 1, according to the following formulas:

$$\text{Schedule Variance} = \text{BCWP} - \text{BCWS}$$

$$\text{Cost variance} = \text{BCWP} - \text{ACWP}$$

$$\text{Schedule variance} = \$750 - \$1{,}000 = \$-250$$

$$\text{Cost variance} = \$750 - \$600 = \$150$$

The schedule variance is a measure of how much work we accomplished in relation to the amount that was planned. Since we had planned $1,000 of work, but accomplished only $750, we are $(250) behind our schedule. This indicates that we need to make relatively more progress in the coming weeks.

The cost variance is a measure of how much we spent in relation to how much we should have spent for the work accomplished. Since we should have spent $750 to accomplish the amount of work we paid $600 for, we are under budget by $150. Since we saved $150, we might spend some of it on extra help to get caught up to our schedule.

Note that negative values indicate poor performance, such as behind schedule or over budget, and positive values indicate good performance, such as ahead of schedule or under budget.

Each week we complete the earned-value-analysis spreadsheet to determine if corrective measures are needed. Let's assume that we have been working

TABLE 14.9   Earned-Value Analysis for End of Week #7

| | Earned-Value Analysis | | | | | | End of Week: | | #7 |
| | Work Task | | | | | | Schedule | | Cost |
Week	A	B	C	Weekly	BCWS	BCWP	Variance	ACWP	Variance
1	1000			1000	1000	750	−250	600	150
2	1000	3500		4500	5500	4800	−700	4500	300
3	1000	3500	2800	7300	12800	10800	−2000	11000	−200
4		3500	2800	6300	19100	17300	−1800	18300	−1000
5			2800	2800	21900	18900	−3000	23500	−4600
6			2800	2800	24700	22700	−2000	25600	−2900
7			2800	2800	27500	26955	−545	29355	−2400
8			2800	2800	30300				
Total	3000	10500	16800	30300					
% Complete	97	93	85						
BCWP	2910	9765	14280	26955					

on the project for seven weeks, and we are finishing the earned-value analysis for the end of week 7, as shown in Table 14.9.

Upon examining the schedule variance we see that the team was behind schedule ever since the project started. Then around week 7 the team began to catch up. Similarly, by examining the cost variance, we see that the team began to overspend around week 4 and especially week 5.

The project manager uses the weekly earned-value analysis results to take corrective actions, such as assigning more or less personnel to the project, or reallocating budget funds among team members or tasks. The schedule variance in week 7 appears to show that the team is getting back on track with regard to the schedule and may finish the project almost on time, but the cost variance indicates they might finish over budget.

A graphical chart of the BCWS, BCWP, and ACWP is shown in Figure 14.5. We see that for weeks 1–4 the BCWP curve was lower than the BCWS, which means that the team was behind schedule. We see that this gap increased, meaning that they got further behind schedule. At week 6 and 7 the gap is narrowing; therefore, they are catching up.

We also see that the ACWP curve is close to the BCWP for the early weeks, meaning that the team was spending their budget in proportion to the work they actually performed. This situation worsened, however, in weeks 4–6 and appears to be improving.

Project control is comparing project progress to the plan so that corrective action can be taken when deviation from planned performance occurs.

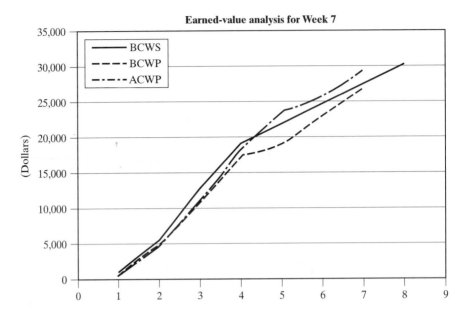

FIGURE 14.5

Graphs of the BCWS, BCWP, and ACWP show the project progress. The gap or space between the BCWP and BCWS lines is the schedule variance. The gap between the BCWP and ACWP is the cost variance.

14.2.3 Closing a Project

At the conclusion of a project we perform a project evaluation. We reflect back on the things that we learned, what we did well, and what we could have done better. We record these findings to be archived for our future use or to be passed on to other project teams so that they do not make the same mistakes.

14.3 TEAMWORK

A **team** is a group of people who have complementary skills and knowledge, work together toward common goals, and hold each other mutually accountable. A product development team, for example, will have members with special skills and knowledge in areas such as sales, marketing, engineering design, purchasing, manufacturing engineering, production, and distribution. They share the goal of developing a successful product. They also are mutually accountable for the performance of the fellow team members.

   Not all teams have **teamwork,** however. In professional sports, for example, we routinely witness examples where teamwork breaks down, as in missed calls, bad timing, or inattention to the game. Engineering design and manufacturing teams are not much different.

14.3.1 Elements of Teamwork

Let's examine a well-executed offensive play in professional football, to identify some of the necessary ingredients for teamwork. The quarterback receives the recommended play from the head coach. During the huddle, the quarterback examines the defensive lineup, spots an opportunity to run a different play, and talks it over with the team. The team simultaneously initiates the new play, in a coordinated fashion, as the quarterback calls for the hike. Individual team members perform separate roles, but with a common purpose of scoring. Blockers protect the quarterback as he drops into the pocket for a pass. A halfback sprints forward into a fake run. As one receiver sprints from the scrimmage line he sees an opening and adapts his running pattern. The quarterback lofts the football in his direction. Successfully catching the ball, the receiver heads toward the goal line with his teammates blocking a path in front of him. Using his well-honed running skills he dodges the last tackler and scores a touchdown. Now that was teamwork!

Successful teamwork depends upon four major behaviors (McGourty and DeMeuse, 2001): communication, decision making, collaboration, and self-management.

Communication Communication is the means by which we exchange information between team members. The quarterback communicated his observation to the offensive line, thereby taking advantage of an opportunity. He also effectively communicated the timing of the hike, preventing an offside penalty. Downfield team members were watching and listening and came to the aid of the receiver.

The principal forms of communication include spoken and written messages. An effective communicator will spend time in preparing better written documents such as e-mails, memoranda, and reports. And an effective communicator will take particular care to make himself or herself understood when speaking to teammates. Listening skills, however, are perhaps even more important. The downfield teammates had their "heads up," watching and listening to the play as it unfolded. Without their "listening" the play might have ended differently. Whether in professional football or business, we can find people who "miss the call" because they are not listening carefully.

Listening is an individual skill, something that we have control over, and something that we can get better at by adopting the following guidelines:

Stop talking. Hear "the play?" Engage in only one discussion at a time.

Show respect. Empathize with the speaker. Show interest in what is being said, respond with appropriate body language, and refrain from interrupting.

Concentrate on what is being said. Do we understand the message completely? Is there something missing?

Stay calm. Keep control of our emotions. React to the ideas or facts not to the speaker presenting them. Try not to jump to any conclusions.

Ask questions. Take responsibility for some of the communication. Paraphrase the speaker's main points until we understand them. We should first try to understand the speaker before trying to have our views understood.

Decision Making   Effective decision making begins with having a clear understanding of the problem. It also involves activities for generating, analyzing, evaluating, and refining alternatives, and then implementing the best one in a timely fashion.

Team decision making takes effort. For example, when we make "individual" decisions we don't necessarily need to communicate with, or understand, others. We can use our own criteria and selection process to decide the best alternative. We rarely have to compromise or consider other options if we don't want to. Team decision making, on the other hand, requires subordinating our own individual desires for the good of the team. For example, during the huddle, the team discussed alternatives and then agreed to commit its entire resources to the new play.

Assuming that the team has gathered the pertinent data and established evaluation criteria and importance weights, they can prepare a weighted rating evaluation to prioritize the alternatives. The final decision, however, comes down to the team making its final selection. Various approaches include a handoff to management, voting, and consensus.

The handoff to management is an abrogation of the team's responsibility to find the best solution for the company. Noneffective teams, however, will pass decisions up the chain of command, especially if the team is concerned about personal retribution.

Voting is a seemingly democratic process, which can lead to some less-than-desirable consequences. Each member may be given one vote and the majority vote wins. Of course, the minority loses. The first question that should be asked is whether each team member should get one vote. If not, how should the number of votes be determined? The second question we could ask is whether ignoring the minority issue may put the company at risk. Another voting approach is that a decision must be by unanimous vote. In this case any team member may veto all choices, such that everyone loses.

A **consensus decision** occurs when the team thoughtfully examinee all of the issues, and agrees upon a course of action that does not compromise any strong convictions of a team member. The approach gives everyone an opportunity to present his case, while recognizing that in some circumstances a decision has to be made. While a team member may not get the choice he favors, at least he knows why, and moreover, agrees that the decision has to be made.

Collaboration   Teams that exhibit effective collaboration have members that are committed to the goals of the team, work cooperatively and constructively, actively participate in team activities, and support fellow team members. The blockers protected the quarterback, for example. And the receiver got blocking help from his teammates when they realized the opportunity for a team score.

Collaboration, in effect, requires subordination of individual desires so that the team can benefit.

Self-Management   Effective teams have members that manage themselves. They monitor their own progress to meet established goals, stay focused on important tasks, use meeting time effectively, solicit constructive feedback on their performance, involve others in all the decision making, and put top priority on getting results.

14.3.2   Stages of Team Development

Teams develop teamwork over time. We don't expect a group of professional football players to exhibit teamwork the first day of practice. Neither can we expect a business team to do so. A professional team will practice their skills until they are ready for the game. Unfortunately, few business teams have the same opportunity. Business team members usually develop their skills of communication, decision making, collaboration, and self-management on-the-job.

One humorous model of team development includes the following sequence of stages:

1. project initiation,
2. wild enthusiasm,
3. disillusionment,
4. chaos,
5. search for the guilty,
6. punishment of the innocent,
7. promotion of the nonparticipants,
8. and definition of the project requirements (Lewis, 2002).

On a more serious note, however, we do find that teams actually experience different stages. Tuckman presented a model describing four stages: **forming**, **storming**, **norming**, and **performing** (1965).

Forming.   At the beginning of a project, participants transition from being individuals to being team members. Somewhat anxiously, members politely interact to learn about the nature of the tasks to be performed, the goals of the project, and the personalities and work styles of fellow members.

Storming.   During this stage, members begin to realize the enormity of the project. They recognize differences in individual abilities, personalities, and work styles. Disagreement and conflict may lead some members to retreat from the group to try and "do it alone." Tension, conflict, and scapegoating often occur.

Norming.   During this stage members begin to cooperate with each other. They begin to understand and respect individual strengths and weaknesses. They begin to focus on common goals and communicate more openly. They evolve acceptable standards of behavior, or norms, for performing their roles and resolving conflict. And slowly a team spirit begins to emerge.

Performing. During this stage the team is productive and satisfaction is high. Team members share accountability for their actions and are strongly united. Project activities are conducted in an atmosphere of trust and mutual support. Conflicts are resolved openly and effectively as members are comfortable with their roles.

The most important point of Tuckman's model is that most groups evolve through these same four stages. They are destined to form, storm, norm, and perform. Therefore, we should recognize that our team interactions will be dynamic and evolve, and more importantly, we need to work through the difficult times.

14.3.3   Effective Team Meetings

To be effective, team meetings need to be well planned and efficiently executed. Planning efforts should concentrate on the agenda. Just like a scope of work and in combination with a schedule, an agenda can lay the framework for how a meeting is to proceed. When possible, it should be sent to team members in advance for review and comment. If an agenda is not prepared, we should use the first five minutes or so of a meeting to prepare one.

Agenda   An **agenda** should list the topics to be discussed along with an estimate of time required and the initials of the person coordinating that item of discussion. Topics should be ordered logically. The topics should be designated as a decision item or an announcement. A facilitator may also be designated for each topic. A facilitator keeps the meeting focused and moving. He or she tactfully intervenes when discussions take tangents and prevents anyone from dominating or from being overlooked. The last item of the agenda should be a brief evaluation of the meeting with regard to its effectiveness and plans to improve future meetings.

Effective Execution   Having an agenda will enhance any meeting. By executing the agenda, the team will accomplish what it expects to accomplish, in the time it expects to take. The following guidelines, however, are recommended to facilitate its execution.

- *Start on time.*   Coming late to a meeting wastes the other participants' time and is disrespectful.
- *Practice effective listening skills.*   Understand first, then try to be understood.
- *Facilitate the facilitator.*   Help keep the meeting on time and on topic.
- *Come prepared.*   We should never "wing it." If we are presenting an agenda topic, we should be ready for our listeners. It may be the only chance to get our points across.
- *Discuss fact not fiction.*   If facts are not available, try to reschedule the topic, such that an informed and factual decision can be made.

- *Take action.*   Encourage a consensus decision and move on.
- *Take minutes.*   Record key topics, decisions made, and actions to be taken.
- *Draft next agenda.*   Before participants leave discuss tentative topics, times, and presenters.
- *Turn off cell phones.*   The other meeting participants are honoring us by attending. We should not let any personal calls interrupt the meeting, thereby wasting their time.

14.3.4   Team Rules

To be an effective team member we need to recognize and accept that certain responsibilities come along with the privileges of being on the team, such as the responsibility to:

- Commit to the goals of team
- Perform assigned tasks completely, accurately and on time
- Respect the contributions of others
- Assist other team members when needed
- Ask for help before we get into trouble
- Follow guidelines for effective meetings
- Actively participate in team deliberations
- Focus on problems not people or personalities
- Constructively resolve conflicts or differences of opinion
- Comment clearly and constructively

14.4   ETHICS AND THE ENGINEERING PROFESSION

As engineering professionals, we make decision that effect our fellow employers, customers, the public, and the profession. As we make product decisions, for example, we strive to design and manufacture each product so that they are both safe and financially successful for our employer. In some cases, however, a safer product may mean less profits for our company. Should we approve engineering changes that endanger customers, even if it means disappointing our managers? Will our management appreciate the long-term benefits of short-term ethical decision making? Who should we be loyal to? Our employer or the public? Are we obligated to follow a course of action? Do we have a professional responsibility to "do the right thing"?

14.4.1   Code of Ethics

Professional **ethics** are standards of conduct to be followed by every member of that profession. Engineering societies such as the American Society of Mechanical Engineers (ASME) and the American Society of Civil Engineers, as well as

---

American Society of Mechanical Engineers

Code of Ethics of Engineers

The Fundamental Principles

Engineers uphold and advance the integrity, honor, and dignity of the engineering profession by:

   I. Using their knowledge and skill for the enhancement of human welfare;

  II. Being honest and impartial, and serving with fidelity the public, their employers, and clients; and

 III. Striving to increase the competence and prestige of the engineering profession.

The Fundamental Canons

1. Engineers shall hold paramount the safety, health, and welfare of the public in the performance of their professional duties.
2. Engineers shall perform services only in areas of their competence.
3. Engineers shall continue their professional development throughout their careers and shall provide opportunities for the professional development of those engineers under their supervision.
4. Engineers shall act in professional matters for each employer or client as faithful agents or trustees, and shall avoid conflicts of interest.
5. Engineers shall build their professional reputation on the merit of their services and shall not compete unfairly with others.
6. Engineers shall associate only with reputable persons or organizations.
7. Engineers shall issue public statements only in an objective and truthful manner.
8. Engineers shall consider environmental impact in the performance of their professional duties.
9. Engineers accepting membership in the American Society of Mechanical Engineers by this action agree to abide by this Society Policy on Ethics and procedures for its implementation.

FIGURE 14.6

ASME Code of Ethics of Engineers. (Reprinted by permission of the American Society of Mechanical Engineers.)

many other societies, embody these ethical standards of conduct in their **code of ethics**. The ASME code of ethics is presented in Figure 14.6.

A code of ethics is important because: (1) a code of ethical standards provides a systematic collection of rules to guide all members of the profession, (2) a code encourages value-laden decisions for the public good, (3) code provides employees a working environment to resist the pressures of unethical business practices, (4) a code protects members in that profession from being harmed by what other engineers may do, and (5) a code helps to sustain the moral reputation the profession.

The fundamental principles of the ASME code emphasize that if we want to maintain and advance the integrity, honor, and dignity of the profession we should use our knowledge and skill for the enhancement of human welfare. Additionally, that this can be accomplished by being honest, impartial, and faithful to our employers, clients, and the public, and by striving to improve our competence and the prestige, in general, of our profession. Note how the principles embody **ethical values** that we can use to guide our decision making. Values such as:

- Integrity—exercising good judgment in the practice of our profession
- Honesty—telling the truth, being sincere
- Fidelity—being loyal to our employer, clients, public, and the profession
- Responsibility—being reliable, dependable, accountable, and trustworthy.

The nine fundamental canons are a set of standards or rules that members of ASME should follow.

1. Engineers shall hold paramount the safety, health, and welfare of the public in the performance of their professional duties. There is nothing more important than the safety, health, and welfare of the public. We are responsible for conducting safety reviews, and for providing users with information for safe operation. We should not approve designs that could endanger others. Further, if our judgment is overruled we are obligated to inform our clients, employers, or other authorities of the possible consequences. We are also obligated to report other individuals or firms that are in violation.

2. Engineers shall perform services only in areas of their competence. We cannot perform engineering work in fields where we are not qualified by education or experience. We have a responsibility to refuse such work.

3. Engineers shall continue their professional development throughout their careers and shall provide opportunities for the professional development of those engineers under their supervision. We are responsible for keeping up with technology, to know what the current standards of engineering practice are. And we should encourage those under our supervision to do so.

4. Engineers shall act in professional matters for each employer or client as faithful agents or trustees, and shall avoid conflicts of interest. We have a responsibility to inform our clients or employers of known conflicts of interest. We should not accept financial or other valuable consideration for specifying materials, equipment, or vendors without disclosure to our clients and employers. We must treat all information received from our clients or employers as confidential. We should not accept outside employment without notifying our employer.

5. Engineers shall build their professional reputation on the merit of their services and shall not compete unfairly with others. We should not misrepresent our professional qualifications. We should not embellish or plagiarize technical reports. We should not injure the professional reputation of other engineers. We should not propose or accept any contracts on a contingency basis such that our judgment might be compromised.

6. Engineers shall associate only with reputable persons or organizations. We should not associate with a dishonest or fraudulent party, nor use that party to engage in unethical practices.

7. Engineers shall issue public statements only in an objective and truthful manner. We should never embellish or misrepresent the facts. We should be objective and present both sides of an argument. We have a responsibility to extend public knowledge and to prevent misunderstandings.

8. Engineers shall consider environmental impact in the performance of their professional duties. We must consider the impact of our plans or designs on the environment and provide for the safety, health, and welfare of the public.

9. Engineers accepting membership in the American Society of Mechanical Engineers by this action agree to abide by this Society Policy on Ethics and procedures for its implementation. If we want to be a member of the profession, we have to abide by the rules of the profession.

When engineers register themselves in a state as professional engineers, they must follow state and local laws relating to professional responsibility. Many of the state codes have sections similar to the professional societies' codes of ethics, such as responsibility to the public, competency for assignments, conflicts of interest, and improper conduct.

14.4.2 Resolving Ethical Dilemmas

Most employees and employers conduct themselves in an ethical manor. However, there may come a time when we are faced with an unethical situation. How should we go about resolving it? Hopefully we can work it out inside the company. If not, we may need to "blow the whistle" on our employers and go public with our concerns.

The following steps are recommended:

Step 1: Obtain the facts of the situation.

Step 2: List the stakeholders, i.e., anyone who has a vested interest in the outcome.

Step 3: Consider the motivations of the stakeholders.

Step 4: Formulate alternative solutions using codes of ethics or basic ethical values.

Step 5: Evaluate the alternatives, reject unethical solutions.

Step 6: Seek assistance from co-workers, supervisors, and ombudsmen.

Step 7: Select the alternative that satisfies the highest ethical values.

Step 8: Implement the selected solution through the chain of command.

Step 9: Monitor the outcome.

Step 10: If unsatisfactory, contact legal counsel, professional society, and the media.

14.5 SUMMARY

- The interdependent elements of a project are the scope of work, time, budget, and quality of the work performed.
- A work breakdown structure is a graphic illustration of the basic work tasks.
- Those who will be working on the project should plan it.
- A project schedule indicates the beginning and ending of project work tasks.

- Earned-value analysis is a method used to determine whether a project is ahead or behind schedule and under or over budget.
- Teamwork involves communication, decision making, collaboration, and self-management.
- Practicing effective listening skills is the first step to being a good communicator.
- Consensus decision making is preferable to voting.
- Teamwork evolves through forming, storming, norming, and performing stages.
- Having a (complete) agenda is the first step to an effective meeting.
- Professional ethics are standards of conduct based on ethical values.
- A code of ethics provides guidelines of ethical conduct that members must follow.

REFERENCES

Badiru, A. B., and P. S. Pulat. 1995. *Comprehensive Project Management.* Englewood Cliffs, NJ: Prentice Hall.

Lewis, J. P. 2002. *Fundamentals of Project Management.* New York: American Management Association.

McGourty, J., and K. P. DeMeuse. 2001. *The Team Developer: An Assessment and Skill Building Program.* New York: John Wiley & Sons.

Stub, A., J. F. Bard, and S. Globerson. 1994. *Project Management: Engineering, Technology and Implementation.* Upper Saddle River, NJ: Prentice Hall.

Tuckman, B. 1965. "Developmental Sequence in Small Groups." *Psychological Bulletin,* no. 63.

KEY TERMS

Actual cost of work performed (ACWP)
Agenda
Budget
Budgeted cost of work performed (BCWP)
Budgeted cost of work scheduled (BCWS)
Code of ethics
Consensus decision
Cost variance
Earned-value analysis
Ethical values

Ethics
Forming
Gantt chart
Milestone
Mission statement
Norming
Organization chart
Percent complete
Performing
Problem statement
Project
Project control
Project objectives

Project plan
Responsibilities table
Risk assessment
Schedule
Schedule variance
Scope of work
Storming
Team
Teamwork
Work breakdown structure
Work task

EXERCISES

Self-Test. Write the letter of the choice that best answers the question.

1. _____ A hierarchical block diagram of major work tasks is called a:
   a. schedule
   b. scope of work
   c. work breakdown structure
   d. budget

2. _____ A project schedule illustrates:
   a. estimated beginning time of tasks
   b. ending time
   c. milestones
   d. all of these

3. _____ A project responsibilities table shows:
   a. person responsible for coordinating a task
   b. total estimated time by person
   c. total estimated time for each task
   d. all of these

4. _____ To illustrate the name and function of each project team member we use a:
   a. task schedule
   b. responsibilities table
   c. design project notebook
   d. organizational chart

5. _____ The four elements of teamwork include the following except:
   a. collaboration
   b. empathy
   c. communication
   d. decision making

6. _____ We can practice effective listening by doing the following except:
   a. asking questions
   b. showing respect
   c. interrupting
   d. stop talking

7. _____ At this stage of team development team standards are established and a team spirit emerges:
   a. forming
   b. storming
   c. norming
   d. performing

8. _____ An effective meeting agenda includes:

   a. topic to be discussed

   b. estimated time for each topic

   c. person assigned to facilitate the topic discussion

   d. all of these

9. _____ A code of ethics:

   a. establishes ethical standards of professional conduct

   b. establishes rules to assist our decision making

   c. encourages value-laden decision making

   d. all of these

10. _____ Values associated with professional ethics include the following except:

   a. honesty

   b. safety

   c. fidelity

   d. integrity

2: d, 4: d, 6: c, 8: d, 10: b

11. What are the fundamental elements of a project and why are they interdependent?

12. What are a project problem statement and mission statement? What purpose do they serve in a project plan?

13. Who should prepare the project plan and why?

14. Use a separate piece of paper to prepare a work breakdown structure diagram for making a batch of cookies from basic ingredients.

15. How does the process of preparing a responsibilities table benefit teamwork?

16. List a few good reasons to keep a complete project notebook.

17. How is the budgeted cost of work scheduled determined?

18. Project team Alpha has accumulated $5,000 of expenses as of the end of week #10. The budgeted cost of work scheduled for week #10 is $6,500. The project has one task, which is about 45 percent complete at the end of week #10. Calculate the schedule and cost variances. Is the project ahead (or behind), and under (over) budget?

19. Team Delta is working on a project that has two work tasks, each worth $5,000, which are 45 percent and 35 percent complete as of week #5. The actual expenses are $2,500 as of the end of week #5. The budgeted cost of work scheduled for week #5 is $6,500. Calculate the schedule and cost variances. Is the project ahead (or behind), and under (over) budget?

20. The project below has three main tasks. Task A lasts for five weeks beginning in week #1, task B lasts for three weeks beginning in week #3, task C lasts for three

weeks beginning in week #5. Assume that each task consumes resources evenly over the weeks that they are scheduled.

Prepare a spreadsheet similar to the one shown to calculate the budgeted cost of work scheduled for the whole schedule. At the end of week #4, the percent complete for task A, B, and C was 90 percent, 55 percent, and 25 percent, respectively. The actual cost of work performed is $11,850 at the end of week #4. Calculate the cost and schedule variances for week #4. Is the project ahead (or behind) and under (or over) budget?

| | Earned Value Analysis | | | | | | End of Week: | | #0 |
| | Work Task | | | | | | Schedule | | Cost |
Week	A	B	C	Weekly	BCWS	BCWP	Variance	ACWP	Variance
1									
2									
3									
4									
5									
6									
7									
8									
Total	5,000	15,000	6,000	26,000					
% Complete									

21. Describe the four principal elements in teamwork.

22. Describe the two forms of voting used in decision making; list the pros and cons of each.

23. Why is a consensus decision preferred?

24. How important is an agenda?

25. Describe ethical values and how they are interwoven into the ASME Code of Ethics.

26. Use the Internet and connect to one of the following Web sites. Briefly describe the fundamental ethical dilemma. What ethical values are involved? What canon for the ASME Code of Ethics could have prevented or helped to correct the situation?

    http://ethics.tamu.edu/pritchar/an-intro.htm

    http://www.onlineethics.org/eng/cases.html

Glossary

activity analysis A procedure used to learn how the customer will use and ultimately retire the product.

actual cost of work performed (ACWP) The accumulated amount actually expended to complete work tasks of a project.

adaptive design The adaptation of a known solution to accomplish a new task.

additive process A manufacturing process that builds a part by adding layers of material.

adhesive bonding The joining of two materials with an adhesive substance.

agenda An ordered list of items to be discussed in a meeting.

aggregate objective function An objective function formed by combining a number of separate objective functions.

alpha prototype A reduced-scale or full-scale part or product prototype having the same geometric features, materials, and layout as the intended final assembly.

analogy Making a comparison to show a similarity in some respect; a method of generating ideas by thinking about the problem in general terms so that the characteristics it has in common with other situations will become apparent.

analysis See "engineering analysis."

anodizing A surface treatment done to some metals to improve appearance and provide better weathering ability.

anthropometrics A field of human factors dealing with the physical measurements of human form such as height and/or reach.

artistic design A type of design that deals with an object's appearance rather than its function.

assembly A collection of two or more parts; to put two or more parts together.

assembly drawings Drawings that show the components that make up a product or subassembly and the location of each part in relation to the other parts.

automatic storage and retrieval system (AS/RS) Integrated materials handling hardware and software used to automatically store and retrieve raw materials or finished goods from inventory locations in the factory.

backtracking Redoing some phase of the design process.

basic dimension A numerical size, location, or orientation of a feature on a part, theoretically, exact value of a geometric characteristic such as size, position, or orientation.

benchmark A standard by which something can be compared; a product used to compare a design.

benchmark studies Assessments of competitive products to identify common or unique features.

bending A manufacturing process that plastically deforms sheet metal using a matched punch-and-die set, or a descending punch to wipe-form the workpiece over the edge of the die.

bending stress A normal stress caused by the application of a bending moment.

beta prototype A full-scale, functional part or product prototype using materials and manufacturing processes that will be used in production.

bill of materials (BOM) A table of product component information organized with column headings for part number, part name, material, quantity used in assembly, and other special notes.

blanking A sheet metalworking process that shears a smaller shaped piece of sheetmetal, called a blank, from the stock sheet; blanks are later used in deep drawing.

blow molding A process used to form polymers wherein a molten parison is injected with air then expands to the shape of the mold.

boring A machining process that increases the diameter of an existing hole by feeding a sharp tool into the rotating workpiece.

brainstorming A group method of creative idea generation in which members suggest as many solutions to a given problem as possible before they analyze the feasibility or worthiness of the suggestions.

brazing A process used to join two metal pieces together with the addition of molten brass or zinc solder.

breakeven point Specific volume of production such that the profit is zero; that is, the revenues equal the total costs.

budget An estimate of manpower and other resources used to complete a set of work tasks.

budgeted cost of work performed (BCWP) The sum of the products of the percent complete for each work task times the amount budgeted each work task.

budgeted cost of work scheduled (BCWS) The amount budgeted for a reporting period to complete portions of scheduled work tasks.

bulk deformation Manufacturing processes that change the shape or form of bulk raw materials by compressive or tensile yielding.

casting Processes in which molten metal is poured into a cast to solidify; used to produce complex geometries within broad tolerances.

cavity The stationary part of mold that forms the external shape of a concave part.

ceramic A family of materials described as strong in compression, weak in tension, brittle, stiff, electrically and thermally insulating, not impact-resistant, medium-weight, very temperature-tolerant, very hard, and corrosion-resistant.

chart An illustration used to portray relationships among numerical data.

check listing A creative method of idea generation that makes use of lists made by others, such as the company, a consultant, or an expert.

choice reaction time The time that occurs when one of several stimuli is presented, each of which requires a different response.

chunk A major physical building block of a product.

civil actions Actions taken by the injured party, called the plaintiff, against the manufacturer or seller, called the defendant, to recover damages for personal injury or loss to property caused by defect(s) in design or manufacture.

clearance fit An intentional space between mating parts such as between a bolt and hole.

closed profile A closed-contour or two-dimensional sketch having a finite area.

cluster Gathering together into a group; e.g., clustering elements into chunks.

code of ethics Ethical standards of conduct established under law or by professional organizations.

codes Documents that prescribe how to design and/or build something; e.g., uniform building codes.

coefficient of friction A relative measure of the amount of friction force between two surfaces; equal to the ratio of the friction force divided by the force normal to the surface.

coefficient of thermal expansion A measure of the amount a material elongates in response to a change in its temperature.

combinations An idea generation method that systematically aggregates sub-solutions to create many alternative product concepts. See "morphological matrix."

company requirements Specifications for the design of a product that originate from the company rather than the customer. Company requirements include marketing, manufacturing, financial, and legal considerations.

component decomposition The process of identifying and separating the components of a product into parts and subassemblies.

composite A heterogeneous mixtures of polyester or epoxy resins and fibers made from materials including: glass, carbon, Kevlar, fibers, and metal. Composites can be stiff, strong, light, nonconducting, and moderately corrosion-resistant, but are somewhat sensitive to temperature.

compound die A complex die that has moving plates that perform two or more metal deformation processes with one stroke of the press.

compound interest Interest amounts which are allowed to accumulate from one period to the next.

compression molding A process that forms a charge of thermoset or elastomer between heated mold halves under pressure while the material cures.

compressive strength A measure of the amount of compressive force per unit area that a material can withstand before it fails.

computer-aided design (CAD) Software used to prepare two-dimensional and three-dimensional drawings of product parts and assemblies.

computer-aided engineering (CAE) Computer packages that aid the engineering of a product including determining loads, stresses, vibrations, motions, heat transfer, fluid mechanics, and thermodynamics.

computer-aided inspection (CAI) Computer software that assists in obtaining and recording raw material and finished goods inspection data.

computer-aided process planning (CAPP) Computer software that establishes the type and sequence of manufacturing processes for each part produced.

computer-aided tolerancing (CAT) Software used to analyze and specify part tolerances with respect to economic production and reliable product function.

computer numerical control (CNC) Computer software and hardware used to automatically generate parts, such as a CNC milling machine or NC lathe.

concept design An abstract embodiment of a working principle, geometry, and material; a phase of design when the physical principles are selected.

concurrent engineering A team approach to product design, in which team members, representing critical business functions, work together in the same office and are coordinated by one senior manager; cross-functional, colocated, and coordinated teamwork; also called simultaneous engineering.

conductivity Physical property of a material that indicates its ability to either conduct electrical current or transfer heat.

configuration design The selection and arrangement of features on a part; or the selection and arrangement of components on a product; a phase of design when geometric features are arranged and connected on a part, or standard components or types are selected for the architecture; see "product architecture."

configuration requirements sketch A sketch drawn to approximate scale showing the essential surroundings of a part, including forces, flows, features of mating parts, support points or areas, adjacent parts, and obstructions or forbidden areas.

configure To consider the number and type of features or components, their arrangement, and relative dimensions.

consensus decision Decision in which the team thoughtfully examines all of the issues, and agrees upon a course of action which does not compromise any strong convictions of a team member.

constraint A physical, economic, or legal restriction on product form or function.

consumer The individual who uses the product; see "customer."

contiguous configuration The condition that all geometric features are connected together as one part.

coordinate measuring machines (CMM) Hardware and software that obtains and records measurements of parts.

copyright A legal right afforded by U.S. law that protects literary works such as books, music, drama, software, and artistic works.

core The moving part of a mold that forms the internal shape of a concave part.

correlation ratings matrix A portion of the house of quality listing values that rate the correlation between qualitative customer requirements and quantitative engineering characteristics.

corrosion resistance The ability of a material to resist chemical or electrochemical attack.

cost variance The difference between the budgeted cost of work performed and the actual cost of work performed.

coupling An interdependent relationship between variables.

coupling matrix The triangular roof of the house of quality that forms a matrix of rating values that estimate the amount of interaction between engineering characteristics.

creep The long-term stretching behavior of a material under load at high temperatures.

creep resistance The ability of a material to resist stretching while under load over long time periods at elevated temperatures.

criminal actions Actions that are directed at individuals who may have been negligent in the performance of their duties and may result in prison sentences.

customer The individual who purchases a product; e.g., a grandparent is the customer for a toddler's toy.

customer requirements Specifications for the design of a product that originate from the customer rather than the company. Customer requirements include functional performance, operating environment, human factors, safety, robustness, maintenance, and repair.

customer satisfaction A measure of how well a product fulfills customer desires.

customer satisfaction curve A graph that illustrates customer satisfaction as a function of an engineering characteristic; e.g., satisfaction versus cost.

customer survey An investigation to obtain customer-needs data.

cycle time The time to produce one part; measured in time per batch or parts per batch.

datum feature Theoretical exact size, axis, or plane from which geometric characteristics of a part are established.

decision A choice among alternatives.

decision-making process The process used to identify alternatives and the outcomes associated with each alternative, and to judge the outcomes to make a selection.

deliverables Items to be completed and given to a customer at specific times during a project such as design reports, drawings, performance simulation summaries, and prototypes.

density The amount of matter per unit volume; two measures for density are mass density and weight density.

design A set of decision-making processes used to determine the form of a product given the functions desired by the customer; processes to prescribe the sizes, shapes, material compositions, and arrangements of parts so that the resulting machine will perform a required task; a package of information such as drawings and specifications sufficient to manufacture a product.

design analysis A portion of the design process that predicts or simulates the performance of each alternative, reiterating to assure that all the candidates are feasible; see "design evaluation."

design artifact/object The item that is being designed; e.g., a product, process, system, or facility.

design concept The abstract embodiment of a physical principle, material, and geometry; same as concept design.

design evaluation The portion of the design process that assesses or weighs results from the analyses to determine which alternative is the best.

design for adjustability An approach used to design products and/or machinery by designing adjustable parts to fit all human beings such as a microphone stand, or car seat.

design for assembly A set of practices that aim to reduce the time and cost required to assemble a product, by improving part handling, insertion, and fastening.

design for close fit A design approach that establishes sets of sizes to match classes of customers, such as shoe sizes, hat sizes, or ring sizes.

design for manufacture A set of practices that aim to improve the fabrication of individual parts.

design for safety A collection of design methods which aim to reduce the risk and severity of personal injury or property damage.

design for the environment A group of design methods that aim to minimize the use of raw materials, increase recyclability, improve remanufacture, increase energy efficiency, and improve the workplace environment.

design for the extreme Design method that strives to fit the smallest or largest person.

design for X A term used to describe any of the various design methods that focus on specific product development concerns.

design method A procedure or set of guidelines for solving a design problem.

design patent A document granting legal monopoly rights to produce, use, sell, or profit from a design; design patents are granted for ornamental aspects of a product such as shape, configuration, and/or any surface decoration, including toys, automotive trim products and household products.

design phase A period or stage in the design of a part or product; concept design phase, configuration design phase, parametric design phase.

design problems Product deficiencies that require resolution; product opportunities that require consideration.

design process The problem-solving process used to formulate a design problem, generate alternative solutions, analyze and evaluate the feasibility and performance of each alternative, and select the best alternative, reiterating if necessary.

design reviews An informal or formal meeting to discuss design project progress to date.

design variable A parameter that can be arbitrarily selected by the designer that influences the behavior of the design candidates; a controllable variable.

design-project report A report that summarizes the work tasks undertaken in a design project and discusses the recommended design in detail; usually includes content on the nature of the design problem, design formulation, concept design, configuration design, parametric design, prototype testing, and detail design description and performance.

detail design A phase of design that results in the preparation of a package of information that includes drawings and specifications sufficient to manufacture a product.

detail drawings Drawings showing orthographic projection views of a part (e.g., front, side, or top); showing geometric features drawn to scale along with full dimensions, tolerances, manufacturing process notes, and title block; detail drawings sometimes include section views and oblique plane views.

detection An assessment of the probability that the controls will detect the cause of the failure mode.

diagram Illustration that is intended to explain how something works or the relationship between the parts.

die Mold used in die casting.

die casting The solidification of molten material after it is injected into a mold under high pressure.

dimensions Quantities that are used to size, locate, or orient a feature of a part; e.g., sizes for length, width, height, diameter, radius.

discounted payback period (DPP) The number of years for the discounted cash flows to equal zero.

discrete design variable A variable that is restricted to specific values, noncontinuous.

drawing The plastic deformation of a bar, rod, or wire as it is pulled through successively smaller dies; also, the plastic deformation of sheet metal into a die, forming cupped, box, or hollow parts.

drilling A machining process that removes material from the workpiece using a rotating bit, thus forming a hole.

ductility The characteristic of a material to deform plastically.

earned value analysis A method used in project engineering to determine whether a project is ahead or behind schedule and either over or under budget.

effective interest rate The interest rate per interest period, equal to the ratio of the nominal rate divided by the number of interest periods per year.

ejection hazards Those that cause injuries as debris particles are flung from moving machine parts.

ejector pin A cylindrical device that pushes a frozen, shrunken part off the core half of a mold.

electrical discharge machining A process of removing metal by means of an electrical discharge spark.

embodiment Manifestation or incarnation of something.

embossing Plastic indentation of surface to form ribs, beads, or lettering on the surface of metal.

empathy A creative idea generation method from Synectics; putting one's self in another's place; identifying physically and personally with the part, product, or process that is to be created.

energy expenditure In human factors, the amount of energy consumed to perform a task.

engineering analysis Using tools such as analytical models or empirical equations to predict the performance or behavior of an object or system.

engineering characteristics Quantities that measure the ability of a product to meet customer and company requirements; e.g., miles/gal is an engineering characteristic of a vehicle that measures whether it would meet customers' fuel economy requirements.

engineering design The set of decision-making processes and activities used to determine the form of an object given the functions desired by the customer.

engineering design specification (EDS) Document containing a comprehensive description of intended uses and functional requirements of a product.

engineering-change notice (ECN) Brief description of changes to be made to a product; detailed on a company-approved form that is subsequently authorized and distributed to all the critical departments in a manufacturing organization.

entanglement hazards Include rotating shafts or workpieces that can catch hair or loose clothing.

entrapment hazards Those that cause injuries when machine parts move towards each other to create pinch points where fingers, hands or other body parts can get pinched or crushed.

equipment Machines or apparatus designed to perform a thermal, chemical, or mechanical process or portion thereof.

ergonomics A synonym for human factors, principally used in Europe.

ethics Standards of conduct.

evaluation The process of rating or assessing the predicted performance of feasible design candidates against established evaluation criteria; used to determine the "best" design alternative.

evaluation criteria Standards by which alternatives may be rated, assessed, or ranked.

external undercut A geometric feature that prevents removal of a part from the cavity half of a mold.

extrusion Bulk deformation process that squeezes heated metal or plastic materials through die producing constant cross section.

facing A finishing operation that removes material from a cylindrical surface; the cutting tool moves radially as the workpiece rotates.

factor of safety A term used to express the ratio of an allowable load (stress) to the applied load (stress); factors of safety greater than one indicate that the applied forces or moments are less than the allowable and are therefore "safe."

fail-safe design principle A principle used to design components such that upon failure of a component, some critical functions are still performed.

failure mode The way in which a part could fail to perform its desired function.

failure modes and effects analysis (FMEA) A systematic method used to identify and correct potential product or process failures before they occur.

fantasy A creative idea generation method from Synectics in which we imagine or wish that something exists or is possible, even though in reality it may be impractical or impossible.

fatigue strength Measure of how much stress a material may sustain given the number of cyclic load applications.

feasible design A design candidate that meets design specifications and/or satisfies design constraints.

feasible region The set of points in the design space that satisfy the constraints.

feature A geometric detail on a part, e.g., wall, hole, notch, rib, louver, boss, and fillet.

figure A diagram that illustrates textual matter.

finishing Preparing the final surface of a part to protect it from the environment or to enhance the visual appearance; e.g., anodizing, chrome plating, and painting.

fit A term used to describe how well parts assemble or match the user's anthropometric limitations.

fixed costs Cost items that do not vary with production such as administrative labor, advertising, legal, and building depreciation.

focus groups A qualitative research technique that brings a group of people together in a controlled environment to discuss the positive attributes or shortcomings of a product.

forging A deformation process that creates a desired shape by plastically compressing material in two halves of a die set by hammer strokes or a hydraulic press.

form A term used to describe a product's shape, configuration, size, materials of construction, and manufacturing processes.

forming An early stage in team development when participants transition from being an individual to being a team member, politely interacting, learning about the nature of the tasks to be performed, the goals of the project, and the personalities and work styles of fellow members.

formulation A phase of design in which customer and company requirements are determined, engineering design specifications are prepared, and a solution plan is prepared.

function A term used to describe what a product does; the function it performs; what a part or product is supposed to do.

function decomposition Product function decomposition is a process of identifying and separating required subfunctions without regard to possible embodiments.

function sharing The ability of a single physical element to perform more than one function.

functional elements The functions or subfunctions that a product performs.

fused deposition modeling (FDM) A rapid prototyping process that deposits a thin filament of melted (fused) material in precise locations on a horizontal layer, using numerically controlled positioners; as the material solidifies, the prototype model is built, layer by layer.

Gantt chart A bar chart showing the timing of project work tasks.

generating (alternatives) The set of activities and decision-making processes used to create alternatives for later analysis and evaluation.

grinding Machining process to remove material from a surface with an abrasive wheel or tool.

guided iteration Logical reasoning of causes and effects to suggest new choices or alternatives.

handling The grasping, moving, orienting, and placing during assembly.

hardness Resistance of a material to surface penetration or deformation.

hazard A source of danger that has the potential to injure people or damage property or the environment.

honing To sharpen or smooth.

house of quality A technique for structuring product information such as customer requirements, engineering characteristics, competitive benchmark data, and design targets.

human factors Abilities, limitations, and other physiological or behavioral characteristics of humans which affect the design and operation of tools, machines, systems, tasks, jobs, and environments.

human senses Sight, hearing, tactile, vestibular, kinesthetic, smell, and taste.

human-machine interface Any part of a machine that a human senses or actuates with a part of his or her body.

hydraulic press A machine tool that uses pressurized fluid to push a piston that compresses material between die halves.

illustrations Graphic documents such as charts, schematics, figures sketches, and diagrams.

impact hazards Those that cause injuries as translating, rotating, or reciprocating machine parts collide with body parts.

impact strength The ability of a material to absorb sudden dynamic shocks or impacts without fracturing.

importance weights Measures that establish how important each customer or company requirement is with respect to the other requirements.

industrial design Decisions or activities to determine essential product aesthetics and basic functions.

infeasible point A point in the design space, or set of design variable values, that does not satisfy constraints.

injection molding Process that melts thermoplastic pellets and injects the molten material into a mold under high pressure.

insertion Moving a part into another in the assembly for fastening or joining.

integral architecture The general structure of a product where a single chunk implements many functional elements, or many chunks implement one function; and the interaction is ill defined.

intellectual property Ideas or concepts that should be protected with contracts, copyrights, trademarks, patents, or trade secrets.

interest The amount paid for the use of money.

interference fit An intentional lack of space between mating parts, such as a gear press-fitted onto a shaft.

internal undercut A geometric feature that prevents removal of a part from the core half of a mold.

inverse analysis Rearranging analysis equations, taking the inverse, to solve for "form"; e.g., given $a = F/m$, a (acceleration) is the function; rearranging we can determine the form (mass) $m = F/a$.

inversion A creative idea generation method that "inverts" the problem or proposed solutions in one or more of several possible ways; e.g., turns things inside out or upside down, noisy instead of quiet.

investment casting A casting process that uses wax patterns dipped in a ceramic slurry to create a mold; wax is removed by melting; lost-wax process.

kinesthetic sense A human sense that uses receptors to feel joint and muscle movement.

laminated-object manufacturing (LOM) Rapid prototyping process that builds each prototype by laminating together thin layers of paper, polymer, or sheet steel, previously cut by a numerically controlled laser.

lapping A finishing process used to polish a surface with a slurry of fine abrasive particles.

layout sketch Illustration drawn roughly to scale to show the geometric or spatial arrangement of the selected components and illustrate their relative shapes and sizes.

loss The amount of total costs that exceed total revenues.

machine Combination of resistant bodies, so arranged that by their means, the mechanical forces of nature can be compelled to do work accompanied by certain determinate motions (Franz Reuleaux, 1876).

machine tool A machine that makes other parts or products.

machining A subtractive process that removes material from a workpiece by a sharp cutting tool that shears away chips of material, to create desired form or features.

manufactured waste Raw materials that become scrap during the making of a part.

masking A term used in human factors to describe how a sound may be concealed or masked by another sound or noise.

material-requirements planning (MRP) Computer software that "explodes" a bill of materials to determine the total amount of materials required for a production run and when they are required.

mechanical press A machine used for sheet metal working and forging operations; an electric motor energizes a flywheel that strokes the hammer or punch.

mechanical property A quantity that characterizes the behavior of a material in response to external or applied forces.

memoranda Written document, three to nine pages in length, sent to a broad audience, and covering more topics in greater depth than a letter.

metals Family of materials that are ductile, strong, stiff, electrically conductive, thermally conductive, fatigue-resistant, creep-resistant, impact-resistant, heavy, temperature-tolerant, medium-hard, but not very corrosion-resistant.

milestone A special event that occurs during a project.

milling A machining process that removes material from a flat surface to form slots, pockets, recesses; a cutting tool rotates as the material is fed.

mission statement A paragraph that states what the project is going to do, for whom, and how the project team will go about doing it.

modular architecture Structure of a product such that chunks implement one or a few functions, and the interactions between chunks are well defined.

modulus of elasticity A mechanical property that measures the stiffness of a material; equal to the ratio of the stress required to produce a unit strain.

mold closure direction The direction in which the core and cavity halves open and close in relation to the part.

morphological matrix A tabular arrangement of subfunctions versus alternative-concept solutions.

multiattribute optimization A method to optimize a problem having more than one objective function.

NC/CNC machining Numerically controlled/computer numerically controlled machining.

nominal interest rate The interest rate per year, typically quoted in interest related loan agreements.

normal probability distribution The most commonly occurring probability distribution, also called a Gaussian distribution.

norming A stage in team development when members begin to cooperate with each other, begin to understand and respect individual strengths and weaknesses, focus on common goals and communicate more openly.

objective function A function of design variables; used to estimate the merit of alternative designs.

occupational safety Protection provided to workers by local, state and federal laws, the most notable being the federally mandated Occupational Health and Safety Act of 1970.

optimize Selecting the best feasible design.

organization chart A diagram illustrating the chain of command; used to establish responsibility and accountability.

original design Development of a new component, assembly, or process that had not existed before (Ullman, 1997).

painting Finishing process used to protect a surface or enhance visual appearance.

parametric design A phase of design that determines specific values for the design variables.

part A piece that requires no assembly.

parting surface/line The surface or line where the core and cavity mold surfaces meet.

patent See "design patent," "utility patent."

payback period The amount of time that it takes for the cash flows to recover the initial investment.

performance A measure of how well a product functions.

performing A stage in team development when the team is productive and satisfaction is high.

physical element An element having an embodiment; clustered with functional elements during the development of product architecture.

physical principles Concepts or relationships from physics that can be selected to perform a function; e.g., spring force, magnetic force, pneumatic force.

physical property A quantity that characterizes a material's response to physical phenomena other than mechanical forces; e.g., melting point, electrical conductivity.

physical prototypes Reduced-scale or full-size models of a part or product.

piece-part A part.

planing Machining process that removes material using a translating cutter as the workpiece feeds.

plating Finishing process that chemically alters the surface of a part.

polishing Finishing process that uses abrasive powders embedded in a rotating leather or felt wheel to remove surface irregularities.

polymers Family of materials that are strong, flexible, electrically and thermally insulating, impact-resistant, lightweight, temperature-sensitive, soft, and corrosion-resistant, but not creep-resistant.

preliminary design Early phases of design, including concept, configuration, and parametric design activities.

preproduction prototype A full-scale part or product made and assembled with final materials and production-like processes.

primary intended purpose The main purpose, function, or service for which the customer buys a product.

primary manufacturing process The manufacturing process that principally alters the material's shape or form.

principal The amount of money borrowed or lent.

problem definition parameter A quantity whose value imposes a specific condition of use or manufacture.

problem statement A simple sentence describing what the problem is, what the desired end state is, what currently prevents it, and how we know when we have achieved the desired outcome.

process plant A combination of systems used to thermally, mechanically, or chemically process energy or materials (both organic and inorganic).

product Designed object or artifact that is purchased and used as a unit (Dixon and Poli, 1995).

product anatomy The fundamental structure of components that compose a product.

product architecture The way in which the physical elements and functional elements of a product are arranged or clustered into chunks and how the chunks interact.

product complexity A qualitative measure of the number and type of components that make up a product.

product development That portion of the product realization process that begins with formulation activities and concludes with production planning and manufacturing engineering activities; excludes activities beginning with production.

product life cycle The cycle of birth and death of a product; starts with the introduction of the product into the market, includes stages of growth, maturity, and decline; ends with removal of product from market.

product-component decomposition The process of identifying and separating the components of a product into parts and subassemblies.

product concept test A market research activity that uses a reduced-scale or full-scale model of a new product or "product concept"; usually nonfunctional but looks like a "finished" product.

product-data management Computer software that manages product development information in a company; usually includes revision control features and security provisions.

product realization process (PRP) Design and manufacturing processes that convert information, materials, and energy into a finished product.

production design Designing and planning the type and arrangement of equipment and the use of labor in a factory to make a product.

products liability A phrase used to describe the legal responsibility to customers under civil action or criminal actions.

profit The amount of total revenues in excess of total costs.

progress report A document sent to clients and other stakeholders to communicate project status in regard to workscope, schedule, and budget.

progressive die A hardened metal tool used to cut or form sheet metal, divided into two or more stations, each performing a separate operation with each stroke of press.

project A unique sequence of work tasks, undertaken once, to achieve a specific set of objectives.

project control Detecting whether a project is on time and within budget, deciding upon and implementing corrective actions if necessary.

project objectives Specific outcomes that a project team expects to produce.

project plan A package of key items to be completed in a project including problem statement, mission statement, project objectives, work breakdown structure, scope of work, responsibilities table, organization chart, budget, schedule, and risk assessment.

Pugh's concept selection method A method used to evaluate alternatives by comparing them to a base case; uses "+," "−," and S symbols to rate them better, worse, or same as.

punch A hardened metal tool that is pushed into a die to cut or form sheet metal.

punching Sheet metalworking process that produces features such as slots, notches, extruded holes, and holes using a punch-and-die.

quality function deployment (QFD) A team-based method that draws upon the expertise of the group members to carefully integrate the voice of the customer in all activities of the company.

quality product One that works as it should, lasts a long time, and is easy to maintain.

range of motion The amount of translation or rotation.

rapid prototyping Processes that use computers and computer-controlled equipment to automatically and rapidly fabricate prototypes.

rapid tooling Use of rapid prototyping to make production tooling or generate tooling inserts.

reaming Refining the diameter of an existing hole.

redesign The process of revising any portion of an existing product's form (i.e., shape, configuration, size, materials, or manufacturing processes); selecting new values for the design variables, reanalyzing, and reevaluating, to obtain better performance and improve customer satisfaction.

redundant design principle A principle used to design components such that additional components or systems, configured in parallel or series, take over the principle function of the failed component or system.

relative dimensions The ratio of one dimension with respect to another; a needle has a small diameter in relation to its length; a sheet is thin compared to its width and height.

research report Document that is similar to a test report, but is longer in length and broader in coverage; includes sections such as an abstract, background, literature review, laboratory/test program description, and bibliography.

respecification Changing design specifications, usually because few, if any, feasible designs are possible.

revenues Amounts received by a company for products or services sold.

reverse engineering The process of physically disassembling an existing product to learn how each component contributes to the product's overall performance (Otto and Wood, 2001).

risk assessment A consideration of the likely risks that face a project and possible contingency plans to overcome such situations.

risk priority number A metric used to assess the risk of a failure mode.

robust design Methods used to design robust products; see "robust product."

robust product One that performs in spite of variations in its material properties, how it was manufactured, the operating environment, or how it is used.

rolling Bulk deformation process used to form sheets, bars, rods, and structural shapes by plastically compressing slabs, billets, and blooms between two rollers.

safe-life design principle A design principle used to design components to operate for their entire predicted useful life without breakdown or malfunction.

safety hierarchy A list of prioritized actions to reduce risk if injury or property damage.

sand casting Molten metal solidifies in a mold made of sand. Standard mold is formed by packing sand around a pattern with the same external shape as the part of the cast.

satisfaction curve Graph relating a solution evaluation parameter to customer satisfaction.

sawing Cutting/dividing material with a toothed blade.

schedule List of work tasks and the dates when they start and finish.

schedule variance The difference between the budgeted cost of work performed and the budgeted cost of work scheduled.

schematic Diagrams of electrical or mechanical systems using abstract symbols.

scope of work A detailed written list of tasks to be completed during the project.

secondary manufacturing process Manufacturing process that adds or removes geometrical features from the basic forms.

sectioned assembly drawing A cutaway portion of the assembly drawing that exposes the details of an interior portion of the assembly.

selection design Decision-making processes used to match the desired functional requirements of a component with the actual performance of standard components listed in vendors' catalogs.

selective laser sintering (SLS) A rapid prototyping process that uses a high-power laser to sinter together fusible materials, such as powdered metals, layer by layer.

sensitivity analysis Analyzing the contribution of a part's variance to the total variance of the assembly.

sensory inputs Human senses used to interact with products, machines, and the environment.

service bureau A third party that fabricates prototypes for a fee.

severity (of a failure mode effect) An assessment of the extent of damage or injury to the product, the user, nearby people, or the environment.

shape The geometrical profile or outline of a part; e.g., circular, triangular, L-shaped, spherical.

shaping Machining process that removes material from a translating workpiece and a stationary cutter.

shear strength Mechanical property indicating the largest stress a material can sustain under shear loading before it yields or fractures.

shearing Cutting or separating sheet metal along a straight line; used to size sheets for subsequent operations.

sheet metalworking Permanent deformation of thin metal sheets produced by bending or shearing forces; often called *stamping*.

side-action feature A geometric feature such as a hole or boss that necessitates an additional press operation after bending.

side core A sliding part of a complex mold used to form undercuts.

simple interest The amount earned per interest period ignoring any amounts earned during prior interest periods.

simple payback period Equal to the investment divided by the annual savings.

simple reaction time The time to initiate a response when only one particular stimulus occurs and the same response is always required.

size Numerical value of a dimension.

sketch Hand-drawn, preliminary, or rough "drawings"; drawn without the use of drawing instruments.

soldering Process used to join two metal pieces together with the addition of molten tin, lead, and silver alloys.

solid modeling Representing a part's geometrical features using topological information such as cylinders, blocks, fillets, slots, holes, and ribs.

solution evaluation parameter Quantity that measures how well a product meets customer or company requirements; engineering characteristic chosen to evaluate candidate designs.

special-purpose assembly An assembly that is designed for a specific application; not a standard component.

special-purpose part Part not typically available or satisfactory for intended purpose.

specific heat The amount of heat required to increase the temperature of a unit mass 1 degree.

stamping Sheet metalworking processes.

stamping die Collection of components, including one or more punches, mounted on a press.

standard part A common interchangeable item, having standard features, typically mass-produced and used in various applications; e.g., nut, bolt, screw, washer, lubricant.

standard (sub)assembly One that is routinely manufactured for general use; e.g., pump, motor, valve, switch.

statistical process control (SPC) Software that obtains and analyzes process characteristic statistics.

statistical tolerance design method A method that uses stochastic measures of each manufacturing process to estimate the probability that parts will fit and/or function based on the actual statistical capability of the processes.

storming A stage of team development when members begin to realize the enormity of the project, recognize differences in individual abilities, personalities, and work styles leading to possible disagreement and conflict.

stereolithographic apparatus (SLA) A material additive process that uses a high-power laser to selectively solidify a liquid photopolymer, layer by layer, into the shape of a finished prototype.

stiffness Characteristic of a material that indicates its resistance to bending, twisting, or stretching.

strain A measure of relative elongation; $\epsilon = \Delta L/L$.

strength Characteristic of a material that indicates how much load is required to deform or break the material.

stress The force intensity per unit area; $\sigma = P/A$.

subassembly An assembly that is included in another assembly or subassembly.

subtractive process A process that removes material from a workpiece; produces manufactured waste.

Synectics A set of four techniques for idea creation: inversion, fantasy, analogy, and empathy (Gordon, 1961).

synthesis Creative processes to generate new ideas or alternatives.

system Two or more pieces of equipment used to perform a set of processes.

team A group of people that have complementary skills and knowledge that work together toward common goals and hold each other mutually accountable.

teamwork A demonstrated attitude and ability to accomplish team goals.

tensile strength See ultimate tensile strength.

tertiary manufacturing process Surface treatment such as polishing, painting, heat-treating, and joining.

test report Document detailing engineering or scientific tests on materials, prototypes, and/or products; variable in length; contents usually include sections on test objectives, procedures, data/results, summary, and recommendations.

theoretical minimum number of parts Minimum number of parts needed for proper functioning of a product.

thermal forming Vacuum forming thin sheets of thermoplastic.

thermoplastic Material that can be repeatedly softened by heating and hardened by cooling.

thermoset Material that permanently sets, or cures, by heating.

through groove Groove that joins a hole to the parting plane.

through hole One that completely penetrates the part.

tinkering Repetitive or iterative cutting and trying, fabrication, and testing; does not use scientific principles or mathematics to predict behavior.

tolerance difference between maximum and minimum size limits of a feature on a part; total permission variation in value of a dimension: the total amount by which a given dimension may vary.

tolerance influencing feature Features that influences the ability of a process to attain tolerance specifications. Ex: undercuts, unsupported walls and unsupported significant projections, supported walls and supported significant projections, primitive connections or intersections, parting plane.

tolerance stack The sum of the tolerances of individual parts assembled as a unit.

tolerance tightness Restrictive demands placed on tolerances, specifying relative variation from standard tolerance abilities.

tolerancing plan Selection of tolerance types, tightness, and quantity, in addition to the designed features and manufacturing processes of a part.

torsion Twisting load; moment applied colinearly with axis of body.

toughness The ability of a material to plastically deform before fracturing; measured by the modulus of toughness.

trade dress Distinctive look of a product or place of business.

trade secret A method used to protect intellectual property.

trademark A symbol, design, word, or combination thereof, used by a manufacturer to distinguish its products from those of its competitors, principally to distinguish its source; e.g., IBM, GE, Xerox, and Coke.

trade-off Compromise made between two favorable outcomes.

turning Machining process that removes material from rotating workpiece; lathes.

ultimate tensile strength Mechanical property of a material that measures the largest load that can be applied to a material.

undercut See "internal undercut" and "external undercut."

utility patent A document granting legal monopoly rights to produce, use, sell, or profit from an invention; e.g., Xerox copying, household appliances, light bulbs, machinery, and cameras.

value The ratio of benefits to costs; the greater a product's benefits in relation to its costs, the more value it has.

variable costs Cost items that depend upon the amount of product manufactured such as raw materials and production labor.

variant design Type of design; modifying the performance of an existing product by varying some of its design variable values or product parameters such as size, or specific material, or manufacturing processes (Pahl and Beitz, 1996).

vestibular sense The inner ear sense that provides human balance.

virtual prototype Nonreal, electronic prototype; modeled inside the memory of a computer.

wear coefficient A mechanical property that measures of the amount of surface removal due to rubbing and sliding.

weighted-rating method The product of an importance weight and an evaluation rating.

welding Fastening process that permanently joins two or more metal parts by controlled melting; fusion of metals.

wire drawing Process that transforms bar stock by pulling it through a set of successively narrowing dies, forming a long strand of wire that is usually wound on a spool as a continuous process.

work breakdown structure A diagram of major work tasks to be completed in a project.

work task A specific activity to be performed during a project.

working geometry The surfaces that a physical principle acts upon along with the motion that results.

working principle The combination of a working geometry and material.

worst-case tolerance design A method that assumes that each manufacturing process will produce parts with the "worst" precision within its capability.

yield strength Mechanical property that measures the tensile stress at which a material yields; often denoted as S_y.

APPENDIX B

Interest Tables

	Interest Factors						i = 0.50%		
n	P/F	P/A	F/P	F/A	A/P	A/F	P/G	A/G	n
1	0.9950	0.9950	1.0050	1.0000	1.0050	1.0000	0.0000	0.0000	1
2	0.9901	1.9851	1.0100	2.0050	0.5038	0.4988	0.9901	0.4988	2
3	0.9851	2.9702	1.0151	3.0150	0.3367	0.3317	2.9604	0.9967	3
4	0.9802	3.9505	1.0202	4.0301	0.2531	0.2481	5.9011	1.4938	4
5	0.9754	4.9259	1.0253	5.0503	0.2030	0.1980	9.8026	1.9900	5
6	0.9705	5.8964	1.0304	6.0755	0.1696	0.1646	14.6552	2.4855	6
7	0.9657	6.8621	1.0355	7.1059	0.1457	0.1407	20.4493	2.9801	7
8	0.9609	7.8230	1.0407	8.1414	0.1278	0.1228	27.1755	3.4738	8
9	0.9561	8.7791	1.0459	9.1821	0.1139	0.1089	34.8244	3.9668	9
10	0.9513	9.7304	1.0511	10.2280	0.1028	0.0978	43.3865	4.4589	10
11	0.9466	10.6770	1.0564	11.2792	0.0937	0.0887	52.8526	4.9501	11
12	0.9419	11.6189	1.0617	12.3356	0.0861	0.0811	63.2136	5.4406	12
13	0.9372	12.5562	1.0670	13.3972	0.0796	0.0746	74.4602	5.9302	13
14	0.9326	13.4887	1.0723	14.4642	0.0741	0.0691	86.5835	6.4190	14
15	0.9279	14.4166	1.0777	15.5365	0.0694	0.0644	99.5743	6.9069	15
16	0.9233	15.3399	1.0831	16.6142	0.0652	0.0602	113.4238	7.3940	16
17	0.9187	16.2586	1.0885	17.6973	0.0615	0.0565	128.1231	7.8803	17
18	0.9141	17.1728	1.0939	18.7858	0.0582	0.0532	143.6634	8.3658	18
19	0.9096	18.0824	1.0994	19.8797	0.0553	0.0503	160.0360	8.8504	19
20	0.9051	18.9874	1.1049	20.9791	0.0527	0.0477	177.2322	9.3342	20
21	0.9006	19.8880	1.1104	22.0840	0.0503	0.0453	195.2434	9.8172	21
22	0.8961	20.7841	1.1160	23.1944	0.0481	0.0431	214.0611	10.2993	22
23	0.8916	21.6757	1.1216	24.3104	0.0461	0.0411	233.6768	10.7806	23
24	0.8872	22.5629	1.1272	25.4320	0.0443	0.0393	254.0820	11.2611	24
25	0.8828	23.4456	1.1328	26.5591	0.0427	0.0377	275.2686	11.7407	25
30	0.8610	27.7941	1.1614	32.2800	0.0360	0.0310	392.6324	14.1265	30
40	0.8191	36.1722	1.2208	44.1588	0.0276	0.0226	681.3347	18.8359	40
50	0.7793	44.1428	1.2832	56.6452	0.0227	0.0177	1035.6966	23.4624	50
60	0.7414	51.7256	1.3489	69.7700	0.0193	0.0143	1448.6458	28.0064	60
70	0.7053	58.9394	1.4178	83.5661	0.0170	0.0120	1913.6427	32.4680	70
80	0.6710	65.8023	1.4903	98.0677	0.0152	0.0102	2424.6455	36.8474	80
90	0.6383	72.3313	1.5666	113.3109	0.0138	0.0088	2976.0769	41.1451	90
100	0.6073	78.5426	1.6467	129.3337	0.0127	0.0077	3562.7934	45.3613	100

	Interest Factors						i = 1.00%		
n	P/F	P/A	F/P	F/A	A/P	A/F	P/G	A/G	n
1	0.9901	0.9901	1.0100	1.0000	1.0100	1.0000	0.0000	0.0000	1
2	0.9803	1.9704	1.0201	2.0100	0.5075	0.4975	0.9803	0.4975	2
3	0.9706	2.9410	1.0303	3.0301	0.3400	0.3300	2.9215	0.9934	3
4	0.9610	3.9020	1.0406	4.0604	0.2563	0.2463	5.8044	1.4876	4
5	0.9515	4.8534	1.0510	5.1010	0.2060	0.1960	9.6103	1.9801	5
6	0.9420	5.7955	1.0615	6.1520	0.1725	0.1625	14.3205	2.4710	6
7	0.9327	6.7282	1.0721	7.2135	0.1486	0.1386	19.9168	2.9602	7
8	0.9235	7.6517	1.0829	8.2857	0.1307	0.1207	26.3812	3.4478	8
9	0.9143	8.5660	1.0937	9.3685	0.1167	0.1067	33.6959	3.9337	9
10	0.9053	9.4713	1.1046	10.4622	0.1056	0.0956	41.8435	4.4179	10
11	0.8963	10.3676	1.1157	11.5668	0.0965	0.0865	50.8067	4.9005	11
12	0.8874	11.2551	1.1268	12.6825	0.0888	0.0788	60.5687	5.3815	12
13	0.8787	12.1337	1.1381	13.8093	0.0824	0.0724	71.1126	5.8607	13
14	0.8700	13.0037	1.1495	14.9474	0.0769	0.0669	82.4221	6.3384	14
15	0.8613	13.8651	1.1610	16.0969	0.0721	0.0621	94.4810	6.8143	15
16	0.8528	14.7179	1.1726	17.2579	0.0679	0.0579	107.2734	7.2886	16
17	0.8444	15.5623	1.1843	18.4304	0.0643	0.0543	120.7834	7.7613	17
18	0.8360	16.3983	1.1961	19.6147	0.0610	0.0510	134.9957	8.2323	18
19	0.8277	17.2260	1.2081	20.8109	0.0581	0.0481	149.8950	8.7017	19
20	0.8195	18.0456	1.2202	22.0190	0.0554	0.0454	165.4664	9.1694	20
21	0.8114	18.8570	1.2324	23.2392	0.0530	0.0430	181.6950	9.6354	21
22	0.8034	19.6604	1.2447	24.4716	0.0509	0.0409	198.5663	10.0998	22
23	0.7954	20.4558	1.2572	25.7163	0.0489	0.0389	216.0660	10.5626	23
24	0.7876	21.2434	1.2697	26.9735	0.0471	0.0371	234.1800	11.0237	24
25	0.7798	22.0232	1.2824	28.2432	0.0454	0.0354	252.8945	11.4831	25
30	0.7419	25.8077	1.3478	34.7849	0.0387	0.0287	355.0021	13.7557	30
40	0.6717	32.8347	1.4889	48.8864	0.0305	0.0205	596.8561	18.1776	40
50	0.6080	39.1961	1.6446	64.4632	0.0255	0.0155	879.4176	22.4363	50
60	0.5504	44.9550	1.8167	81.6697	0.0222	0.0122	1192.8061	26.5333	60
70	0.4983	50.1685	2.0068	100.6763	0.0199	0.0099	1528.6474	30.4703	70
80	0.4511	54.8882	2.2167	121.6715	0.0182	0.0082	1879.8771	34.2492	80
90	0.4084	59.1609	2.4486	144.8633	0.0169	0.0069	2240.5675	37.8724	90
100	0.3697	63.0289	2.7048	170.4814	0.0159	0.0059	2605.7758	41.3426	100

	Interest Factors					i = 2.00%			
n	P/F	P/A	F/P	F/A	A/P	A/F	P/G	A/G	n
1	0.9804	0.9804	1.0200	1.0000	1.0200	1.0000	0.0000	0.0000	1
2	0.9612	1.9416	1.0404	2.0200	0.5150	0.4950	0.9612	0.4950	2
3	0.9423	2.8839	1.0612	3.0604	0.3468	0.3268	2.8458	0.9868	3
4	0.9238	3.8077	1.0824	4.1216	0.2626	0.2426	5.6173	1.4752	4
5	0.9057	4.7135	1.1041	5.2040	0.2122	0.1922	9.2403	1.9604	5
6	0.8880	5.6014	1.1262	6.3081	0.1785	0.1585	13.6801	2.4423	6
7	0.8706	6.4720	1.1487	7.4343	0.1545	0.1345	18.9035	2.9208	7
8	0.8535	7.3255	1.1717	8.5830	0.1365	0.1165	24.8779	3.3961	8
9	0.8368	8.1622	1.1951	9.7546	0.1225	0.1025	31.5720	3.8681	9
10	0.8203	8.9826	1.2190	10.9497	0.1113	0.0913	38.9551	4.3367	10
11	0.8043	9.7868	1.2434	12.1687	0.1022	0.0822	46.9977	4.8021	11
12	0.7885	10.5753	1.2682	13.4121	0.0946	0.0746	55.6712	5.2642	12
13	0.7730	11.3484	1.2936	14.6803	0.0881	0.0681	64.9475	5.7231	13
14	0.7579	12.1062	1.3195	15.9739	0.0826	0.0626	74.7999	6.1786	14
15	0.7430	12.8493	1.3459	17.2934	0.0778	0.0578	85.2021	6.6309	15
16	0.7284	13.5777	1.3728	18.6393	0.0737	0.0537	96.1288	7.0799	16
17	0.7142	14.2919	1.4002	20.0121	0.0700	0.0500	107.5554	7.5256	17
18	0.7002	14.9920	1.4282	21.4123	0.0667	0.0467	119.4581	7.9681	18
19	0.6864	15.6785	1.4568	22.8406	0.0638	0.0438	131.8139	8.4073	19
20	0.6730	16.3514	1.4859	24.2974	0.0612	0.0412	144.6003	8.8433	20
21	0.6598	17.0112	1.5157	25.7833	0.0588	0.0388	157.7959	9.2760	21
22	0.6468	17.6580	1.5460	27.2990	0.0566	0.0366	171.3795	9.7055	22
23	0.6342	18.2922	1.5769	28.8450	0.0547	0.0347	185.3309	10.1317	23
24	0.6217	18.9139	1.6084	30.4219	0.0529	0.0329	199.6305	10.5547	24
25	0.6095	19.5235	1.6406	32.0303	0.0512	0.0312	214.2592	10.9745	25
30	0.5521	22.3965	1.8114	40.5681	0.0446	0.0246	291.7164	13.0251	30
40	0.4529	27.3555	2.2080	60.4020	0.0366	0.0166	461.9931	16.8885	40
50	0.3715	31.4236	2.6916	84.5794	0.0318	0.0118	642.3606	20.4420	50
60	0.3048	34.7609	3.2810	114.0515	0.0288	0.0088	823.6975	23.6961	60
70	0.2500	37.4986	3.9996	149.9779	0.0267	0.0067	999.8343	26.6632	70
80	0.2051	39.7445	4.8754	193.7720	0.0252	0.0052	1166.7868	29.3572	80
90	0.1683	41.5869	5.9431	247.1567	0.0240	0.0040	1322.1701	31.7929	90
100	0.1380	43.0984	7.2446	312.2323	0.0232	0.0032	1464.7527	33.9863	100

	Interest Factors						i = 4.00%		
n	P/F	P/A	F/P	F/A	A/P	A/F	P/G	A/G	n
1	0.9615	0.9615	1.0400	1.0000	1.0400	1.0000	0.0000	0.0000	1
2	0.9246	1.8861	1.0816	2.0400	0.5302	0.4902	0.9246	0.4902	2
3	0.8890	2.7751	1.1249	3.1216	0.3603	0.3203	2.7025	0.9739	3
4	0.8548	3.6299	1.1699	4.2465	0.2755	0.2355	5.2670	1.4510	4
5	0.8219	4.4518	1.2167	5.4163	0.2246	0.1846	8.5547	1.9216	5
6	0.7903	5.2421	1.2653	6.6330	0.1908	0.1508	12.5062	2.3857	6
7	0.7599	6.0021	1.3159	7.8983	0.1666	0.1266	17.0657	2.8433	7
8	0.7307	6.7327	1.3686	9.2142	0.1485	0.1085	22.1806	3.2944	8
9	0.7026	7.4353	1.4233	10.5828	0.1345	0.0945	27.8013	3.7391	9
10	0.6756	8.1109	1.4802	12.0061	0.1233	0.0833	33.8814	4.1773	10
11	0.6496	8.7605	1.5395	13.4864	0.1141	0.0741	40.3772	4.6090	11
12	0.6246	9.3851	1.6010	15.0258	0.1066	0.0666	47.2477	5.0343	12
13	0.6006	9.9856	1.6651	16.6268	0.1001	0.0601	54.4546	5.4533	13
14	0.5775	10.5631	1.7317	18.2919	0.0947	0.0547	61.9618	5.8659	14
15	0.5553	11.1184	1.8009	20.0236	0.0899	0.0499	69.7355	6.2721	15
16	0.5339	11.6523	1.8730	21.8245	0.0858	0.0458	77.7441	6.6720	16
17	0.5134	12.1657	1.9479	23.6975	0.0822	0.0422	85.9581	7.0656	17
18	0.4936	12.6593	2.0258	25.6454	0.0790	0.0390	94.3498	7.4530	18
19	0.4746	13.1339	2.1068	27.6712	0.0761	0.0361	102.8933	7.8342	19
20	0.4564	13.5903	2.1911	29.7781	0.0736	0.0336	111.5647	8.2091	20
21	0.4388	14.0292	2.2788	31.9692	0.0713	0.0313	120.3414	8.5779	21
22	0.4220	14.4511	2.3699	34.2480	0.0692	0.0292	129.2024	8.9407	22
23	0.4057	14.8568	2.4647	36.6179	0.0673	0.0273	138.1284	9.2973	23
24	0.3901	15.2470	2.5633	39.0826	0.0656	0.0256	147.1012	9.6479	24
25	0.3751	15.6221	2.6658	41.6459	0.0640	0.0240	156.1040	9.9925	25
30	0.3083	17.2920	3.2434	56.0849	0.0578	0.0178	201.0618	11.6274	30
40	0.2083	19.7928	4.8010	95.0255	0.0505	0.0105	286.5303	14.4765	40
50	0.1407	21.4822	7.1067	152.6671	0.0466	0.0066	361.1638	16.8122	50
60	0.0951	22.6235	10.5196	237.9907	0.0442	0.0042	422.9966	18.6972	60
70	0.0642	23.3945	15.5716	364.2905	0.0427	0.0027	472.4789	20.1961	70
80	0.0434	23.9154	23.0498	551.2450	0.0418	0.0018	511.1161	21.3718	80
90	0.0293	24.2673	34.1193	827.9833	0.0412	0.0012	540.7369	22.2826	90
100	0.0198	24.5050	50.5049	1237.6237	0.0408	0.0008	563.1249	22.9800	100

	Interest Factors					i = 6.00%			
n	P/F	P/A	F/P	F/A	A/P	A/F	P/G	A/G	n
1	0.9434	0.9434	1.0600	1.0000	1.0600	1.0000	0.0000	0.0000	1
2	0.8900	1.8334	1.1236	2.0600	0.5454	0.4854	0.8900	0.4854	2
3	0.8396	2.6730	1.1910	3.1836	0.3741	0.3141	2.5692	0.9612	3
4	0.7921	3.4651	1.2625	4.3746	0.2886	0.2286	4.9455	1.4272	4
5	0.7473	4.2124	1.3382	5.6371	0.2374	0.1774	7.9345	1.8836	5
6	0.7050	4.9173	1.4185	6.9753	0.2034	0.1434	11.4594	2.3304	6
7	0.6651	5.5824	1.5036	8.3938	0.1791	0.1191	15.4497	2.7676	7
8	0.6274	6.2098	1.5938	9.8975	0.1610	0.1010	19.8416	3.1952	8
9	0.5919	6.8017	1.6895	11.4913	0.1470	0.0870	24.5768	3.6133	9
10	0.5584	7.3601	1.7908	13.1808	0.1359	0.0759	29.6023	4.0220	10
11	0.5268	7.8869	1.8983	14.9716	0.1268	0.0668	34.8702	4.4213	11
12	0.4970	8.3838	2.0122	16.8699	0.1193	0.0593	40.3369	4.8113	12
13	0.4688	8.8527	2.1329	18.8821	0.1130	0.0530	45.9629	5.1920	13
14	0.4423	9.2950	2.2609	21.0151	0.1076	0.0476	51.7128	5.5635	14
15	0.4173	9.7122	2.3966	23.2760	0.1030	0.0430	57.5546	5.9260	15
16	0.3936	10.1059	2.5404	25.6725	0.0990	0.0390	63.4592	6.2794	16
17	0.3714	10.4773	2.6928	28.2129	0.0954	0.0354	69.4011	6.6240	17
18	0.3503	10.8276	2.8543	30.9057	0.0924	0.0324	75.3569	6.9597	18
19	0.3305	11.1581	3.0256	33.7600	0.0896	0.0296	81.3062	7.2867	19
20	0.3118	11.4699	3.2071	36.7856	0.0872	0.0272	87.2304	7.6051	20
21	0.2942	11.7641	3.3996	39.9927	0.0850	0.0250	93.1136	7.9151	21
22	0.2775	12.0416	3.6035	43.3923	0.0830	0.0230	98.9412	8.2166	22
23	0.2618	12.3034	3.8197	46.9958	0.0813	0.0213	104.7007	8.5099	23
24	0.2470	12.5504	4.0489	50.8156	0.0797	0.0197	110.3812	8.7951	24
25	0.2330	12.7834	4.2919	54.8645	0.0782	0.0182	115.9732	9.0722	25
30	0.1741	13.7648	5.7435	79.0582	0.0726	0.0126	142.3588	10.3422	30
40	0.0972	15.0463	10.2857	154.7620	0.0665	0.0065	185.9568	12.3590	40
50	0.0543	15.7619	18.4202	290.3359	0.0634	0.0034	217.4574	13.7964	50
60	0.0303	16.1614	32.9877	533.1282	0.0619	0.0019	239.0428	14.7909	60
70	0.0169	16.3845	59.0759	967.9322	0.0610	0.0010	253.3271	15.4613	70
80	0.0095	16.5091	105.7960	1746.5999	0.0606	0.0006	262.5493	15.9033	80
90	0.0053	16.5787	189.4645	3141.0752	0.0603	0.0003	268.3946	16.1891	90
100	0.0029	16.6175	339.3021	5638.3681	0.0602	0.0002	272.0471	16.3711	100

		Interest Factors					i = 8.00%		
n	P/F	P/A	F/P	F/A	A/P	A/F	P/G	A/G	n
1	0.9259	0.9259	1.0800	1.0000	1.0800	1.0000	0.0000	0.0000	1
2	0.8573	1.7833	1.1664	2.0800	0.5608	0.4808	0.8573	0.4808	2
3	0.7938	2.5771	1.2597	3.2464	0.3880	0.3080	2.4450	0.9487	3
4	0.7350	3.3121	1.3605	4.5061	0.3019	0.2219	4.6501	1.4040	4
5	0.6806	3.9927	1.4693	5.8666	0.2505	0.1705	7.3724	1.8465	5
6	0.6302	4.6229	1.5869	7.3359	0.2163	0.1363	10.5233	2.2763	6
7	0.5835	5.2064	1.7138	8.9228	0.1921	0.1121	14.0242	2.6937	7
8	0.5403	5.7466	1.8509	10.6366	0.1740	0.0940	17.8061	3.0985	8
9	0.5002	6.2469	1.9990	12.4876	0.1601	0.0801	21.8081	3.4910	9
10	0.4632	6.7101	2.1589	14.4866	0.1490	0.0690	25.9768	3.8713	10
11	0.4289	7.1390	2.3316	16.6455	0.1401	0.0601	30.2657	4.2395	11
12	0.3971	7.5361	2.5182	18.9771	0.1327	0.0527	34.6339	4.5957	12
13	0.3677	7.9038	2.7196	21.4953	0.1265	0.0465	39.0463	4.9402	13
14	0.3405	8.2442	2.9372	24.2149	0.1213	0.0413	43.4723	5.2731	14
15	0.3152	8.5595	3.1722	27.1521	0.1168	0.0368	47.8857	5.5945	15
16	0.2919	8.8514	3.4259	30.3243	0.1130	0.0330	52.2640	5.9046	16
17	0.2703	9.1216	3.7000	33.7502	0.1096	0.0296	56.5883	6.2037	17
18	0.2502	9.3719	3.9960	37.4502	0.1067	0.0267	60.8426	6.4920	18
19	0.2317	9.6036	4.3157	41.4463	0.1041	0.0241	65.0134	6.7697	19
20	0.2145	9.8181	4.6610	45.7620	0.1019	0.0219	69.0898	7.0369	20
21	0.1987	10.0168	5.0338	50.4229	0.0998	0.0198	73.0629	7.2940	21
22	0.1839	10.2007	5.4365	55.4568	0.0980	0.0180	76.9257	7.5412	22
23	0.1703	10.3711	5.8715	60.8933	0.0964	0.0164	80.6726	7.7786	23
24	0.1577	10.5288	6.3412	66.7648	0.0950	0.0150	84.2997	8.0066	24
25	0.1460	10.6748	6.8485	73.1059	0.0937	0.0137	87.8041	8.2254	25
30	0.0994	11.2578	10.0627	113.2832	0.0888	0.0088	103.4558	9.1897	30
40	0.0460	11.9246	21.7245	259.0565	0.0839	0.0039	126.0422	10.5699	40
50	0.0213	12.2335	46.9016	573.7702	0.0817	0.0017	139.5928	11.4107	50
60	0.0099	12.3766	101.2571	1253.2133	0.0808	0.0008	147.3000	11.9015	60
70	0.0046	12.4428	218.6064	2720.0801	0.0804	0.0004	151.5326	12.1783	70
80	0.0021	12.4735	471.9548	5886.9354	0.0802	0.0002	153.8001	12.3301	80
90	0.0010	12.4877	1018.9151	12723.9386	0.0801	0.0001	154.9925	12.4116	90
100	0.0005	12.4943	2199.7613	27484.5157	0.0800	0.0000	155.6107	12.4545	100

					Interest Factors			i = 10.00%				
n	P/F	P/A	F/P	F/A	A/P	A/F	P/G	A/G	n
1	0.9091	0.9091	1.1000	1.0000	1.1000	1.0000	0.0000	0.0000	1
2	0.8264	1.7355	1.2100	2.1000	0.5762	0.4762	0.8264	0.4762	2
3	0.7513	2.4869	1.3310	3.3100	0.4021	0.3021	2.3291	0.9366	3
4	0.6830	3.1699	1.4641	4.6410	0.3155	0.2155	4.3781	1.3812	4
5	0.6209	3.7908	1.6105	6.1051	0.2638	0.1638	6.8618	1.8101	5
6	0.5645	4.3553	1.7716	7.7156	0.2296	0.1296	9.6842	2.2236	6
7	0.5132	4.8684	1.9487	9.4872	0.2054	0.1054	12.7631	2.6216	7
8	0.4665	5.3349	2.1436	11.4359	0.1874	0.0874	16.0287	3.0045	8
9	0.4241	5.7590	2.3579	13.5795	0.1736	0.0736	19.4215	3.3724	9
10	0.3855	6.1446	2.5937	15.9374	0.1627	0.0627	22.8913	3.7255	10
11	0.3505	6.4951	2.8531	18.5312	0.1540	0.0540	26.3963	4.0641	11
12	0.3186	6.8137	3.1384	21.3843	0.1468	0.0468	29.9012	4.3884	12
13	0.2897	7.1034	3.4523	24.5227	0.1408	0.0408	33.3772	4.6988	13
14	0.2633	7.3667	3.7975	27.9750	0.1357	0.0357	36.8005	4.9955	14
15	0.2394	7.6061	4.1772	31.7725	0.1315	0.0315	40.1520	5.2789	15
16	0.2176	7.8237	4.5950	35.9497	0.1278	0.0278	43.4164	5.5493	16
17	0.1978	8.0216	5.0545	40.5447	0.1247	0.0247	46.5819	5.8071	17
18	0.1799	8.2014	5.5599	45.5992	0.1219	0.0219	49.6395	6.0526	18
19	0.1635	8.3649	6.1159	51.1591	0.1195	0.0195	52.5827	6.2861	19
20	0.1486	8.5136	6.7275	57.2750	0.1175	0.0175	55.4069	6.5081	20
21	0.1351	8.6487	7.4002	64.0025	0.1156	0.0156	58.1095	6.7189	21
22	0.1228	8.7715	8.1403	71.4027	0.1140	0.0140	60.6893	6.9189	22
23	0.1117	8.8832	8.9543	79.5430	0.1126	0.0126	63.1462	7.1085	23
24	0.1015	8.9847	9.8497	88.4973	0.1113	0.0113	65.4813	7.2881	24
25	0.0923	9.0770	10.8347	98.3471	0.1102	0.0102	67.6964	7.4580	25
30	0.0573	9.4269	17.4494	164.4940	0.1061	0.0061	77.0766	8.1762	30
40	0.0221	9.7791	45.2593	442.5926	0.1023	0.0023	88.9525	9.0962	40
50	0.0085	9.9148	117.3909	1163.9085	0.1009	0.0009	94.8889	9.5704	50
60	0.0033	9.9672	304.4816	3034.8164	0.1003	0.0003	97.7010	9.8023	60
70	0.0013	9.9873	789.7470	7887.4696	0.1001	0.0001	98.9870	9.9113	70
80	0.0005	9.9951	2048.4002	20474.0021	0.1000	0.0000	99.5606	9.9609	80
90	0.0002	9.9981	5313.0226	53120.2261	0.1000	0.0000	99.8118	9.9831	90
100	0.0001	9.9993	13780.6123	137796.1234	0.1000	0.0000	99.9202	9.9927	100

		Interest Factors				i = 12.00%			
n	P/F	P/A	F/P	F/A	A/P	A/F	P/G	A/G	n
1	0.8929	0.8929	1.1200	1.0000	1.1200	1.0000	0.0000	0.0000	1
2	0.7972	1.6901	1.2544	2.1200	0.5917	0.4717	0.7972	0.4717	2
3	0.7118	2.4018	1.4049	3.3744	0.4163	0.2963	2.2208	0.9246	3
4	0.6355	3.0373	1.5735	4.7793	0.3292	0.2092	4.1273	1.3589	4
5	0.5674	3.6048	1.7623	6.3528	0.2774	0.1574	6.3970	1.7746	5
6	0.5066	4.1114	1.9738	8.1152	0.2432	0.1232	8.9302	2.1720	6
7	0.4523	4.5638	2.2107	10.0890	0.2191	0.0991	11.6443	2.5515	7
8	0.4039	4.9676	2.4760	12.2997	0.2013	0.0813	14.4714	2.9131	8
9	0.3606	5.3282	2.7731	14.7757	0.1877	0.0677	17.3563	3.2574	9
10	0.3220	5.6502	3.1058	17.5487	0.1770	0.0570	20.2541	3.5847	10
11	0.2875	5.9377	3.4785	20.6546	0.1684	0.0484	23.1288	3.8953	11
12	0.2567	6.1944	3.8960	24.1331	0.1614	0.0414	25.9523	4.1897	12
13	0.2292	6.4235	4.3635	28.0291	0.1557	0.0357	28.7024	4.4683	13
14	0.2046	6.6282	4.8871	32.3926	0.1509	0.0309	31.3624	4.7317	14
15	0.1827	6.8109	5.4736	37.2797	0.1468	0.0268	33.9202	4.9803	15
16	0.1631	6.9740	6.1304	42.7533	0.1434	0.0234	36.3670	5.2147	16
17	0.1456	7.1196	6.8660	48.8837	0.1405	0.0205	38.6973	5.4353	17
18	0.1300	7.2497	7.6900	55.7497	0.1379	0.0179	40.9080	5.6427	18
19	0.1161	7.3658	8.6128	63.4397	0.1358	0.0158	42.9979	5.8375	19
20	0.1037	7.4694	9.6463	72.0524	0.1339	0.0139	44.9676	6.0202	20
21	0.0926	7.5620	10.8038	81.6987	0.1322	0.0122	46.8188	6.1913	21
22	0.0826	7.6446	12.1003	92.5026	0.1308	0.0108	48.5543	6.3514	22
23	0.0738	7.7184	13.5523	104.6029	0.1296	0.0096	50.1776	6.5010	23
24	0.0659	7.7843	15.1786	118.1552	0.1285	0.0085	51.6929	6.6406	24
25	0.0588	7.8431	17.0001	133.3339	0.1275	0.0075	53.1046	6.7708	25
30	0.0334	8.0552	29.9599	241.3327	0.1241	0.0041	58.7821	7.2974	30
40	0.0107	8.2438	93.0510	767.0914	0.1213	0.0013	65.1159	7.8988	40
50	0.0035	8.3045	289.0022	2400.0182	0.1204	0.0004	67.7624	8.1597	50
60	0.0011	8.3240	897.5969	7471.6411	0.1201	0.0001	68.8100	8.2664	60
70	0.0004	8.3303	2787.7998	23223.3319	0.1200	0.0000	69.2103	8.3082	70
80	0.0001	8.3324	8658.4831	72145.6925	0.1200	0.0000	69.3594	8.3241	80
90	0.0000	8.3330	26891.9342	224091.1185	0.1200	0.0000	69.4140	8.3300	90
100	0.0000	8.3332	83522.2657	696010.5477	0.1200	0.0000	69.4336	8.3321	100

	Interest Factors				i = 14.00%				
n	P/F	P/A	F/P	F/A	A/P	A/F	P/G	A/G	n
1	0.8772	0.8772	1.1400	1.0000	1.1400	1.0000	0.0000	0.0000	1
2	0.7695	1.6467	1.2996	2.1400	0.6073	0.4673	0.7695	0.4673	2
3	0.6750	2.3216	1.4815	3.4396	0.4307	0.2907	2.1194	0.9129	3
4	0.5921	2.9137	1.6890	4.9211	0.3432	0.2032	3.8957	1.3370	4
5	0.5194	3.4331	1.9254	6.6101	0.2913	0.1513	5.9731	1.7399	5
6	0.4556	3.8887	2.1950	8.5355	0.2572	0.1172	8.2511	2.1218	6
7	0.3996	4.2883	2.5023	10.7305	0.2332	0.0932	10.6489	2.4832	7
8	0.3506	4.6389	2.8526	13.2328	0.2156	0.0756	13.1028	2.8246	8
9	0.3075	4.9464	3.2519	16.0853	0.2022	0.0622	15.5629	3.1463	9
10	0.2697	5.2161	3.7072	19.3373	0.1917	0.0517	17.9906	3.4490	10
11	0.2366	5.4527	4.2262	23.0445	0.1834	0.0434	20.3567	3.7333	11
12	0.2076	5.6603	4.8179	27.2707	0.1767	0.0367	22.6399	3.9998	12
13	0.1821	5.8424	5.4924	32.0887	0.1712	0.0312	24.8247	4.2491	13
14	0.1597	6.0021	6.2613	37.5811	0.1666	0.0266	26.9009	4.4819	14
15	0.1401	6.1422	7.1379	43.8424	0.1628	0.0228	28.8623	4.6990	15
16	0.1229	6.2651	8.1372	50.9804	0.1596	0.0196	30.7057	4.9011	16
17	0.1078	6.3729	9.2765	59.1176	0.1569	0.0169	32.4305	5.0888	17
18	0.0946	6.4674	10.5752	68.3941	0.1546	0.0146	34.0380	5.2630	18
19	0.0829	6.5504	12.0557	78.9692	0.1527	0.0127	35.5311	5.4243	19
20	0.0728	6.6231	13.7435	91.0249	0.1510	0.0110	36.9135	5.5734	20
21	0.0638	6.6870	15.6676	104.7684	0.1495	0.0095	38.1901	5.7111	21
22	0.0560	6.7429	17.8610	120.4360	0.1483	0.0083	39.3658	5.8381	22
23	0.0491	6.7921	20.3616	138.2970	0.1472	0.0072	40.4463	5.9549	23
24	0.0431	6.8351	23.2122	158.6586	0.1463	0.0063	41.4371	6.0624	24
25	0.0378	6.8729	26.4619	181.8708	0.1455	0.0055	42.3441	6.1610	25
30	0.0196	7.0027	50.9502	356.7868	0.1428	0.0028	45.8132	6.5423	30
40	0.0053	7.1050	188.8835	1342.0251	0.1407	0.0007	49.2376	6.9300	40
50	0.0014	7.1327	700.2330	4994.5213	0.1402	0.0002	50.4375	7.0714	50
60	0.0004	7.1401	2595.9187	18535.1333	0.1401	0.0001	50.8357	7.1197	60
70	0.0001	7.1421	9623.64	68733.18	0.1400	0.0000	50.9632	7.1356	70
80	0.0000	7.1427	35676.98	254828.44	0.1400	0.0000	51.0030	7.1406	80
90	0.0000	7.1428	132262.47	944724.77	0.1400	0.0000	51.0152	7.1422	90
100	0.0000	7.1428	490326.24	3502323.13	0.1400	0.0000	51.0188	7.1427	100

	Interest Factors						i = 16.00%		
n	P/F	P/A	F/P	F/A	A/P	A/F	P/G	A/G	n
1	0.8621	0.8621	1.1600	1.0000	1.1600	1.0000	0.0000	0.0000	1
2	0.7432	1.6052	1.3456	2.1600	0.6230	0.4630	0.7432	0.4630	2
3	0.6407	2.2459	1.5609	3.5056	0.4453	0.2853	2.0245	0.9014	3
4	0.5523	2.7982	1.8106	5.0665	0.3574	0.1974	3.6814	1.3156	4
5	0.4761	3.2743	2.1003	6.8771	0.3054	0.1454	5.5858	1.7060	5
6	0.4104	3.6847	2.4364	8.9775	0.2714	0.1114	7.6380	2.0729	6
7	0.3538	4.0386	2.8262	11.4139	0.2476	0.0876	9.7610	2.4169	7
8	0.3050	4.3436	3.2784	14.2401	0.2302	0.0702	11.8962	2.7388	8
9	0.2630	4.6065	3.8030	17.5185	0.2171	0.0571	13.9998	3.0391	9
10	0.2267	4.8332	4.4114	21.3215	0.2069	0.0469	16.0399	3.3187	10
11	0.1954	5.0286	5.1173	25.7329	0.1989	0.0389	17.9941	3.5783	11
12	0.1685	5.1971	5.9360	30.8502	0.1924	0.0324	19.8472	3.8189	12
13	0.1452	5.3423	6.8858	36.7862	0.1872	0.0272	21.5899	4.0413	13
14	0.1252	5.4675	7.9875	43.6720	0.1829	0.0229	23.2175	4.2464	14
15	0.1079	5.5755	9.2655	51.6595	0.1794	0.0194	24.7284	4.4352	15
16	0.0930	5.6685	10.7480	60.9250	0.1764	0.0164	26.1241	4.6086	16
17	0.0802	5.7487	12.4677	71.6730	0.1740	0.0140	27.4074	4.7676	17
18	0.0691	5.8178	14.4625	84.1407	0.1719	0.0119	28.5828	4.9130	18
19	0.0596	5.8775	16.7765	98.6032	0.1701	0.0101	29.6557	5.0457	19
20	0.0514	5.9288	19.4608	115.3797	0.1687	0.0087	30.6321	5.1666	20
21	0.0443	5.9731	22.5745	134.8405	0.1674	0.0074	31.5180	5.2766	21
22	0.0382	6.0113	26.1864	157.4150	0.1664	0.0064	32.3200	5.3765	22
23	0.0329	6.0442	30.3762	183.6014	0.1654	0.0054	33.0442	5.4671	23
24	0.0284	6.0726	35.2364	213.9776	0.1647	0.0047	33.6970	5.5490	24
25	0.0245	6.0971	40.8742	249.2140	0.1640	0.0040	34.2841	5.6230	25
30	0.0116	6.1772	85.8499	530.3117	0.1619	0.0019	36.4234	5.8964	30
40	0.0026	6.2335	378.7212	2360.7572	0.1604	0.0004	38.2992	6.1441	40
50	0.0006	6.2463	1670.7038	10435.6488	0.1601	0.0001	38.8521	6.2201	50
60	0.0001	6.2492	7370.2014	46057.5085	0.1600	0.0000	39.0063	6.2419	60
70	0.0000	6.2498	32513.1648	203201.0302	0.1600	0.0000	39.0478	6.2478	70
80	0.0000	6.2500	143429.72	896429.47	0.1600	0.0000	39.0587	6.2494	80
90	0.0000	6.2500	632730.88	3954561.75	0.1600	0.0000	39.0615	6.2499	90
100	0.0000	6.2500	2791251.20	17445313.75	0.1600	0.0000	39.0623	6.2500	100

		Interest Factors				i = 18.00%			
n	P/F	P/A	F/P	F/A	A/P	A/F	P/G	A/G	n
1	0.8475	0.8475	1.1800	1.0000	1.1800	1.0000	0.0000	0.0000	1
2	0.7182	1.5656	1.3924	2.1800	0.6387	0.4587	0.7182	0.4587	2
3	0.6086	2.1743	1.6430	3.5724	0.4599	0.2799	1.9354	0.8902	3
4	0.5158	2.6901	1.9388	5.2154	0.3717	0.1917	3.4828	1.2947	4
5	0.4371	3.1272	2.2878	7.1542	0.3198	0.1398	5.2312	1.6728	5
6	0.3704	3.4976	2.6996	9.4420	0.2859	0.1059	7.0834	2.0252	6
7	0.3139	3.8115	3.1855	12.1415	0.2624	0.0824	8.9670	2.3526	7
8	0.2660	4.0776	3.7589	15.3270	0.2452	0.0652	10.8292	2.6558	8
9	0.2255	4.3030	4.4355	19.0859	0.2324	0.0524	12.6329	2.9358	9
10	0.1911	4.4941	5.2338	23.5213	0.2225	0.0425	14.3525	3.1936	10
11	0.1619	4.6560	6.1759	28.7551	0.2148	0.0348	15.9716	3.4303	11
12	0.1372	4.7932	7.2876	34.9311	0.2086	0.0286	17.4811	3.6470	12
13	0.1163	4.9095	8.5994	42.2187	0.2037	0.0237	18.8765	3.8449	13
14	0.0985	5.0081	10.1472	50.8180	0.1997	0.0197	20.1576	4.0250	14
15	0.0835	5.0916	11.9737	60.9653	0.1964	0.0164	21.3269	4.1887	15
16	0.0708	5.1624	14.1290	72.9390	0.1937	0.0137	22.3885	4.3369	16
17	0.0600	5.2223	16.6722	87.0680	0.1915	0.0115	23.3482	4.4708	17
18	0.0508	5.2732	19.6733	103.7403	0.1896	0.0096	24.2123	4.5916	18
19	0.0431	5.3162	23.2144	123.4135	0.1881	0.0081	24.9877	4.7003	19
20	0.0365	5.3527	27.3930	146.6280	0.1868	0.0068	25.6813	4.7978	20
21	0.0309	5.3837	32.3238	174.0210	0.1857	0.0057	26.3000	4.8851	21
22	0.0262	5.4099	38.1421	206.3448	0.1848	0.0048	26.8506	4.9632	22
23	0.0222	5.4321	45.0076	244.4868	0.1841	0.0041	27.3394	5.0329	23
24	0.0188	5.4509	53.1090	289.4945	0.1835	0.0035	27.7725	5.0950	24
25	0.0160	5.4669	62.6686	342.6035	0.1829	0.0029	28.1555	5.1502	25
30	0.0070	5.5168	143.3706	790.9480	0.1813	0.0013	29.4864	5.3448	30
40	0.0013	5.5482	750.3783	4163.2130	0.1802	0.0002	30.5269	5.5022	40
50	0.0003	5.5541	3927.36	21813.09	0.1800	0.0000	30.7856	5.5428	50
60	0.0000	5.5553	20555.14	114189.67	0.1800	0.0000	30.8465	5.5526	60
70	0.0000	5.5555	107582.22	597673.46	0.1800	0.0000	30.8603	5.5549	70
80	0.0000	5.5555	563067.66	3128148.11	0.1800	0.0000	30.8634	5.5554	80
90	0.0000	5.5556	2947003.54	16372236.33	0.1800	0.0000	30.8640	5.5555	90
100	0.0000	5.5556	15424131.91	85689616.14	0.1800	0.0000	30.8642	5.5555	100

		Interest Factors			i = 20.00%				
n	P/F	P/A	F/P	F/A	A/P	A/F	P/G	A/G	n
1	0.8333	0.8333	1.2000	1.0000	1.2000	1.0000	0.0000	0.0000	1
2	0.6944	1.5278	1.4400	2.2000	0.6545	0.4545	0.6944	0.4545	2
3	0.5787	2.1065	1.7280	3.6400	0.4747	0.2747	1.8519	0.8791	3
4	0.4823	2.5887	2.0736	5.3680	0.3863	0.1863	3.2986	1.2742	4
5	0.4019	2.9906	2.4883	7.4416	0.3344	0.1344	4.9061	1.6405	5
6	0.3349	3.3255	2.9860	9.9299	0.3007	0.1007	6.5806	1.9788	6
7	0.2791	3.6046	3.5832	12.9159	0.2774	0.0774	8.2551	2.2902	7
8	0.2326	3.8372	4.2998	16.4991	0.2606	0.0606	9.8831	2.5756	8
9	0.1938	4.0310	5.1598	20.7989	0.2481	0.0481	11.4335	2.8364	9
10	0.1615	4.1925	6.1917	25.9587	0.2385	0.0385	12.8871	3.0739	10
11	0.1346	4.3271	7.4301	32.1504	0.2311	0.0311	14.2330	3.2893	11
12	0.1122	4.4392	8.9161	39.5805	0.2253	0.0253	15.4667	3.4841	12
13	0.0935	4.5327	10.6993	48.4966	0.2206	0.0206	16.5883	3.6597	13
14	0.0779	4.6106	12.8392	59.1959	0.2169	0.0169	17.6008	3.8175	14
15	0.0649	4.6755	15.4070	72.0351	0.2139	0.0139	18.5095	3.9588	15
16	0.0541	4.7296	18.4884	87.4421	0.2114	0.0114	19.3208	4.0851	16
17	0.0451	4.7746	22.1861	105.9306	0.2094	0.0094	20.0419	4.1976	17
18	0.0376	4.8122	26.6233	128.1167	0.2078	0.0078	20.6805	4.2975	18
19	0.0313	4.8435	31.9480	154.7400	0.2065	0.0065	21.2439	4.3861	19
20	0.0261	4.8696	38.3376	186.6880	0.2054	0.0054	21.7395	4.4643	20
21	0.0217	4.8913	46.0051	225.0256	0.2044	0.0044	22.1742	4.5334	21
22	0.0181	4.9094	55.2061	271.0307	0.2037	0.0037	22.5546	4.5941	22
23	0.0151	4.9245	66.2474	326.2369	0.2031	0.0031	22.8867	4.6475	23
24	0.0126	4.9371	79.4968	392.4842	0.2025	0.0025	23.1760	4.6943	24
25	0.0105	4.9476	95.3962	471.9811	0.2021	0.0021	23.4276	4.7352	25
30	0.0042	4.9789	237.3763	1181.8816	0.2008	0.0008	24.2628	4.8731	30
40	0.0007	4.9966	1469.7716	7343.8578	0.2001	0.0001	24.8469	4.9728	40
50	0.0001	4.9995	9100.44	45497.19	0.2000	0.0000	24.9698	4.9945	50
60	0.0000	4.9999	56347.51	281732.57	0.2000	0.0000	24.9942	4.9989	60
70	0.0000	5.0000	348888.96	1744439.78	0.2000	0.0000	24.9989	4.9998	70
80	0.0000	5.0000	2160228	10801137	0.2000	0.0000	24.9998	5.0000	80
90	0.0000	5.0000	13375565	66877821	0.2000	0.0000	25.0000	5.0000	90
100	0.0000	5.0000	82817975	414089868	0.2000	0.0000	25.0000	5.0000	100

			Interest Factors				i = 25.00%			
n	P/F	P/A	F/P	F/A	A/P	A/F	P/G	A/G	n
1	0.8000	0.8000	1.2500	1.0000	1.2500	1.0000	0.0000	0.0000	1
2	0.6400	1.4400	1.5625	2.2500	0.6944	0.4444	0.6400	0.4444	2
3	0.5120	1.9520	1.9531	3.8125	0.5123	0.2623	1.6640	0.8525	3
4	0.4096	2.3616	2.4414	5.7656	0.4234	0.1734	2.8928	1.2249	4
5	0.3277	2.6893	3.0518	8.2070	0.3718	0.1218	4.2035	1.5631	5
6	0.2621	2.9514	3.8147	11.2588	0.3388	0.0888	5.5142	1.8683	6
7	0.2097	3.1611	4.7684	15.0735	0.3163	0.0663	6.7725	2.1424	7
8	0.1678	3.3289	5.9605	19.8419	0.3004	0.0504	7.9469	2.3872	8
9	0.1342	3.4631	7.4506	25.8023	0.2888	0.0388	9.0207	2.6048	9
10	0.1074	3.5705	9.3132	33.2529	0.2801	0.0301	9.9870	2.7971	10
11	0.0859	3.6564	11.6415	42.5661	0.2735	0.0235	10.8460	2.9663	11
12	0.0687	3.7251	14.5519	54.2077	0.2684	0.0184	11.6020	3.1145	12
13	0.0550	3.7801	18.1899	68.7596	0.2645	0.0145	12.2617	3.2437	13
14	0.0440	3.8241	22.7374	86.9495	0.2615	0.0115	12.8334	3.3559	14
15	0.0352	3.8593	28.4217	109.6868	0.2591	0.0091	13.3260	3.4530	15
16	0.0281	3.8874	35.5271	138.1085	0.2572	0.0072	13.7482	3.5366	16
17	0.0225	3.9099	44.4089	173.6357	0.2558	0.0058	14.1085	3.6084	17
18	0.0180	3.9279	55.5112	218.0446	0.2546	0.0046	14.4147	3.6698	18
19	0.0144	3.9424	69.3889	273.5558	0.2537	0.0037	14.6741	3.7222	19
20	0.0115	3.9539	86.7362	342.9447	0.2529	0.0029	14.8932	3.7667	20
21	0.0092	3.9631	108.4202	429.6809	0.2523	0.0023	15.0777	3.8045	21
22	0.0074	3.9705	135.5253	538.1011	0.2519	0.0019	15.2326	3.8365	22
23	0.0059	3.9764	169.4066	673.6264	0.2515	0.0015	15.3625	3.8634	23
24	0.0047	3.9811	211.7582	843.0329	0.2512	0.0012	15.4711	3.8861	24
25	0.0038	3.9849	264.6978	1054.7912	0.2509	0.0009	15.5618	3.9052	25
30	0.0012	3.9950	807.7936	3227.1743	0.2503	0.0003	15.8316	3.9628	30
40	0.0001	3.9995	7523.1638	30088.6554	0.2500	0.0000	15.9766	3.9947	40
50	0.0000	3.9999	70064.92	280255.69	0.2500	0.0000	15.9969	3.9993	50
60	0.0000	4.0000	652530	2610118	0.2500	0.0000	15.9996	3.9999	60
70	0.0000	4.0000	6077163	24308649	0.2500	0.0000	16.0000	4.0000	70
80	0.0000	4.0000	56597994	226391973	0.2500	0.0000	16.0000	4.0000	80
90	0.0000	4.0000	527109897	2108439585	0.2500	0.0000	16.0000	4.0000	90
100	0.0000	4.0000	4909093465	19636373857	0.2500	0.0000	16.0000	4.0000	100

	Interest Factors				i = 30.00%				
n	P/F	P/A	F/P	F/A	A/P	A/F	P/G	A/G	n
1	0.7692	0.7692	1.3000	1.0000	1.3000	1.0000	0.0000	0.0000	1
2	0.5917	1.3609	1.6900	2.3000	0.7348	0.4348	0.5917	0.4348	2
3	0.4552	1.8161	2.1970	3.9900	0.5506	0.2506	1.5020	0.8271	3
4	0.3501	2.1662	2.8561	6.1870	0.4616	0.1616	2.5524	1.1783	4
5	0.2693	2.4356	3.7129	9.0431	0.4106	0.1106	3.6297	1.4903	5
6	0.2072	2.6427	4.8268	12.7560	0.3784	0.0784	4.6656	1.7654	6
7	0.1594	2.8021	6.2749	17.5828	0.3569	0.0569	5.6218	2.0063	7
8	0.1226	2.9247	8.1573	23.8577	0.3419	0.0419	6.4800	2.2156	8
9	0.0943	3.0190	10.6045	32.0150	0.3312	0.0312	7.2343	2.3963	9
10	0.0725	3.0915	13.7858	42.6195	0.3235	0.0235	7.8872	2.5512	10
11	0.0558	3.1473	17.9216	56.4053	0.3177	0.0177	8.4452	2.6833	11
12	0.0429	3.1903	23.2981	74.3270	0.3135	0.0135	8.9173	2.7952	12
13	0.0330	3.2233	30.2875	97.6250	0.3102	0.0102	9.3135	2.8895	13
14	0.0254	3.2487	39.3738	127.9125	0.3078	0.0078	9.6437	2.9685	14
15	0.0195	3.2682	51.1859	167.2863	0.3060	0.0060	9.9172	3.0344	15
16	0.0150	3.2832	66.5417	218.4722	0.3046	0.0046	10.1426	3.0892	16
17	0.0116	3.2948	86.5042	285.0139	0.3035	0.0035	10.3276	3.1345	17
18	0.0089	3.3037	112.4554	371.5180	0.3027	0.0027	10.4788	3.1718	18
19	0.0068	3.3105	146.1920	483.9734	0.3021	0.0021	10.6019	3.2025	19
20	0.0053	3.3158	190.0496	630.1655	0.3016	0.0016	10.7019	3.2275	20
21	0.0040	3.3198	247.0645	820.2151	0.3012	0.0012	10.7828	3.2480	21
22	0.0031	3.3230	321.1839	1067.2796	0.3009	0.0009	10.8482	3.2646	22
23	0.0024	3.3254	417.5391	1388.4635	0.3007	0.0007	10.9009	3.2781	23
24	0.0018	3.3272	542.8008	1806.0026	0.3006	0.0006	10.9433	3.2890	24
25	0.0014	3.3286	705.6410	2348.8033	0.3004	0.0004	10.9773	3.2979	25

	Interest Factors				i = 35.00%			
n	P/F	P/A	F/P	F/A	A/P	A/F	P/G	A/G
1	0.7407	0.7407	1.3500	1.0000	1.3500	1.0000	0.0000	0.0000
2	0.5487	1.2894	1.8225	2.3500	0.7755	0.4255	0.5487	0.4255
3	0.4064	1.6959	2.4604	4.1725	0.5897	0.2397	1.3616	0.8029
4	0.3011	1.9969	3.3215	6.6329	0.5008	0.1508	2.2648	1.1341
5	0.2230	2.2200	4.4840	9.9544	0.4505	0.1005	3.1568	1.4220
6	0.1652	2.3852	6.0534	14.4384	0.4193	0.0693	3.9828	1.6698
7	0.1224	2.5075	8.1722	20.4919	0.3988	0.0488	4.7170	1.8811
8	0.0906	2.5982	11.0324	28.6640	0.3849	0.0349	5.3515	2.0597
9	0.0671	2.6653	14.8937	39.6964	0.3752	0.0252	5.8886	2.2094
10	0.0497	2.7150	20.1066	54.5902	0.3683	0.0183	6.3363	2.3338
11	0.0368	2.7519	27.1439	74.6967	0.3634	0.0134	6.7047	2.4364
12	0.0273	2.7792	36.6442	101.8406	0.3598	0.0098	7.0049	2.5205
13	0.0202	2.7994	49.4697	138.4848	0.3572	0.0072	7.2474	2.5889
14	0.0150	2.8144	66.7841	187.9544	0.3553	0.0053	7.4421	2.6443
15	0.0111	2.8255	90.1585	254.7385	0.3539	0.0039	7.5974	2.6889
16	0.0082	2.8337	121.7139	344.8970	0.3529	0.0029	7.7206	2.7246
17	0.0061	2.8398	164.3138	466.6109	0.3521	0.0021	7.8180	2.7530
18	0.0045	2.8443	221.8236	630.9247	0.3516	0.0016	7.8946	2.7756
19	0.0033	2.8476	299.4619	852.7483	0.3512	0.0012	7.9547	2.7935
20	0.0025	2.8501	404.2736	1152.2103	0.3509	0.0009	8.0017	2.8075
21	0.0018	2.8519	545.7693	1556.4838	0.3506	0.0006	8.0384	2.8186
22	0.0014	2.8533	736.7886	2102.2532	0.3505	0.0005	8.0669	2.8272
23	0.0010	2.8543	994.6646	2839.0418	0.3504	0.0004	8.0890	2.8340
24	0.0007	2.8550	1342.7973	3833.7064	0.3503	0.0003	8.1061	2.8393
25	0.0006	2.8556	1812.7763	5176.5037	0.3502	0.0002	8.1194	2.8433

	Interest Factors						i = 40.00%		
n	P/F	P/A	F/P	F/A	A/P	A/F	P/G	A/G	n
1	0.7143	0.7143	1.4000	1.0000	1.4000	1.0000	0.0000	0.0000	1
2	0.5102	1.2245	1.9600	2.4000	0.8167	0.4167	0.5102	0.4167	2
3	0.3644	1.5889	2.7440	4.3600	0.6294	0.2294	1.2391	0.7798	3
4	0.2603	1.8492	3.8416	7.1040	0.5408	0.1408	2.0200	1.0923	4
5	0.1859	2.0352	5.3782	10.9456	0.4914	0.0914	2.7637	1.3580	5
6	0.1328	2.1680	7.5295	16.3238	0.4613	0.0613	3.4278	1.5811	6
7	0.0949	2.2628	10.5414	23.8534	0.4419	0.0419	3.9970	1.7664	7
8	0.0678	2.3306	14.7579	34.3947	0.4291	0.0291	4.4713	1.9185	8
9	0.0484	2.3790	20.6610	49.1526	0.4203	0.0203	4.8585	2.0422	9
10	0.0346	2.4136	28.9255	69.8137	0.4143	0.0143	5.1696	2.1419	10
11	0.0247	2.4383	40.4957	98.7391	0.4101	0.0101	5.4166	2.2215	11
12	0.0176	2.4559	56.6939	139.2348	0.4072	0.0072	5.6106	2.2845	12
13	0.0126	2.4685	79.3715	195.9287	0.4051	0.0051	5.7618	2.3341	13
14	0.0090	2.4775	111.1201	275.3002	0.4036	0.0036	5.8788	2.3729	14
15	0.0064	2.4839	155.5681	386.4202	0.4026	0.0026	5.9688	2.4030	15
16	0.0046	2.4885	217.7953	541.9883	0.4018	0.0018	6.0376	2.4262	16
17	0.0033	2.4918	304.9135	759.7837	0.4013	0.0013	6.0901	2.4441	17
18	0.0023	2.4941	426.8789	1064.6971	0.4009	0.0009	6.1299	2.4577	18
19	0.0017	2.4958	597.6304	1491.5760	0.4007	0.0007	6.1601	2.4682	19
20	0.0012	2.4970	836.6826	2089.2064	0.4005	0.0005	6.1828	2.4761	20
21	0.0009	2.4979	1171.3556	2925.8889	0.4003	0.0003	6.1998	2.4821	21
22	0.0006	2.4985	1639.8978	4097.2445	0.4002	0.0002	6.2127	2.4866	22
23	0.0004	2.4989	2295.8569	5737.1423	0.4002	0.0002	6.2222	2.4900	23
24	0.0003	2.4992	3214.1997	8032.9993	0.4001	0.0001	6.2294	2.4925	24
25	0.0002	2.4994	4499.8796	11247.1990	0.4001	0.0001	6.2347	2.4944	25

Index